Military textiles

The Textile Institute and Woodhead Publishing

The Textile Institute is a unique organisation in textiles, clothing and footwear. Incorporated in England by a Royal Charter granted in 1925, the Institute has individual and corporate members in over 90 countries. The aim of the Institute is to facilitate learning, recognise achievement, reward excellence and disseminate information within the global textiles, clothing and footwear industries.

Historically, The Textile Institute has published books of interest to its members and the textile industry. To maintain this policy, the Institute has entered into partnership with Woodhead Publishing Limited to ensure that Institute members and the textile industry continue to have access to high calibre titles on textile science and technology.

Most Woodhead titles on textiles are now published in collaboration with The Textile Institute. Through this arrangement, the Institute provides an Editorial Board which advises Woodhead on appropriate titles for future publication and suggests possible editors and authors for these books. Each book published under this arrangement carries the Institute's logo.

Woodhead books published in collaboration with The Textile Institute are offered to Textile Institute members at a substantial discount. These books, together with those published by The Textile Institute that are still in print, are offered on the Woodhead web site at: www.woodheadpublishing.com. Textile Institute books still in print are also available directly from the Institute's website at: www.textileinstitutebooks.com.

A list of Woodhead books on textile science and technology, most of which have been published in collaboration with The Textile Institute, can be found at the end of the contents pages.

Woodhead Publishing in Textiles: Number 73

Military textiles

Edited by
Eugene Wilusz

The Textile Institute

CRC Press
Boca Raton Boston New York Washington, DC

WOODHEAD PUBLISHING LIMITED

Cambridge, England

Published by Woodhead Publishing Limited in association with
The Textile Institute
Woodhead Publishing Limited, Abington Hall, Granta Park,
Great Abington
Cambridge CB21 6AH, England
www.woodheadpublishing.com

Published in North America by CRC Press LLC, 6000 Broken Sound Parkway, NW,
Suite 300, Boca Raton, FL 33487, USA

First published 2008, Woodhead Publishing Limited and CRC Press LLC
© Woodhead Publishing Limited, 2008
The authors have asserted their moral rights.

British Library Cataloguing in Publication Data
A catalogue record for this book is available from the British Library.

Library of Congress Cataloging in Publication Data
A catalog record for this book is available from the Library of Congress.

Woodhead Publishing ISBN 978-1-84569-206-3 (book)
Woodhead Publishing ISBN 978-1-84569-451-7 (e-book)
CRC Press ISBN 978-1-4200-7960-9
CRC Press order number WP7960

The publishers' policy is to use permanent paper from mills that operate a
sustainable forestry policy, and which has been manufactured from pulp which is
processed using acid-free and elementary chlorine-free practices. Furthermore,
the publishers ensure that the text paper and cover board used have met
acceptable environmental accreditation standards.

Typeset by SNP Best-set Typesetter Ltd., Hong Kong
Printed by TJ International Limited, Padstow, Cornwall, England

Contents

G. SUN, University of California, USA and
S. D. WORLEY and R. M. BROUGHTON Jr, Auburn
University, USA

P. SUDHAKAR and N. GOBI, K. S. Rangasamy
College of Technology, India and M. SENTHILKUMAR,
PSG Polytechnic College, India

Contributor contact details

(* = main contact)

Editor

Dr E. Wilusz
US Army Natick Soldier Research,
  Development and Engineering
  Center
Kansas Street
Natick
MA 01760-5020
USA

Email: eugene.wilusz@us.army.mil

Chapter 1

Dr E. Sparks
Engineering Systems Department
Defence College of Management
  and Technology
Cranfield University
Shrivenham
Swindon
SN6 8LA
UK

Email: e.sparks@cranfield.ac.uk

Chapter 2

Professor G. A. Thomas
115 Textile Engineering
  Department
Auburn University
Auburn
AL 36849
USA

Email: gwynedd_thomas@
auburn.edu

Chapter 3

Professor N. Pan
Textiles and Clothing
129 Everson Hall
University of California
One Shields Avenue
Davis
CA 95616
USA

Email: npan@ucdavis.edu

Chapter 4

A. V. Cardello
US Army Natick Soldier Research,
  Development and Engineering
  Center
Kansas Street
Natick
MA 01760-5020
USA

Email: armand.cardello@
us.army.mil

Chapter 5

Dr F. S. Kilinc-Balci*
Polymer and Fiber Engineering
  Department
115 Textile Building
Auburn University
Auburn
AL 36849-5327
USA

Email: fselcen@gmail.com

Dr Y. Elmogahzy
Polymer and Fiber Engineering
  Department
115 Textile Building
Auburn University
Auburn
AL 36849-5327
USA

Email: elmogye@auburn.edu

Chapter 6

Professor N. Pan
Textiles and Clothing
129 Everson Hall
University of California
One Shields Avenue
Davis, CA 95616
USA

Email: npan@ucdavis.edu

Chapter 7

Dr C. Thwaites
W. L. Gore and Associates UK Ltd
Kirkton Campus
Livingston
West Lothian
EH54 7BH
UK

Email: cthwaite@wlgore.com

Chapter 8

C. A. Gomes
Foster-Miller, Inc
350 Second Ave
Waltham
MA 02451-1196
USA

Email: cgomes@foster-miller.com

Chapter 9

Dr T. Tam* and A. Bhatnagar
Honeywell International Inc
15801 Woods Edge Rd
Colonial Heights
VA 23834
USA

Email: Thomas.tam@honeywell.
com

Chapter 10

D. R. Dunn
H. P. White Laboratory, Inc
3114 Scarboro Road
Street
Maryland
MD 21154
USA

Email: info@hpwhite.com

Chapter 11

Q. Truong and E. Wilusz*
US Army Natick Soldier
    Research, Development and
    Engineering Center
Individual Protection Directorate
Chemical Technology Team
Kansas Street
Natick
MA 01760-5019
USA

Email: Quoc.Truong@us.army.mil
eugene.wilusz@us.army.mil

Chapter 12

Prof. G. Sun*
Division of Textiles and Clothing
University of California
Davis
CA 95616
USA

Email: gysun@ucdavis.edu

S. D. Worley
Department of Chemistry and
    Biochemistry
Auburn University
Auburn
AL 36849
USA

R. M. Broughton Jr
Department of Polymer and Fiber
    Engineering
Auburn University
Auburn
AL 36849
USA

Chapter 13

P. Sudhakar* and N. Gobi
Department of Textile Technology
K. S. Rangasamy College of
    Technology
KSR Kalvi Nagar
Thokavadi Post
Tiruchengode-637215
Namakkal district
Tamilnadu
India

Email: sudhakaren_p@
rediffmail.com
gobsn@yahoo.com

M. Senthilkumar
Department of Textile Technology
PSG Polytechnic College
Coimbatore
Tamilnadu
637209
India

Chapter 14

P. Sudhakar,* S. Krishnaramesh
    and D. Brightlivingstone
Department of Textile Technology
K. S. Rangasamy College of
    Technology
KSR Kalvi Nagar
Thokavadi Post
Tiruchengode-637215
Namakkal district
Tamilnadu
India

Email: sudhakaren_p@rediffmail.
com
krishtextech@yahoo.co.in

Chapter 15

C. Winterhalter
US Army Natick Soldier Research,
    Development and Engineering
    Center
Kansas Street
Natick
MA 01760-5020
USA

Email: carole.winterhalter@
us.army.mil

Woodhead Publishing in Textiles

In memory of S/Sgt Eugene Wilusz

Introduction

The purpose of this book is to provide an update on the considerable advances that have occurred in the field of military textiles in recent years. Because developments by the military are often adopted in the civilian sector, this book is expected to be of interest to a wide range of individuals, including scientists and engineers working in the various disciplines of textiles and materials science. For a detailed overview of the subject, the reader is referred to the excellent chapters by Richard A. Scott on 'Textiles in Defence' (Chapter 16) and by David A. Holmes on 'Textiles for Survival' (Chapter 17) in the *Handbook of Technical Textiles* (Woodhead Publishing Limited, 2000). The reader is also referred to the comprehensive volume edited by Dr Scott on *Textiles for Protection* (Woodhead Publishing Limited, 2005).

Textiles are a very important class of materials used by the military and civilians alike. Since we all wear clothing, each and every one of us has developed a certain knowledge, and even expertise, at least with regard to the clothing we wear and textile items we use on a daily basis. We know which type of clothing we like and which we don't. We know if the clothing fits or if it doesn't. We know if the clothing is comfortable, or if it is not. We know which towels we like to use, perhaps because they are soft and plushy or because they absorb a lot of water. We encounter these textiles on a daily basis and generally don't give them a second thought.

Individuals in the military, and many others, rely on clothing and items made from textiles for protection and even life support. Police officers, firefighters, and those who work in various industrial settings rely on safety clothing for protection against bullets, flames, hazardous chemical splashes, or punctures by sharp objects. Physically active individuals involved in various sports and other outdoor activities, such as hiking and camping, depend on their clothing for more than comfort. Hikers depend on their backpacks. Campers depend on their tents and sleeping bags. Individuals who live in cold climates depend on their clothing for protection against the cold. Parachutists depend on the textiles in their parachutes.

Despite textiles having been around and in use for so long, advances and improvements continue to be made. Some recent advances have come from the world of nanotechnology. Surfaces of fibers have been chemically modified through surface grafting chemistry. Fibers have been extruded through unique dies resulting in fibers with a variety of cross-sectional shapes. Nanoparticles have been incorporated into fibers as well as adsorbed on fiber surfaces. Nanofibers have been prepared by electrospinning from solution. These developments have resulted in novel improvements to fiber and fabric properties, and many more advances are expected in the near future.

One of the exciting areas which continues to develop is that of electronic textiles. The incorporation of conducting fibers and electronic components into textiles has opened a new world of possibilities. Shirts already exist that are capable of monitoring physiological parameters, such as heart rate, blood pressure, and temperature. Numerous other sensors can also be incorporated. It is anticipated that circuit boards and even batteries will be woven directly into fabrics.

In the military, some uses of textiles are dress clothing, combat clothing, ballistic protective vests, chemical biological protective clothing, cold weather clothing, sleeping bags, tents and parachutes. While all of these items do the job for which they are intended, all of them, and many more, are constantly under improvement. It is imperative that the latest advances be incorporated into these items to make them more effective, lighter in weight or less costly.

In this volume the effort has been made to capture the most recent developments in military textiles. Contributors to this volume are well-known experts in their respective subject matters and represent a truly global perspective on the subject.

The book is divided into two parts. Part I covers general requirements for military textiles, and Part II is about protection. Each of the chapters reports the latest developments in specialty areas and also future trends. It is hoped that this book will be a useful contribution toward providing tomorrow's military servicemen and servicewomen with the best possible protective clothing and other textile items.

The editor wishes to extend his sincere thanks to all of the experts who devoted considerable time and effort in contributing chapters to this volume. He also wishes to thank Ms Lucy Cornwell of Woodhead Publishing Limited for her many efforts in helping to make this book a reality.

Dr E. Wilusz
US Army Natick Soldier Research,
Development and Engineering Center, USA

Part I

General requirements for military textiles

1
Future soldier requirements: Dealing with complexity

E. SPARKS, Cranfield University, UK

1.1   Introduction

Underpinning the development of military textiles and the use of these textiles in creating soldier clothing and equipment, are requirements providing performance measurements against which success is determined (Sommerville and Sawyer, 1997). Requirements form part of the discipline of systems engineering, which is concerned with the management of complexity, the behaviour that is exhibited when elements have a high number of inter-relationships or dependencies. But why are future soldier requirements complex, and what relationship do the requirements have to military textiles and the creation of soldier clothing and equipment? The latter question is the driver for future research as well as the creation of physical concept demonstrators. Collections of requirements form the basis of specifications for clothing and/or equipment and provide industry with a blueprint against which they design and test textiles; these in turn are manufactured into clothing and equipment which is again tested for suitability.

The question of complexity is demonstrated in Fig. 1.1, which diagrammatically represents the many facets that must be thought about when considering future soldier requirements. It is in no way complete, with the ability to add far greater fidelity using subsequent iterations. It does, however, serve the purpose of highlighting the inherent complexity of the activity, with dependencies between elements that may be outside our control, but may impact overall success (Stevens et al., 1998). Therefore, the title of 'Future soldier requirements: Dealing with complexity' recognises that many factors must be considered when making decisions about the best mix of clothing and equipment with which to supply the soldier, and that making these may require consideration of parameters that are outside our direct control because of the critical relationships that exist with other entities.

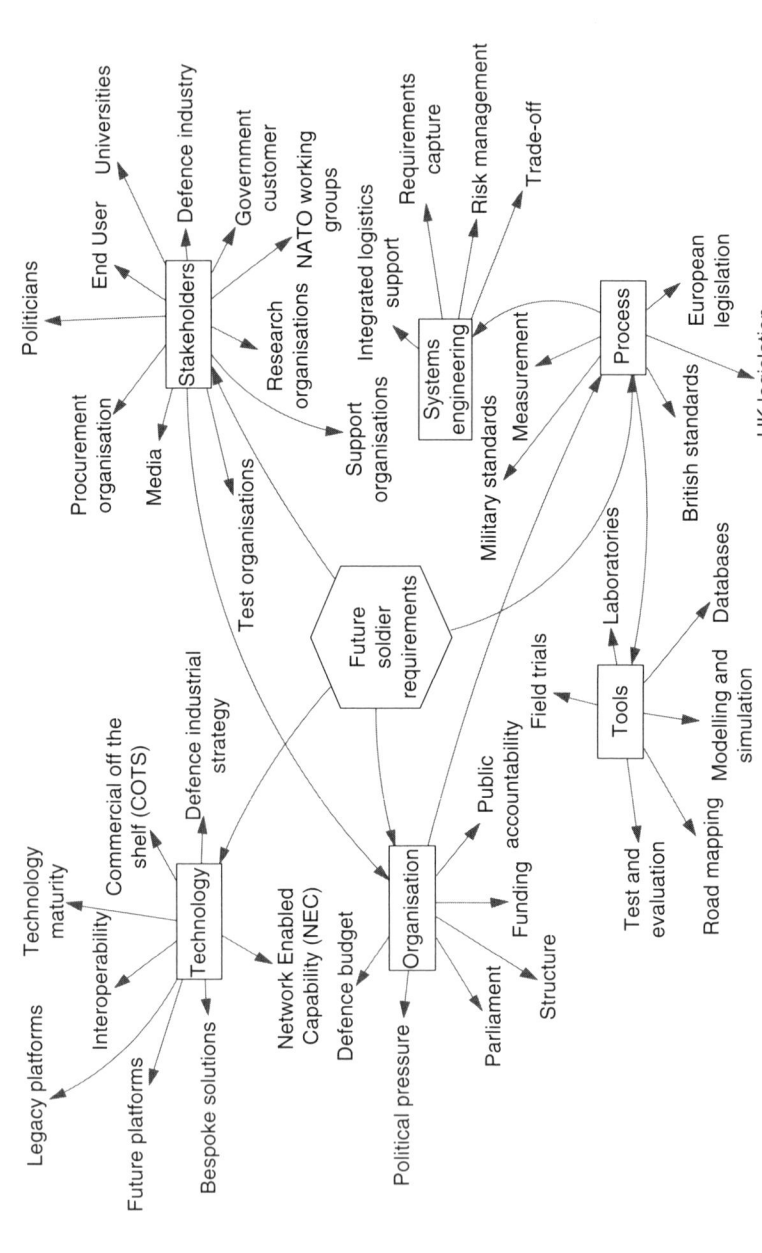

1.1 Future soldier requirements.

Figure 1.1 provides broad categories/areas of interest that will be revisited in later discussions to address future soldier requirements. They include stakeholders, process, organisation, technology and tools. All of these form components of systems engineering, which provides applied theory for dealing with systems from conception through to design and disposal, or from 'cradle to grave' (Forsberg and Mooz, 1992).

The adoption of this discipline within UK defence has come as a result of the Strategic Defence Review (HM Stationary Office, 1998), commissioned by Government to investigate time and cost overruns for management of defence procurement projects. This was also an opportunity to assess UK procurement processes and the types and quantities of military platforms required in the light of significant changes to the threats and theatres of operation post Cold War (Armstrong and Goldstein, 1990). The outcome was a realisation that greater flexibility and adaptability were required, moving from the traditional procurement of specific pieces of equipment to the term 'capability', bringing about effect to prosecute defence aims. Capability includes wider service-related issues, such as logistics, doctrine and manpower, as well as equipment. It represented a step change in business and a required process to support it, leading to the introduction of systems engineering (Controller and Auditor General, 2003). Optimisation of individual systems became no longer acceptable; the new age was about things working together towards a common aim, capitalising on synergy and underpinning the new manoeuvrist approach to warfare.

The following sections define the drivers for this change and then focus specifically on the soldier and the complexity associated with derivation of future requirements for clothing and equipment. Principles from the domain of systems thinking and systems engineering are utilised within the discussion but are not explicitly introduced on the grounds that they are sufficiently intuitive to provide benefit to the reader without necessitating specialist knowledge. In all instances, discussion is primarily focused from a UK perspective.

1.2   The current and future challenges faced by the soldier

The domain of defence is changing, partly as a consequence of the wider environment (finance, society and politics) but also due to shifting threats and changes in strategic level military doctrine. Uncertainty in the geographic location of the 'front line', and the nature of operations that we may be engaged in and with whom, present significant challenges, not only to the Armed Forces but also to the developers and researchers supporting equipment procurement. Buzz words for 21st century warfare include 'integrated', 'high-tempo', 'combined', 'joint', 'multi-national', 'inter-agency' and 'full spectrum'. Effectiveness is expected to be increased with the 'ability to move at short notice and with endurance, adapting through a seamless spectrum of conflict prevention, conflict and post-conflict activities' (Director Infantry, 2000). Information superiority through Network Enabled Capability (NEC) will provide near real-time data from the sensor to shooter, supporting prosecution of high-level defence aims, protecting UK interests (Secretary of State for Defence, 2005). When distilled, these statements recognise a number of key factors that can be summarised as follows:

- We need our equipment to work more effectively together due to a greatly increased number of commitments at geographically dispersed locations.
- We are unlikely to deploy on large-scale operations on our own, meaning that our forces and their equipment must be capable of working with other nations.
- We need to exploit technological advancement rapidly with flexible, adaptable systems in order to counter agile adversaries with unorthodox doctrine.
- We need to ensure better value for money to counter years of time and cost overruns with equipment that has failed to meet stakeholder requirements.

This significant shift in future vision, and delivery of this vision, has come as a result of the Strategic Defence Review conducted in 1998, which offered an opportunity to look at the entire defence procurement situation from first principles, and determine how a future process could support the fundamental restructuring of the armed forces in line with the emerging global threat picture. The study was *a foreign policy led strategic defence review to reassess Britain's security interests and defence needs and consider how the roles, missions and capabilities of our armed forces should be adjusted to meet new strategic realities*' (HM Stationery Office, 1998).

The mechanism to achieve the above statement was identified as the application of tools and techniques from the discipline of systems engineering complemented by systems thinking, the former having been developed in its current state by the defence industry after the Second World War, when military technology was at the forefront of many nations' development agendas (Bud and Gummett, 2002). The two, in conjunction, help in the scoping and understanding of problems, with systems thinking providing the more abstract views of systems, and systems engineering focusing on process and management of complexity using activities such as requirements definition to capture stakeholder needs and their transition through to systems that can be designed, developed and tested (Buede, 2000).

In the context of the soldier, the big picture 'system' approach is a shift from the more traditional research-led development, where requirements have generally been driven by the output from previous research, centred on one element of clothing or equipment. An example is personal protection, where changing weapon threats have been investigated in order to determine required changes to body armour materials and design (Couldrick, 2005). Traditionally, optimisation would focus on a specific piece of equipment (in this case body armour), which would potentially be to the detriment of the whole soldier system (i.e. in conjunction with all other clothing and equipment that is worn and carried) leading to integration issues including inability to sight the personal weapon, and helmet obstruction when lying in the prone position (Vang, 1991). Recognition of the importance of wider issues in the design and development process requires that more time is spent at the front end of projects, understanding the real problem to be addressed before considering potential solution options.

The consideration of problems more abstractly, without specifying what the solution will look like from the outset, has led to the creation of capability domains when scoping and creating future systems. Adopted by NATO, they are defined as survivability, mobility, sustainability, C4I (command, control, communications, computing and intelligence) and lethality (NATO LG3, 1999). These groupings try to aggregate many different elements into more nebulous terms, providing a solution-independent approach. For example: 'We want to increase the level of survivability of the individual', rather than 'We want to increase the level of protection afforded by body armour'. In making the first statement, there may be more than one potential way of tackling the problem; for instance one might increase the soldier's speed over ground by reducing the weight that the soldier carries and, in so doing, reduce the time they are a target, making them therefore more survivable.

When considering the soldier and his or her clothing and equipment as an abstract problem, there are a number of tools and techniques within the discipline of systems engineering to enhance and refine our understanding in order that we can write requirements for the future. One of the techniques used to achieve this is mind mapping, or brainstorming, (Rawlinson, 1981) providing the opportunity for a number of subject matter experts to discuss and map different potential influences when describing systems. Figure 1.2 is an example of the different parameters that might be considered when looking at soldier effectiveness in the context of clothing and equipment. Although in no way complete, it provides a high-level view of the wider parameters that will impact upon soldier effectiveness as our key future driver in the context of wider future defence aims. The intent of including this diagram, as with Fig. 1.1, is to express visually the complexity of the environment when the soldier and his or her requirements are

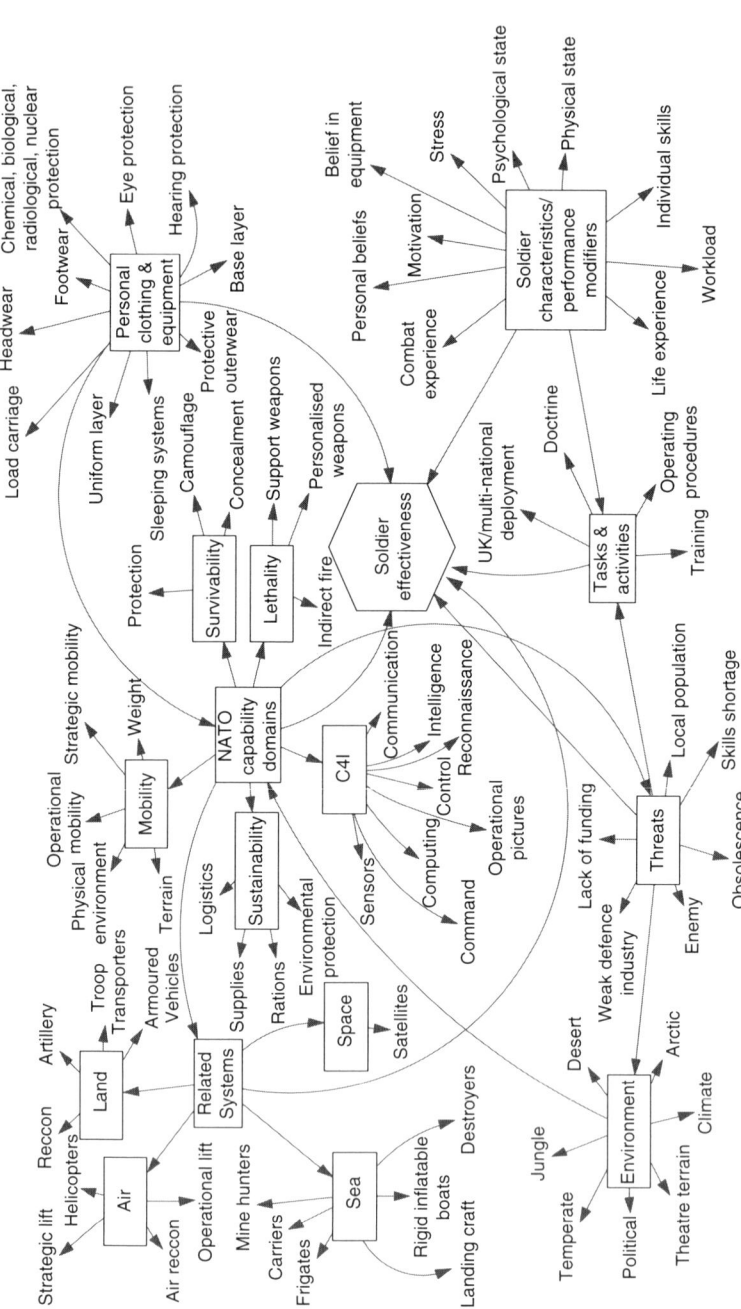

1.2 Soldier effectiveness.

considered more widely. It is no longer just about their clothing and equipment; it is about their personal characteristics, the platforms and equipment with which they interface and the influences of threats, physical environment and politics. Less tangible parameters are identified such as user acceptance, which is critical in the adoption of equipment, as well as parameters that can have performance measures associated with them, for example environmental protection. Each of the areas identified could be expanded to numerous levels of detail, depending on the criticality to success and available resources for modelling. What Fig. 1.2 identifies is that elements outside one's control may impact upon the success of the system. The soldier and his or her attributes is a prime example of a modifier to overall effectiveness and is a specific example of complexity that will be discussed in the next section, due to the impact that it has on future soldier requirements.

1.3   Dynamic complexity: The impact of the human

Complexity arises when there are mutual dependencies and interwoven components, which, as discussed in Section 1.2 increases in likelihood if one is trying to achieve integration. In terms of soldier system requirements, there is complexity due to the number of interfaces that exist with other pieces of clothing and equipment, but also because the human that wears the clothing and equipment is complex in his/her own right. The difficulty with people, whether soldiers or members of any other profession, is their unpredictability; they are dynamically complex (Checkland, 1981). Whereas the performance, or required performance, of a machine can be measured or estimated based on predicted behaviour, with a human this is not possible with the same degree of accuracy (Wilson et al., 2000). It is this fundamental difficulty that on the one hand creates a remarkably adaptable system, and on the other a significant requirements challenge. How does one measure and 'design in' non-tangible characteristics such as those shown in Fig. 1.1, and what are the implications if one ignores them?

The United States has recognised for a number of years that the soldier is the key component within wider battlefield effectiveness, as the soldier's ability impacts upon the use of other systems critical for mission success. MANPRINT (Booher, 1990) looked at the impact of the soldier on the use of other pieces of military hardware and concluded that insufficient consideration had been given to human characteristics within the design cycle. This has led to the failure of a number of highly valuable pieces of equipment, in some cases with catastrophic effect (Wheatley, 1991). Therefore, soldier requirements sit at the very heart of military capability. Designing systems for protection of soldiers, whether from environmental or battlefield threats, whilst considering the tasks and activities that they must undertake and how we can enhance their effectiveness, in turn ensures that their performance can be enhanced as another platform within the bigger defence picture.

Having justified the existence of numerous research organisations and national soldier system programmes, the challenge is in defining the balance of future soldier requirements and the way in which they should be written. Underpinning this activity is a fundamental understanding of dynamic complexity, its impact and its association with the activity of requirements definition. Measurement is a primary function within the domain of systems engineering, and specifically requirements documentation; how can you ascertain if something does what you want it to do if you cannot quantify it? Contracts are written specifying the performance of clothing and equipment, which are legally binding. It is possible to measure Tog values for protection against the cold; it is possible to measure ballistic protection against specified threat systems; it is even possible to measure speed over ground when wearing the clothing and equipment that has been designed if user trials are carried out; but how can one incorporate parameters such as user acceptance with anything other than subjective input? Wearable computers for the soldier are an applied example. This field of technology is under investigation by a number of countries as part of the move towards network enabled forces.

Wearable computers have many potential benefits when used for situational awareness, as well as wider roles such as health monitoring (Jovanov *et al.*, 2001). When writing requirements for this system, it is possible to specify the technical performance in terms of durability and bandwidth for information received, but what about comfort when worn by the individual? Other than subjective scale assessments (Bodine and Gemperle, 2003), how can we determine the perceived benefit of an item to the wearer? It may appear to be a secondary issue, but it cascades to the battlefield where soldiers make decisions not to use pieces of equipment based on their personal perceptions; which in turn may affect their military effectiveness. Acceptance may appear a non-critical issue, but perceived comfort cuts across a myriad of soldier clothing and equipment with numerous cases of soldiers purchasing commercial equipment because it is believed to be superior to that which is issued. This can be detrimental to the individual, as the specification for commercial equipment will be different to that for military equipment, particularly with respect to characteristics such as infra-red signature, but also it may jeopardise that individual, as part of a larger team, which in turn may impact on mission success or failure. All of that from someone not believing that their kit is up to the job! Process application of systems thinking and systems engineering to the problem tries to answer the question on a macro and micro level. Two key questions are considered: whether the right system is being built (from a more abstract perspective) and whether the system has been built right (a more process-driven perspective). The end user of the equipment is more concerned by the former, whereas the customer paying for the equipment is more concerned by the latter. Systems engineering provides a great deal of guidance for requirements capture and management, but, as the next section will discuss, we are still not at the bottom of our soldier requirements' challenge.

1.4   Provision of capability and how to make trade-off decisions

The previous sections have described the changes within defence procurement, shifting from equipment to capability as a consequence of emerging world threats, and providing the justification for application of systems engineering tools and techniques to scope activities such as requirements capture, as a result of the discipline's track record in management of complex inter-related problems. The focus then returns to the challenges of defining soldier requirements caused by the dynamic and complex nature of humans as systems and the difficulty in measuring their characteristics and those of the clothing and equipment that they use.

The crux of future soldier requirements can ultimately be reduced to the two elements within the title of this section and the activities associated with them. Ultimately, the challenge of writing requirements and the balance of those requirements in producing clothing and equipment for the soldier are driven by achieving capability enhancement, and in doing so making trade-off decisions between different system options. This leads us back to the issue of measurement, as stakeholders generally rely on some form of quantification of parameters when making their decisions.

When defining need based on threats, tasks and activities to be carried out and the capability of current military platforms, the stakeholder community provides a wish list that is unlikely to be completely fulfilled. This may be as a result of insufficient funds, or it may be that what they are asking for is not technologically feasible at this time. This leads to the activity of trading off, where one option is chosen over another (Buede, 2004). Often considered as a design level activity (Ashby *et al.*, 2004), it is in fact used at numerous points within the development cycle, from capability to system, to sub-system and component; any time where one statement or performance measure is chosen over another. Systems engineering links requirements to the process of trading off, with measurement of system performance identifying those concepts or options that most closely meet the defined need (Daniels *et al.*, 2001), although there is no standardised method for trade-off activities (Felix, 2004).

At all stages and levels of fidelity within the development cycle, the key is understanding the relationship between decisions made at the top of the chain (capability level) and the impact this will have on the systems produced at the bottom of the chain (physical concepts). The highest capability level is one where generic questions are considered such as: Do we need more mobility, or more survivability? If we have more sustainability, what impact does this have on mobility? At the more specific level we may have questions such as: I have this fabric with this performance and another fabric with a higher level of performance but greater associated cost; which will provide the greatest benefit within the defined constraints?

Requirements choices have to be made based on the constraints of the platform, (in this case the human, as they can only carry so much and have only limited processing power, depending on natural ability and training) as well as on the relationship that the platform has with other related systems within the environment. This reiterates the optimisation issue, where sub-optimisation of components may lead to overall optimisation of the system. Creating behaviour that is greater than the sum of the parts (Smuts, 1973) is central to delivery of capability.

Test and measurement form the foundations of requirements choices and trade-off activities. At some point it is necessary to make decisions about systems and their performance and in doing so we need to provide a body of evidence to support why these decisions are justified, not only to withstand the scrutiny that is associated with expenditure of public funds, but to provide an enduring catalogue of documentation for development of future systems. If and when things change, there is then a clear understanding of previous decisions, their rationale and therefore evidence to decide the most appropriate way forward. When addressing capability, it is difficult to make decisions in the early stages of development as it might not be clear what success looks like. The things that we need to enhance effectiveness may not exist, or may be services rather than physical components and so the performance that they provide cannot be clearly understood. This is where modelling is used to understand different possible futures through creation of virtual worlds where ideas can be tested without a single bullet being fired. In utilising these tools for decision making and trade-off, it is important to remember that models are only as good as the information put into them, with the phrase often used 'rubbish in leads to rubbish out'. They are only representations of the real world, and as such must be relevant to the problem one is trying to answer (Wilson, 1993). Therefore, the assumptions upon which modelling is carried out are vital to the level of confidence that can be associated with the output (Wang, 2001).

Modelling and simulation are used extensively in the UK to scrutinise requirements and trade off different options prior to full-scale development. Although costly to produce, once created, models can provide significant cost reductions over the lifecycle of a project (Cropley and Campbell, 2004) enabling greater and greater levels of detail to be explored without the need for field or laboratory trials. The relevance of modelling to soldier requirements lies in the ability to test human characteristics, with difficulties in representing human attributes due to the aggregative nature of many of the variables. An example is fatigue, which can seriously affect the ability of the individual to carry out their role within a battle. Fatigue is a function of a number of things including hydration, and physical and mental workload, with further modifiers such as fear. Furthermore, fatigue is not universally quantifiable, as individual attributes such as fitness and motivation will affect humans differently. This is in contrast to modelling a weapons system that has a defined rate of fire, and known accuracy and trajectory, with a pretty accurate measure of the mean time between failures. Therefore, how does one model the human with any certainty if they are so unpredictable and complex (Curtis, 1996)? If you cannot model what the requirements are to enhance soldier effectiveness, how do you know what performance soldier clothing and equipment should have and therefore which of a number of options is most appropriate? Doing nothing is not an option!

Laboratory testing of human attributes creates both positive and negative implications for modelling activities. Empirical data create a body of evidence that can enhance validity of assumptions but, conversely, can create issues when trying to aggregate output. This links back to the reductionist nature of scientific testing to ensure that cause and effect can be attributed (Okasha, 2002), but in breaking down the problem to such a low level of detail there is a tendency to lose the type of behaviour that is exhibited by the dynamic complexity of the system. Because the procurement stakeholders are interested in gross measures of effectiveness such as mission success or failure as indicators of system suitability, it becomes difficult to aggregate or in some instances extrapolate information that has been generated in a laboratory as there is no empirical evidence to support it, with the result that confidence in output is reduced. Therefore, the techniques considered for requirements and trade-off activities of dynamically complex systems need to bring together multiple strands of information, in both quantitative and qualitative form, to provide a clear audit trail of decisions made and the ability to look at the impact of changes at varying levels of detail with confidence in the data quality (Pipino et al., 2002). Work has been carried out on future soldier requirements and the fusion of quantitative and qualitative methods in derivation of soldier systems, specifically focused on clothing and equipment (Sparks, 2006). The approach, although directly applied within the UK defence domain, is still in its infancy, with further development required through application to future systems development. What it provides is a generic process for achieving the elements discussed within the sections above, linking high-level capability needs through to lower-level design issues and underpinning data from diverse sources in order that subjective and objective considerations are incorporated in the prioritisation of research and physical concept generation. Rather than ignoring subjective or intangible parameters, it is recognised that without due consideration, overall success can be significantly impacted (Booher, 1990). Just as modelling and measurement can be based on assumptions, fused data can apply more rigorous techniques for expression of uncertainty (Grainger, 1997) providing the data integrity to satisfy the formalised systems engineering processes.

Often intangible benefits, not quantitative input, drive the final decision on a system, as what will or will not be developed is decided by people, exhibiting all of the dynamic complexity and unpredictability that has been discussed.

1.5   Summary

The domain of defence is changing and, with it, the approach that is taken in the derivation of requirements for development and procurement of future systems. Straight replacement of equipment has been superseded by provision of capability, creating a need to understand the impact of interactions between numerous parameters. Dependencies of this nature increase the level of complexity, and create problems, the solution of which may include areas outside of our direct control. Increased complexity heightens uncertainty, which in turn impacts on validation and verification of whether the right system has been built and subsequently whether it has been built right. Dynamic complexity exhibits the highest level of uncertainty due to the unpredictable and potentially aggregated effect of variables, with the human as a key example.

Time, cost and performance drivers, coupled with an emerging threat picture and increased inter-dependencies between systems, have led to the adoption of systems engineering as a discipline that will provide the necessary tools and techniques to ensure future procurement success. Central to its application are the activities of requirements generation and trade-off, with modelling and measurement underpinning the analysis of data.

The soldier and the requirements pertaining to his or her clothing and equipment exhibit dynamic complexity and, as such, are difficult to associate with quantifiable measures. The solution to this problem is either to carry on with development of equipment in isolation, which may or may not enhance effectiveness, or to suggest ways of introducing more intangible parameters into the requirements and trade-off activities whilst managing risk and uncertainty. Although the first option reduces risk in 'building the system right', it may significantly impact upon whether the 'right system has been built'. As this latter question is central to the premise of capability, it would appear that the second option is better. With this in mind, development of generic processes to fuse objective and subjective data enables researchers and developers to consider the soldier and the required capability at a number of levels of resolution, cascading relationships from high-level defence doctrine through to detailed design (Sparks, 2006). Of central importance to this process are the views of the stakeholders, particularly the customers, whose decisions will shape final concept generation based on the information presented to them.

In essence, future requirements include art, science, engineering and an element of gut feeling. The art provides the abstract system views, the science provides the analysis, the engineering provides the technology, but it is the people that put it all together and make it work.

1.6   References

ARMSTRONG, D. and GOLDSTEIN, E. (1990), *The End of the Cold War*, Frank Cass, London.

ASHBY, P., IREMONGER, M. and GOTTS, P. (2004), The Trade-off Between Protection and Performance for Dismounted Infantry in the Assault. *Proceedings of the Personal Armour Systems Symposium 2004*, The Hague, The Netherlands, 6–10 September.

BODINE, K. and GEMPERLE, F. (2003), Effects of Functionality on Perceived Comfort of Wearables, *Proceedings of the Seventh IEEE International Symposium on Wearable Computers*, White Plains, New York, 21–23 October.

BOOHER, H. R. (1990), *MANPRINT: An Approach to Systems Integration*, Van Nostrand Reinhold, New York.

BUD, R. and GUMMETT, P. (eds.) (2002), *Cold War, Hot Science: Applied Research in Britain's Defence Laboratories 1945–1990*, NMSI Trading Ltd, Science Museum, London.

BUEDE, D. (2000), *The Engineering Design of Systems. Models and Methods*, John Wiley & Sons, Chichester, UK.

BUEDE, D. (2004), On Trade Studies. *Managing Complexity and Change! INCOSE 2004 – 14th Annual International Symposium Proceedings*, Toulouse. 20–24 June.

CHECKLAND, P. (1981), *Systems Thinking, Systems Practice*, John Wiley and Sons, Chichester, UK.

Controller and Auditor General (2003), *Through Life Management*, HM Stationary Office. HC 698 Session 2002–2003.

COULDRICK, C. (2005), *Assessment of Personal Armour Using CASPER*, Cranfield University, Shrivenham. DCMT/ESD/CAC/1151/05.

CROPLEY, D. and CAMPBELL, P. (2004), The Role of Modelling and Simulation in Military and Systems Engineering, *Systems Engineering Test and Evaluation Conference, SETE 2004*, Adelaide, 8–10 November.

CURTIS, N. (1996), Possible Methodologies for Analysis of the Soldier Combat System: Operations Research Support to Project Wundurra. *DSTO-TR-0148*.

DANIELS, J., WERNER, P. and BAHILL, T. (2001), Quantitative Methods for Trade Off Analysis, *Systems Engineering*, 4 (3) pp. 190–212.

Director Infantry (2000), *Future Infantry . . . The Route to 2020.* 118/00/00.

FELIX, A. (2004), Standard Approach to Trade Studies. *Managing Complexity and Change! INCOSE 2004 – 14th Annual International Symposium Proceedings*, Toulouse. 20–24 June.

FORSBERG, K. and MOOZ, H. (1992), The Relationship of Systems Engineering to the Project Cycle, *Engineering Management Journal*, 4 (3).

GRAINGER, P. (1997), Principles of Cost Effectiveness Analysis, *Journal of Defence Science*, 2 (4).

HM Stationary Office (1998), *Strategic Defence Review*. CM3999.

JOVANOV, E., RASKOVIC, D., PRICE, J., CHAPMAN, J., MOORE, A. and KRISHNAMURTHY, A. (2001), Patient Monitoring Using Personal Area Networks of Wireless Intelligent Sensors, *Proceedings 38th Annual Rocky Mountain Bio-engineering Symposium*, Copper Mountain, CO, April.

NATO LG3 (1999), *NATO Measurement Framework*. WG3.

OKASHA, S. (2002), *Philosophy of Science: A Very Short Introduction*, Oxford University Press, Oxford, UK.

PIPINO, L., LEE, Y. and WANG, R. (2002), Data Quality Assessment, *Communications of the ACM,* 45 (4), pp. 211–218.

RAWLINSON, J. (1981), *Creative Thinking and Brainstorming*, Gower, London.

Secretary of State for Defence (2005), *Network Enabled Capability*, HM Stationery Office. JSP 777.

SMUTS, J. (1973), *Holism and Evolution* (reprint), Greenwood Press, Connecticut.

SOMMERVILLE, I. and SAWYER, P. (1997), Viewpoints: Principles, Problems and a Practical Approach to Requirements Engineering, *Annals of Software Engineering*, 3, pp. 101–130.

SPARKS, E. (2006), *From Capability to Concept: Fusion of Systems Analysis Techniques for Derivation of Future Soldier Systems*, PhD thesis, Cranfield University, Engineering Systems Department, Shrivenham.

STEVENS, R., BROOK, P., JACKSON, K. and ARNOLD, S. (1998), *Systems Engineering: Coping with Complexity*, Prentice Hall. 130950858.

VANG, L. (1991), *Handbook on Clothing. Biomedical Effects of Military Clothing and Equipment Systems*. Panel 8 on the Defence Applications of Human and Bio-medical Sciences.

WANG, C. (2001), *Measuring the Quality of Mission Oriented Research*, Airframes and Engines Division and Aeronautical and Maritime Research Laboratory. DSTO-GD-0276.

WHEATLEY, E. (1991), *MANPRINT: Human Factors in Land Systems Procurement*. Army Staff Duties.

WILSON, A., BUNTING, A. and WHEATLEY, A. (2000), *FIST Technology Options and Infantry Performance*. DERA/CHS/PPD/TR000151.

WILSON, B. (1993), *Systems: Concepts, Methodologies and Applications*, John Wiley & Sons, Chichester, UK.

2

Non-woven fabrics for military applications

G. A. THOMAS, Auburn University, USA

2.1   Introduction

Humans have used forms of protective armor in combat for at least five millennia. At first animal skins and furs were the only protection both in combat and in cold weather. Ancient civilizations used leather as a form of protection beginning in roughly 3000 BC. The use of leather has continued as a means of various types of body protection. Some 700 years later, ancient cultures such as those in Egypt learned to alter leather by boiling and tanning it. Leather was very effective in warding off blows from bludgeoning weapons and can be found serving this role in some cultures and subcultures up to the present day.[1]

The first fabricated weapons of note in warfare were swords and spears, so more advanced armor was at first designed specifically to address these threats. The Egyptians were using armor to protect from slashing and cutting weapons as early as 1500 BC. The first forms of armor were probably cloth garments with bronze scales or plates sewn mounted on them. The Assyrians apparently developed lamellar armor between 900 and 600 BC by mounting small rectangular plates upon a garment in parallel rows. Later, the Greeks made armor from bronze plates that not only fitted over the individual parts of the body, but were shaped to fit over the part of the body where it would be carried. Chain mail seems to have been invented by the Celts in Europe, but it was quickly adopted by the Romans and many subsequent civilizations afterward.[1]

By the end of the sixteenth century and with the advent of firearms, armor had to withstand and absorb impact from large caliber projectiles. The weight of armor increased up to about 50 kg, which was a burden on the wearer. The leather garment originally created to be worn under armor was used alone, because it gave the wearer mobility. A debate began then about what was more important, optimum protection or comfort and mobility.

> As early as the 14th century, armor was given a proof rating which guaranteed its protective qualities against weapons of the time. By the 17th century, ballistic testing was required for proofing protective gear. Some surviving armor shows marks of ballistic testing.
>
> During all of the armor developments of the ancient and medieval cultures, the greatest threats to soldiers and their armor were the ballistic weapons. Of these, the bow and the crossbow first posed the most dangerous challenges to survival on the battlefield.
>
> Standard bows were able to penetrate many armors at ranges of 30–50 yards in early warfare, but the wooden or metal overlaid wooden shield was able to effectively defeat most of these weapons. The Celts apparently were the military technologists who again changed warfare by introducing the longbow by the 13th century A.D. This devastating projectile weapon was, in a sense, the first hint of the effectiveness of the later repeating rifles on battlefields. The longbow could put up to six arrows in the air simultaneously and accurately at targets 200 yards away before the first arrow in the volley hit. In continental Europe, crossbows became so effective against armor that the Church actually banned their use in warfare for a time.
>
> Eventually, armor for nobles became thick and heavy enough to withstand most hits by even longbows or crossbows, so a further development in lethality was needed. This step came in the form of the gun.
>
> Guns and gunpowder were introduced to Europe from China, where such weapons were in widespread use by the 12th century. Early guns were no more effective against royal armor than bows, but they eventually became powerful enough to render the use of any armor of the times ineffective. Thus it seemed that the struggle between weapons and armor had been won by the weapons until the reappearance of a new and practical concept in the Second World War.[2]

2.1.1 Modern armor

The British Royal Air Force and the US Army Air Corps created and issued protective vests to flight personnel beginning early in the Second World War. These early ballistic resistant armors were known as 'flak' jackets because German Anti-Aircraft Artillery was known as FLAK (Fliegerabwehrkanonen). Thus, flak jackets are ballistic-resistant garments intended solely for the purpose of defending a body from shrapnel, or explosion fragments, and not from bullets. These first flak vests contained steel plates carried in multiple plies of nylon fabric that protected against relatively low velocity shrapnel.[3]

During the period of the 1950s through early 1960s, the various military branches began to define levels of protection they believed would represent the real threats to service personnel from combat weapons.[4] (See Fig. 2.1)

Early Military Standards

◆ US Army
— range = 5 feet
— witness plate 6 inches behind armor target
— penetration of armor + plate = fail
— no plate penetration = pass
— determine max velocity at which pass occurs

◆ US Navy
— range = 5 feet
— no witness plate
— target penetration = fail
— no penetration by a projectile = pass
— fragment penetration without projectile penetration = pass

2.1 Test protocols from early military standard determinations.

2.1.2 Scientific armor studies begin

By the Vietnam War, combat infantrymen were wearing ceramic and/or ballistic nylon vests to protect themselves against both fragment and lower speed projectile threats.

Today it is common practice for both combat personnel and military police to use ballistic protection fabrics and plates to defend themselves against fragments and some small arms threats. The military standards which were used to rate the effectiveness of these materials varied according to end use and even according to the military service branch which was testing them, but in general, the stopping power of the material was evaluated based on its ability to completely stop a penetrating projectile (see Fig. 2.1). Some military standards also evaluated the material deformation and target deformation after impact.

Despite its obvious lack of sophistication by present standards, it quickly became apparent in such testing that no material or combination of materials could withstand the entire spectrum of ballistic objects or magnitudes of velocity of such objects and remain intact or protect the wearer/user.

The most common major standards for civilian and police ballistic threats that are used by the market's suppliers of fabrics and fibers to compare performance of products are those in the USA and in the European Union. The US Standard is from the National Institute of Justice (NIJ) and identifies four levels of threat plus two subparts. These levels range from rather low velocity or low mass projectiles at Level I to very high velocity, high mass projectiles at Level IV. The NIJ standard in current use is 0101.04, although the older 0101.03 standard can still be found in application for body armors produced when that part was in effect and will be in use until their lifespan has been exceeded (see Tables 2.1 and 2.2).

In both of these NIJ standards, armor is tested using Roma Plastilina #1 modeling clay as a test backing to determine how much impact is

Table 2.1 NIJ Standard 0101.03 for protection classes

Threat level	Caliber	Projectile description	Mass (g)	Velocity (m/s)
I	.22 Long rifle	Lead	2.6	320
I	.38 Special	Rounded, lead	10.2	259
IIA	9 mm	Full metal jacket	8.0	332
IIA	.357 Magnum	Jacketed soft point	10.2	381
II	9 mm	Full metal jacket	8.0	358
II	.357 Magnum	Jacketed soft point	10.2	425
IIIA	9 mm	Full metal jacket	8.0	426
IIIA	.44 Magnum	Lead semi-wadcutter	15.55	426
III	7.62 mm Winchester	Full metal jacket	9.7	838
IV	.30–06	Armor piercing	10.8	868

Table 2.2 NIJ 0101.04 (http://www.nlectc.org/pdffiles/0101.04RevA.pdf)

Threat level	Caliber	Projectile description	Weight g (gr)	Velocity m/s (ft/s)
I	.22 Long rifle	Lead	2.6 (40)	329 (1080)
I	.380 ACP	Full metal jacket	6.2 (95)	322 (1055)
IIA	9 mm	Full metal jacket	8.0 (124)	341 (1120)
IIA	.40 S&W	Full metal jacket	11.7 (180)	322 (1055)
II	9 mm	Full metal jacket	8.0 (124)	367 (1205)
II	.357 Magnum	Jacketed soft point	10.2 (158)	436 (1430)
IIIA	9 mm	Full metal jacket	8.0 (124)	436 (1430)
IIIA	.44 Magnum	Jacketed hollow point	15.6 (240)	436 (1430)
III	7.62 mm NATO	Full metal jacket	9.6 (148)	847 (2780)
IV	.30–06	Armor piercing	10.8 (166)	878 (2880)

transferred to the body after the bullet is stopped. The US standard is 44 mm (1.73 inches) of indenting into the clay after bullet stop (Fig. 2.2).

The various classes within the standard represent the energy threats and penetration power of various bullets and bullet types. If armor is present, the total energy a bullet delivers to its target is not as important as how well it penetrates the target. The smaller the bullet, the more energy per square inch (or square centimeter), and the greater the penetrating power exists (Fig. 2.3).

Scheme of recommended target strikes

- Level I, IIA, II and IIIA require two shots at 30 degrees and four at 90 degrees
- Level III ('high powered' rifle tests) require six shots at 90°
- Level IV (armor piercing rifle) tests require one shot at 90°
- All targets are tested for deformation against Roma Plastilena modeling clay backing, 24″ × 24″ × 4″ in dimension

Exhibit 5: Test ammunition shot series

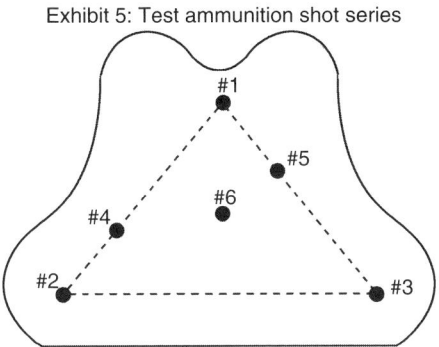

2.2 Recommended ballistic testing procedure for National Institute of Justice standards. Graphic courtesy of National Institute of Justice (NIJ Standard 0101.03, p. 10, method 'A').

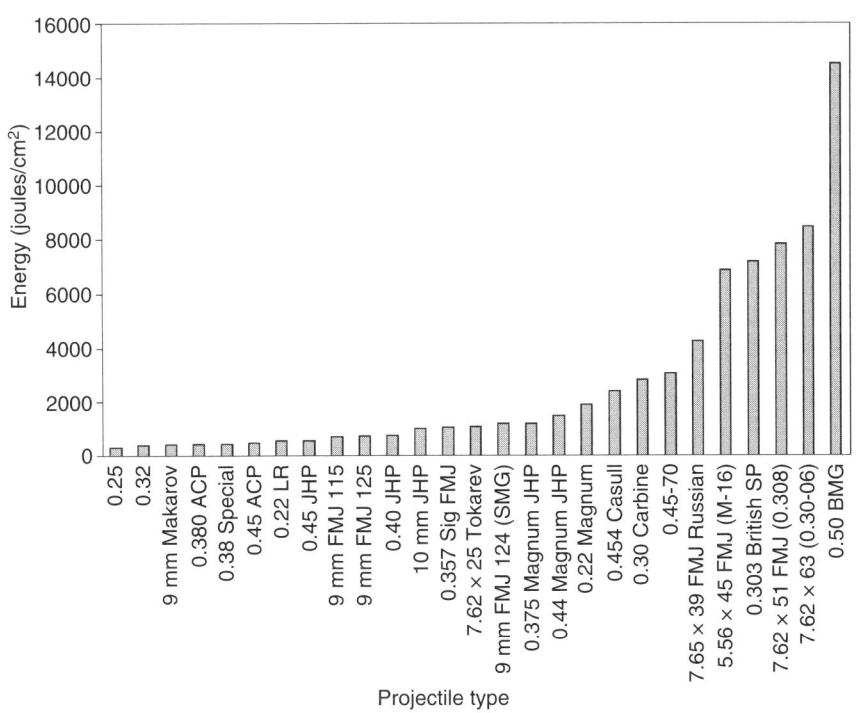

2.3 Energy delivered to a target by various ammunitions.

When blunt tipped, hollow point or lead nose bullets are the projectile threat, the energy is released much more quickly than when ballistically optimized bullets are present. Metal jacketed bullets stay together longer and penetrate farther than soft lead or hollow point bullets do, therefore they are a greater threat to ballistic resistant armors. Large quantities of energy, released quickly onto an armor that successfully stops the bullet, can still be very serious unless the armor system can absorb the energy of impact.

As an example of what this means, the following illustration is offered: if a police officer is on duty and a criminal shoots at him/her, the officer wants the best possible armor to stop the penetration of the bullet first. After that, the officer wants low impact to the body from the bullet's energy when it is stopped. Some highly touted bullet types, like the .38 Special JHP, the .45 ACP, and the .40 S&W deliver a lot of energy quickly into a soft target (a body) to bring it down. For this very same reason, they are very ineffective against body armor, and they are the easiest bullets to stop with modern armor. On the other hand, the 9 mm FMJ, the .357 magnum JHP and the .44 magnum JHP/SJHP are very dangerous. The magnum rounds deliver penetrating power followed instantly by a massive blow even if the bullet is stopped.

Worse yet for blunt impact force than the magnum handguns are the shotguns. Although most soft body armor above Level IIA can stop the pellets, or even a slug, the energy of impact from a slug or from 00 or 000 buckshot can still permanently injure or kill the wearer. For this reason, new efforts are being made to reduce the impact from weapons after bullet termination.

Bullets like the 7.62×25 mm and the 5.7 mm FN are very small, but they can go through most soft body armor without problem. They have what may be described as a 'high energy density', that is, high velocity, notable mass and a very small area of impact into which they concentrate all their deadly penetrative energy. These weapons are far more dangerous than the slower, thicker .45 ACP or .40 S&W projectiles for this reason. Rifles above .22 magnum caliber require rigid, or 'hard', armor. Such armor types may consist of either metal, ceramic, pressed hard plastic or combinations thereof to stop anything from a .30 caliber M-1 carbine, .30–30 rifle or more energetic projectiles. The term 'bulletproof' has been discarded by both armor testers and armor producers in favor of the more descriptive term 'ballistic resistant' shortly after a rational testing scheme for these materials was adapted.

The grim reality of the race between protection and lethality is that no matter how assiduously the designer attempts to protect a user from death and injury, there is always something that can deliver a fatal or disabling wound through any given armor. Until humans so radically change their nature that they cease their desire to murder or maim their fellow creatures, this will remain true.

Categories of military armor

Ballistic-resistant materials for military purposes presently fall into three general categories:

1. garments, such as vests.
2. helmets.
3. vehicle and structural reinforcement.

Ballistic-resistant vests, jackets, and similar garments are often mainly for protection against shrapnel and bomb fragments. Protection from military caliber small arms is quite challenging in most cases because of the high velocities, low aspect ratios and hard surfaces of the projectiles. Although such high-level protection is vital, it is cumbersome for long-term use in field situations.

Law enforcement armor needs

Police protective equipment is usually designed for handgun threats and sharp instrument threats such as one would encounter from ordinary criminals. Higher level protection is available for protection from more organized criminal threats, terrorism and riots, but it is not normally issued for daily use. Police equipment is ideally designed for constant use against the most commonly expected threat.

Police departments usually have to rely on city budget managers, city councils and mayors to receive whatever protective products they can get, and most such people are not sufficiently educated about ballistic protection to decide these life and death issues. The real dangers of daily situations in the life of a law enforcement officer are poorly understood by buyers, the press and the public. Even the end users are often ill-informed about what protective materials can and cannot do.

It seems appropriate, therefore, to discuss what levels of protections are provided by various products and categories, and how the products are defined for specific end-uses.

2.2   Protective materials, devices and end-use requirements

All ballistic resistant materials have certain common characteristics. The use of polymer materials has made the protection to weight ratio very favorable for their use over metals or ceramics. Lower weight also permits greater mobility and better capability for police or military personnel to perform their assignments with reduced threats from attackers.

In addition to the desired characteristic of low weight, there are also important demands for flexibility and thermal transport. Stiff, inflexible ballistic garments inhibit performance even at low weight. Garments or materials that trap body heat and moisture are unpleasant for intended wearers and are cited as one of the main reasons such garments are not worn in the line of duty.

2.2.1 Conventional approaches

Regardless of any individual fiber capabilities, all fibers must be formed into a structure to be useful as armor. Conventional devices for protecting police and military personnel from ballistic threats are now at least peripherally known in both the professional and the civilian community, albeit within previously discussed boundaries of understanding. It is still not uncommon for both the users of these products and the news media to refer to such products as 'bulletproof vests' or even 'Kevlar® vests'. Most of those who lightly use these phrases do not know the material is not universally 'bulletproof', nor do they realize that not all such materials are made of Kevlar®, a fiber produced only by DuPont.

If an expert were to tell the lay person that flexible body and structural armor products are actually textiles, they would often be met with astonishment and even disbelief. Yet all but a few such products are made of fiber, and anything produced from or with fiber is a textile. Most of the products designed to protect the wearer from ballistic threats are now made of woven filament materials produced by technologies that originated in far ancient times. Other, newer, types of products are also appearing both on the market and in research labs that bypass the ancient techniques of weaving, are faster to produce and offer unique capabilities that woven materials do not have.

Of the significant technologies available for consideration – weaving, knitting, non-wovens and resin fortified, filament lay-up composites – only knitting seems to be inappropriate for use in the ballistic resistant materials area at present.

Weaving is by definition the interlacing of at least two sets of yarns with each other and conventionally at approximately right angles to each other. For the weaving process to occur, the set of warp yarns must be parted in some desired order for a pattern, weft must be inserted through the opening, the warp yarns must exchange positions, trapping the weft between them, and the weft must be pushed into place in the cloth. Once these operations have been performed, there is a fabric which has been manufactured on the loom. This fabric must be taken away from the loom and more unwoven yarn moved forward to make more fabric as a result.

The style specifications describe a desired look or function of a fabric. What they mean is how the fabric should be made.

'End' is the common mill expression for a warp yarn in a woven fabric. In the USA, textile specifications are still given in avoirdupois units (inches, pounds, etc.). Ends per inch (EPI for short) refers to the warp yarns per inch in the fabric off the loom. 'Pick' is yet another term for weft or filling, but it applies to weft yarns in the fabric after it has been woven. Picks per inch are normally abbreviated PPI.

All weaving processes have certain characteristics in common and all require certain processing steps. Yarn is the basic building component of woven fabric structure. Yarns must be prepared for presentation to the weaving machine at least in so far as requiring an assembly of some useful length and organization of the yarns are concerned. All weaving processes require at least two different sets of yarns for the process to be accomplished, and all present weaving processes need to have one set of yarns presented simultaneously to the weaving machine.

2.2.2 Fiber components

Almost all ballistic resistant structures require the use of yarns rather than fibers as their primary components. Yarn is the correct textile term for unitary or conglomerate assemblies of fiber materials which are used to make fabrics by weaving. It is not sufficient simply to state that yarns are the basic product materials which compose woven or knitted goods. Modern textile manufacturing has offered the weaver a choice among types of yarns which could be applied to the production of a fabric simply by virtue of several distinct yarn production methods. These methods are not free from consideration of the fiber material to be applied, but the production methods themselves do determine subsequent processing steps which are required.

Yarns in a fabric can be described in several ways, most of which depend on the type of fibers which compose the yarns. All methods used for yarn size descriptions use ratios of mass (or weight) and length.

There are many forms of yarn counts which exist in textile science. These include direct yarn counts (mass/unit length) and indirect (length/unit weight or mass). In the synthetic yarns industry such as is encountered in ballistic resistant armor, direct yarn counts are preferred. The most common direct yarn counts are:

- Denier, the number of grams of mass in a yarn per 9000 meters, is the measure used by man-made fiber producers to describe their products.

• Tex, the number of grams of mass in a yarn per 1000 meters, is a measure employed by the scientific community in textiles.

2.2.3 Unconventional non-wovens approaches

Needle-punching is a simpler operation than weaving by which a variety of properties can be obtained in the fabric by varying the process components. Continuous ballistic fibers are chopped into smaller fibers, carded and (usually) randomly oriented by cross-lapping to form an isotropic mat or sheet. This sheet is subsequently consolidated by a set of barbed needles. The needles push a limited amount of fibers at 90° through the sheet of randomly oriented fiber felt. The felt material engages fragments much better than traditional woven fabrics.

A 1966 US Department of Defense study found that a needle-punched structure containing ballistic resistant nylon could be produced at one third the weight of a woven duck fabric while retaining 80% of its ballistic resistance.[5] The process is still being used with success today in special applications.

2.3   Proper selection of fibers

Nylon became the ballistic resistant fiber of choice (i.e. 'ballistic nylon') for many years because it had a high strength-to-weight ratio and could be fashioned in sufficient layers to capture shrapnel fragments from some explosive projectiles and devices.

According to one source,[4]

> Reports received by the Office of the Surgeon General of the Army on the combat testing of the new Army nylon vest showed that the armor deflected approximately 65 per cent of all types of missiles, 75 per cent of all fragments, and 25 per cent of all small-arms fire. The reports also stated that the armor reduced torso wounds by 60 to 70 per cent, while those inflicted in spite of the armor's protection were reduced in severity by 25 to 35 per cent.

As polymer science progressed, fibers such as high tenacity polyamides, aramids, and linear, high density polyethylene (HPPE) were developed for ballistic resistant applications. The protection offered per unit weight of the material increased greatly. Such structures provide higher comfort, and less conspicuous means of providing protection against a ballistic threat. Ballistic nylons are no longer used because modern fibers offer superior performance.

2.3.1 Aramid types

Aramid fibers are condensation polymers belonging to the polyamide family of fibers, but their amide links are formed at aromatic ring structures

2.4 Chemical structure of para-aramid fibers (Stouffer, J., http://web.umr.edu/~wlf/Synthesis/kevlar.html).

(Fig. 2.4). This chemistry allows the fiber to form very rigid, long chain structures with high modulus, high tensile strength and high temperature resistance. Unlike nylons, aramid fibers are not thermoplastic and must be solution spun into sulfuric acid or similar oxidative solvents for formation.

Two typical aramids used in ballistic-resistant fabrics are DuPont Kevlar® and Teijin–Twaron®. DuPont introduced Kevlar® 29 aramid in the early 1970s for vests and helmets. This fiber's name has become synonymous with ballistic-resistant material in the popular media. Kevlar® 129 was introduced in the late 1980s and was offered in smaller denier per filament for increased flexibility and comfort. It was designed to defeat rounds such as the 9 mm full metal jacket (FMJ) handgun projectile.

The most current Kevlar® fiber for military use in both fragment and bullet defeat roles is KM2. This venerable contender in the military armor role is the preferred type for use in the US military's 'Interceptor' body armor.

Teijin–Twaron produces several types of Twaron® for ballistic-resistant garments. The first generation, Twaron® Standard, was introduced in 1986. The latest generation of Twaron® is CT Microfilament. This product contains up to 50% more individual filaments than other equivalent weight aramid yarns. The 930 dtex Twaron® CT Microfilament yarn has a 1000 filament content. The result of this new technology is a weight reduction of 41% from Twaron® standard with equivalent performance.

2.3.2 Linear polyethylene types

A totally different technology from aramid fibers is used to produce the extremely lightweight polyethylene ballistic-resistant fibers. Polyethylene is an additive polymer, which requires a special withdrawal procedure called gel spinning for its formation as a ballistic-resistant material (Fig. 2.5). The fibers have extremely linear molecular chains, resulting in very high parallel orientation and crystallinity. This fiber type has very low specific gravity and tensile strength 15 times greater than steel. This family of fibers includes the Dyneema® products from DSM and the Spectra® products from

Polyethylene 'mer' formula

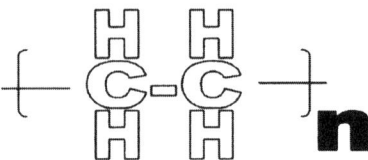

Very long chains, high crystallinity

2.5 Chemical structure of HPPE/ECPE/UHMWPE fibers.

2.6 Poly(*p*-phenylene-2,6-benzobisoxazole), or PBO structure (Toyobo Company, Ltd., http://www.toyobo.co.jp/e/seihin/kc/pbo).

Honeywell. They are variously known as high performance polyethylenes (HPPE), extended chain polyethylenes (ECPE) or ultra-high molecular weight polyethylenes (UHMWPE).

One important concern in the use of polyethylene fiber in high temperature environments is its sensitive thermoplastic nature. Tests by both Honeywell and DSM have shown little influence on the fiber performance in room temperature conditions after they were stored at elevated temperatures.

2.3.3 PBO types

One of the more newsworthy candidates in the ballistic-resistant fibers market is PBO. This fiber is marketed by Toyobo of Japan under the trade name 'Zylon'. PBO is the abbreviation for poly(*p*-phenylene-2,6-benzobisoxazole), a rigid-rod, isotropic, crystal polymer (Fig. 2.6).

Data from Toyobo indicates that the tensile modulus of PBO is greater than carbon, HPPE or aramid fiber types. The fiber is chemically more similar to aramid than to HPPE and therefore has great resistance to heat. Its specific gravity is higher than HPPE, however, so the sonic modulus of the fiber is lower than the linear polyethylenes.

2.3.4 Liquid crystal polymers

Vectran is a high-performance thermoplastic multifilament yarn spun from Vectran® liquid crystal polymer (LCP). Vectran® is the only commercially

2.7 Chemical structure of M5, PIPD fiber (Magellan Systems International, LLC, http://www.m5fiber.com/magellan/m5_fiber.htm).

available melt spun LCP fiber in the world. It is not yet a player in the ballistic-resistant fibers market, but modifications to this fiber may permit it to become a contender in the future.

2.3.5 M5 fiber

PIPD or poly{2,6-diimidazo[4,5-b4′,5′-e]pyridinylene-1,4(2,5-dihydroxy)-phenylene} is a much anticipated and apparent likely contender in the ballistic protection market (Fig. 2.7). The fiber is being developed and marketed by Magellan Systems International, but it is not yet commercially available. Tests by the US Army at the Natick Soldier Center labs have indicated a very promising likelihood of success with this new high-strength polymer.

2.4   Variations of fiber forms

The characteristics of any fabric or fiber-based material structure are most dependent at the outset on whether yarns are continuous filament or staple fiber types (Figs 2.8 and 2.9). The two varieties are easily distinguished by the length of fibers which make up the yarns. In continuous filament yarns, each individual fiber has a length equal to that of the entire yarn being processed. With the exception of silk, all yarns of this type are man-made. Interestingly, silk is the only natural fiber that has been successfully used in forms of ballistic-resistant armor.

The man-made yarns may be further distinguished between regenerated types such as rayon, acetate, glass, etc., or purely synthetic types including polyesters, polyamides, polyolefins, etc. (Fig. 2.10). In all cases, the continuous filament yarns are delivered wound in very great lengths onto a surface such as a tube or a spool.

Staple fiber yarns have measurable, discrete lengths and are easily recognizable as shorter than filaments. They are the common types of fibers we have been accustomed to seeing from our youth such as cotton, wool and pillow or quilt battings of synthetic fibers. Although the synthetics and regenerated fibers are produced in continuous filament form as either yarns

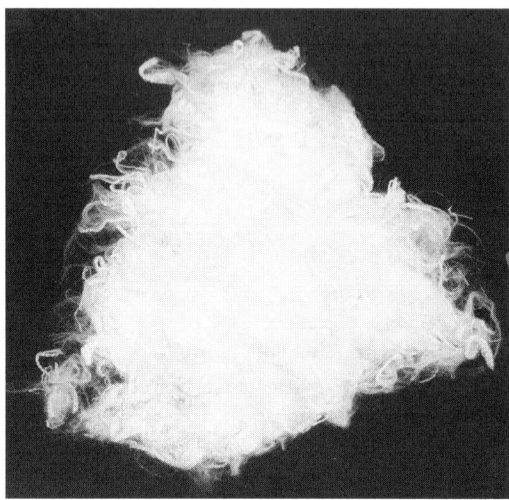

2.8 Aramid fiber in staple form (photo by the author).

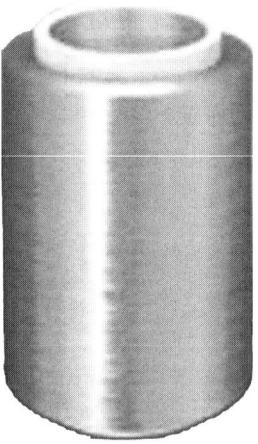

2.9 Continuous filament form (Toyobo Company, Ltd., http://www.toyobo.co.jp/e/seihin/kc/pbo).

or tow, they can be cut into determinate discrete lengths as required by a manufacturer of fiber-based goods.

2.4.1 Methods of creating non-wovens

Although numerous methods have existed for decades to produce fiber-based material structures within the broad category known as 'non-wovens',

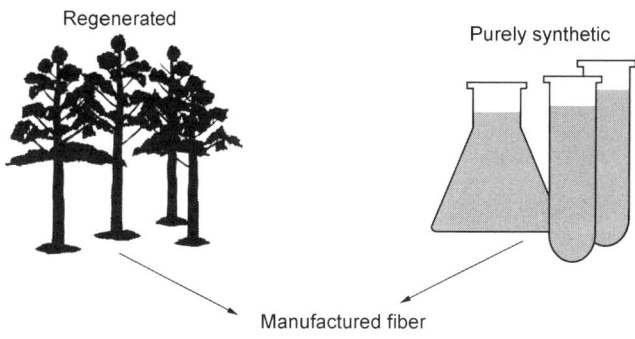

Regenerated

Purely synthetic

Manufactured fiber

2.10 Manufactured (man-made or artificial) fiber sources.

not all of these are of practical use for ballistic-resistant structures. Indeed the definition of a non-woven is in itself a difficulty, since there is disagreement among professionals about what constitutes a member of this category of fabric.

Certainly wovens are not non-wovens, but are non-woven felts needled into woven fabrics both or neither? Certainly knits are not wovens, yet they are not non-wovens. And if a knit incorporates non-woven into it, does it become a non-woven? While such questions are comical, they are also the subject of serious debate because large corporate investments in marketing and customer outreach depend on what at first appears to be a trivial and fun semantic.

INDA, the Association of the Nonwoven Fabrics Industry, should perhaps wield some considerable authority in this arena to help define what a non-woven is. According to INDA's *The Nonwovens Handbook*,[6] 'Nonwoven fabrics are flat, porous sheets that are made directly from separate fibers or from molten plastic or from plastic film. They are not made by weaving or knitting and do not require converting of fibers to yarn'. Even with this definition, some experts disagree with the restrictions inherent in the wording.

For the sake of convenience, a ballistic-resistant non-woven structure is defined herein as one that is fiber based, not exclusively woven, not exclusively knitted and not exclusively a fiber–matrix composite in construction. But some will disagree.

2.4.2 Filament

In conventional ballistic-resistant structures, filament yarns are used to absorb projectile impact force. The logic behind the use of filaments is to present a network of high-modulus, high-strength fiber structure components that individually extend the entire breadth or length of the structure

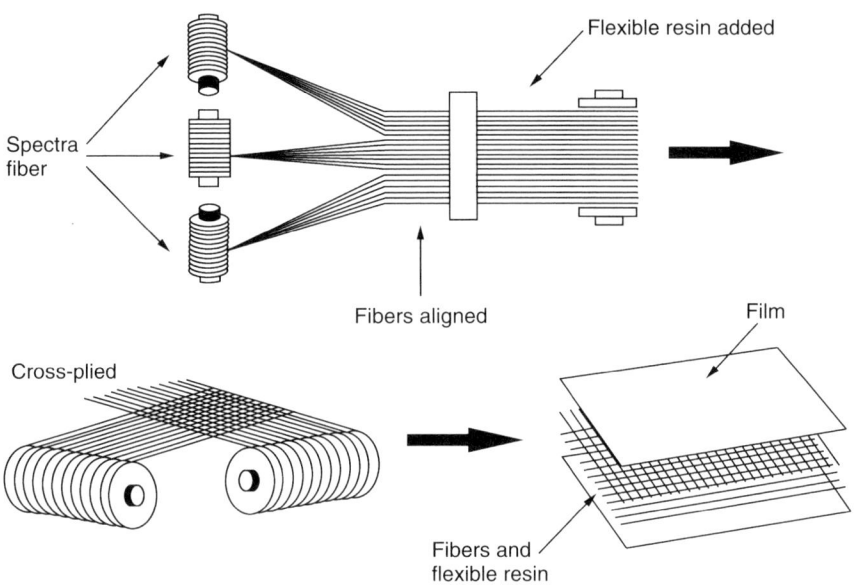

2.11 Spectra Shield™ manufacturing process.

into which a ballistic impact is directed. Such filament structures do not depend on frictional forces among themselves to hold themselves into a physical continuum and thereby avoid inherent weak places within themselves to resist penetrative impacts.

Parallel filament lay-up with resin reinforcement

A very significant type of ballistic resistant structure is encompassed by those that may be described as filament lay-up composites. Although these structures are neither woven nor knitted, and they are sometimes marketed as non-wovens, they also fit the definition of a fiber–matrix composite.

In the filament lay-up structure, all of the fibers are lined parallel to each other as in the beaming operation for woven fabric. A binder is then applied to form the structure into a continuous resin-fixed web of aligned fibers. The resin holds the fibers' spacing for further processing. A web of similarly constructed filaments is aligned at 90° to form a continuous roll. The 0-degree and 90-degree webs are further consolidated to form a cross-plied unidirectional roll product (Fig. 2.11).

The roll product developed by this technology is a patented process; commonly this material is referred to as 'shield'. The shield technology is applicable to all types of continuous ballistic fibers including HPPE/ECPE fibers, aramid and PBO fibers.[7]

Stitchbonding

The stitchbonding process is best described as a warp knitting process that is modified to use far fewer filaments, often of a much coarser type than is typical of warp knitting, and often also involving the use of felts or loose fiber mats. Although this process in not presently used for any commercial ballistic resistant products, there is clearly reason to believe that it could offer some significant advantages by combining the lateral and transverse stability of a warp knit type structure while utilizing the isotropic impact absorbing power of a fiber mat or needled non-woven.

Stitchbonding machines were initially introduced in eastern European countries during the Cold War era, and they managed to make incursions into Western textile production facilities despite the politics of the time. Krčma[8] distinguishes between what he calls a 'true' stitchbonding system and a knitting through system for thread systems only. The former system would mimic closely a triaxial weaving system with the corresponding advantages of an additional two translational energy vectors available to divert impact forces. At the same time the disadvantages of warp knit loop overshot and undershot geometries would create numerous opportunities for high impact forces to stress brittle high-modulus ballistic-resistant fibers beyond their breaking strain limits. Thus, the advantages of such a 'knit through' structure may likely be cancelled out before they come into play.

True stitchbonded structures include those formed by machines such as the Maliwatt and Arachne types.[9] Although these types of fabrics are conventionally used for insulations, there is considerable promise for their application in the market niches for needled non-wovens as well.

2.4.3 Staple fiber

Staple fibers have not traditionally been used in ballistic-resistant non-woven structures because they have the exact limitation of discrete, discontinuous character that the use of filaments seeks to overcome. On the other hand, if formed together correctly, these tiny, particulate materials can offer potential advantages of structural isotropy that filaments specifically cannot offer. They can also be consolidated and compressed so that the fiber population density in such structures is greater than that which can be achieved with woven or composite structures.

The disadvantage to the use of staple fibers is that they are presented to the manufacturing process in a random, unconsolidated, and non-uniform mass. Most commonly staple fibers are packed in 'bale' form. They must be mechanically processed through several stages before they can be made ready for use.

Opening and blending

In one classical definition of the opening process, 'The term opening origi-nates with compact baled fibers being separated into small loose pieces or tufts'.[10] Because of the immense pressures required to compress a loosely arranged mass of fibers into a tight, dense bale of roughly 225 kilograms, fiber-to-fiber interfaces are increased and thus large groupings of fibers will form themselves into tufts.

Blending is included in preparing staple fiber for use because it is the most logical place for this step to occur. Even in the case of modern, high-modulus ballistic-resistant fibers, there are slight variations in the physical characteristics of the fibers from one lot to another. These varia-tions are reduced with blending of various lots of fibers. The most advanced method of preparing staple fibers for conversion into ballistic-resistant non-wovens is blending of two or more fiber types together at this step. The manufacturer must determine whether the customer needs the blend to be expressed as a ratio of percentages of fiber types present by weight ratios or by actual fiber populations. The most common terminology refers to weight ratios.

Opening machines today fall into two major categories:

1. Those using a spiked apron conveyor feed with rotating beaters posi-tioned at the ends of the conveyors (Fig. 2.12).
2. Those designed accurately and delicately to remove small layers of fibers from bale surfaces in a series of feeder bales known as a 'lay down' (Fig. 2.13).

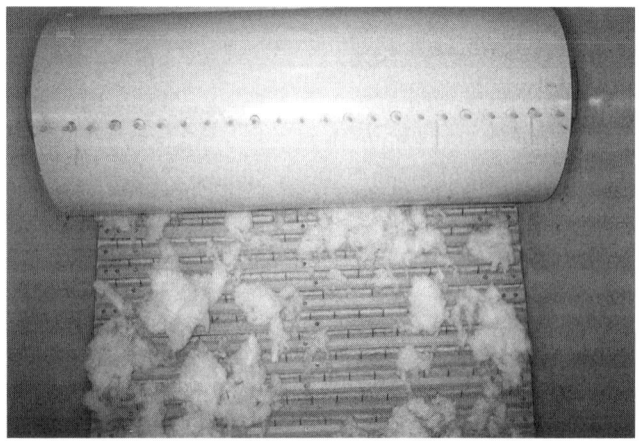

2.12 Spiked 'apron' feed lifts partially separated fiber tufts to a rotary beater (photo by the author).

2.13 Metered layer removal by modern bale opener (Marzoli spa, *Marzoli Spinning Solutions Blowroom Machines*, 2001).

2.14 Opened, blended staple fibers being fed in mat form into a card (photo by the author).

Both of these methods may be used in modern facilities, but the extremely high strength of ballistic resistant fibers and the range of useful fiber lengths for such a specification make the spiked conveyor and beater arrangement the more flexible alternative for non-wovens plants.

Mat formation methods

Once the staple fibers have been opened and blended together, they must be metered out into a form that approaches the final desired density or volume (Fig. 2.14).

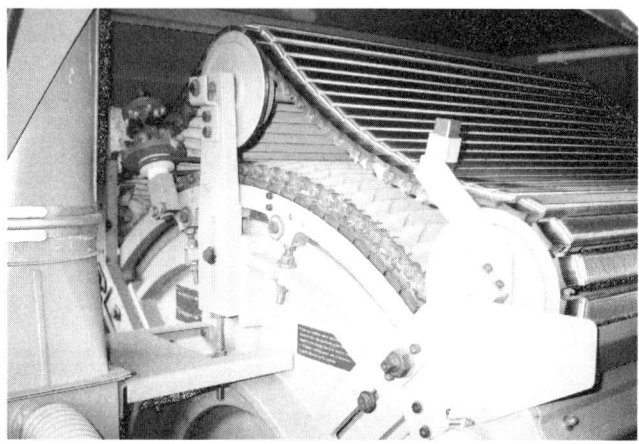

2.15 A standard 'flat top' card, showing wire clothing on flats (photo by author).

The fibers must also be arranged in a desired orientation, or machine limitations that restrict the orientation of the fibers to a single direction or small range of directions must be recognized so that other manufacturing methods may be applied to achieve the desired result.

The earliest, and still most prevalent, method of forming staple fibers into a mat is the card. The card was originally designed to create a thick strand of paralleled fibers from cleaned, blended, opened fibers in preparation for converting those fibers into yarns. To accomplish this task, it is constructed of at least three large rotating cylinders, each of which is covered with a fine, angled and chisel-pointed wire 'clothing' (Fig. 2.15).

The modern card is actually not ideally suited to the formation of non-woven webs for ballistic-resistant fabrics because it is designed to produce a stream of nearly perfectly paralleled fibers to eventually form into a staple fiber yarn. A ballistic-resistant material must be able to engage an incoming projectile – bullet, shrapnel fragment or energetically propelled rubble from an explosion – from any angle, under any spins or tumble condition and in any geometry. Yet, this basic, long pedigreed piece of traditional textile equipment was the first to be applied to the formation of useful mats for non-woven fabrics. It is, in fact, one of the most commonly applied technologies for the manufacture of non-woven ballistic resistant materials.

One step in this manufacturing process was still lacking. Converting a thin, paralleled mat of fibers into a useful, ballistic-resistant structure requires a technology that was unknown to textiles before the successful advent of non-woven fabrics. That technology is known as cross-lapping.

Cross-lapping (cross-plying)

Webs delivered from a card are only two to four fibers thick. Such a fine, gossamer-like structure may be useful for adhesive-bonded non-wovens like dryer sheets with fabric softeners, but they certainly have far too little ballistic resistance to be useful. In order to create a structure with sufficient fiber population and varied orientation to engage various projectile shapes, a new way of combining fiber layers was required.

The functions of the cross-lapper are:

1. To fold a desired number of multiple layers of carded webs together to form a final web or fiber mat of desired weight per unit area.
2. While layering the carded webs together, to lay them onto each other at varying angles that are different from the original carding machine delivery direction.

The cross-lapper can perform this function by picking up the carded web on a moving conveyor, laying it onto a conveyor that is moving perpendicular to that conveyor and at a slower speed from the first conveyor. This scheme of delivery allows the webs to be stacked on each other in various thicknesses and average angles of fiber orientation, depending on conveyor speed differences (Fig. 2.16).

Further control of the final web thicknesses, orientations and uniformity can come from total frictional contact and pressure between conveyors and individual speed controls of the driving rolls. This latter scheme is becoming the most favored and common among needlepunchers.

Needlepunching

Needlepunching is a simpler operation than weaving by which a variety of properties can be obtained in the fabric by varying the process components.

2.16 A modern type of cross-lapper (http://www.nonwovens.net/photo26.htm).

Continuous ballistic fibers are chopped into smaller fibers, carded and (usually) randomly oriented by cross-lapping to form an isotropic mat or sheet. This sheet is subsequently consolidated by a set of barbed needles. The needles push a limited amount of fibers at 90° through the sheet of randomly oriented fiber felt. The felt material engages fragments much better than traditional woven fabrics. Needlepunching is a rather simple operation, but a variety of properties can be realized in a needled web structure by varying different parameters of the process.

One of the most important parameters that can be controlled in the process is the shape of the individual needles used to consolidate the felted structure. Needles are designed for a variety of purposes, including relief structuring, creating density gradients in the fabric and for simple, uniform consolidation (Fig. 2.17). For ballistic resistant structures, the most common needle type is the simple barbed, triangular or four-pointed star-shaped cross-section types.

Needle barbs may be varied in shape, number and orientation along the axis of the needle. Additional control of fiber entanglement angles, depth, extent and frictional contact lengths are provided by the barb throat depth and barb angle ('kick-up').

The next considerations are those of needle population in the fixing structure, known as the needleboard, the rate of feed of the fiber mat and the punch frequency. The foregoing factors combine to create the critical defining characteristic of a needled non-woven fabric known as punches per square inch.

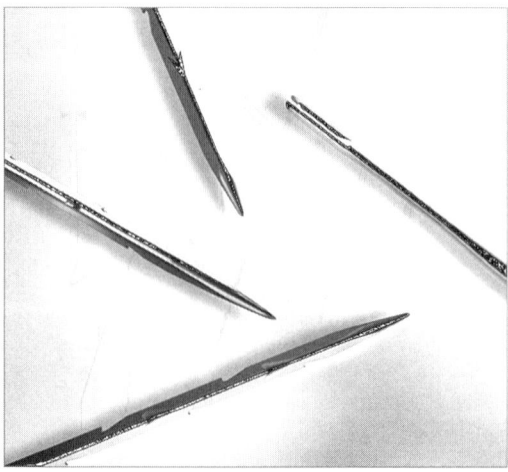

2.17 Examples of various types of felting needles (Groz-Beckert, http://gbu.groz-beckert.com/website/gbu/en/fn_innovations.html).

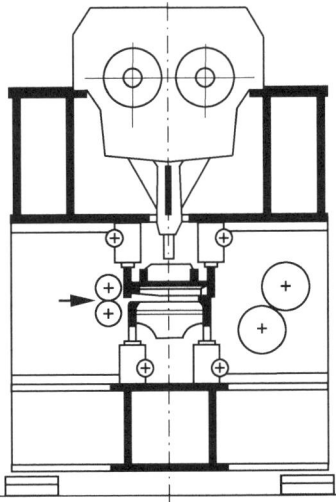

2.18 Schematic of a 'top punch' needlepunch machine or 'needle loom' (Fehrer AG, http://www.fehrerag.com/Fehrer/frame.htm).

Finally, needlepunch machines, or needle looms, as some companies call them, may have their needleboards arranged to punch from the top down, from the bottom up, or in both directions simultaneously (Fig. 2.18).

While some ballistic-resistant and ballistic-assisting non-wovens may be formed directly on one pass through a needlepunch machine, most require a lighter needling step known as pre-needling.

The final fabric product from the above process is actually only a network of randomly arranged fibers, held together only by frictional contact among its constituent fibers.

2.5   Filament lay-up composites

The filament lay-up composite, or those structures made by parallel lay and resin reinforcement as described in the section 'Parallel filament lay-up with resin reinforcement' in Section 2.4.2, occupy an increasingly important and, ironically, traditional sector of the ballistic-resistant materials spectrum. These unique structures are designed to engage an incoming projectile with a much larger population of high strength fibers than can be brought against such a threat with a woven or knitted fabric. The presence of a reinforcing resin also assists in the energy dissipation and the composite structure together quickly acts to strip a bullet of its casing and flatten it upon impact. Two major products in the present market that use this same principle are Honeywell Spectra Shield and DSM Dyneema UD armors. Both products depend on the same ballistic resistance principles to defeat incoming threats.

Energy absorption and dissipation energy is the secret to ballistic resistance. A ballistic-resistant fiber's strength must be utilized in the most effective manner for such a fabric or structure to be effective. The principle has been expressed in the following manner[11]:

A woven fabric dissipates energy at yarn interlacings. When a projectile strikes the surface of a fabric, energy is distributed along the yarn axis to each interlacing point. Most woven fabrics exhibit yarn strength translational efficiencies between 60 and 80%. Only about one-third of the strength loss can be attributed to degradation during weaving. The remaining strength reduction is caused by mechanical interaction between warp and filling yarns during tensile loading. High warp crimp in a woven structure is accompanied by low strength translation efficiency. A compromise must be reached in fabric construction between weave density and fabric strength where neither is at an optimum level.

Spectra Shield fabric forces the projectile to engage many more fibers upon initial impact than a woven fabric because of the wide dispersion of filaments in the untwisted yarn. Resin prevents the projectile shock wave from pushing the fibers out of the projectile's path; the fiber strength has higher translation efficiency in the structure.

Ideally, a structure should dissipate impact energy rather than obstructing it. Fiber friction is one property which may assist in absorbing energy while utilizing the strain wave velocity of a fibrous system. This theory is of interest when considering a non-woven structure, because large numbers of fibers are present in a non-woven, oriented in many different directions.

Strain wave velocity is the speed at which a fiber or structure can absorb and disperse strain energy. It can be expressed as

$$v = \sqrt{F / \mu}$$

where
v = strain wave velocity
F = force applied to the fiber (from projectile)
μ = linear density expressed as kg/m
At the same time, one can also express v as

$$v = \sqrt{E / \rho}$$

where
E = material Young's modulus
ρ = specific gravity of material
By combining the equations, an expression for optimum dissipation of impact energy can be found.

$$F = E\mu/\rho$$

The more impact energy a structure disperses, the more efficient the energy absorption mechanism is. Three reactions occur in a needlepunched structure when a projectile strikes it. These reactions are fiber elongation, fiber slippage, and fiber breakage. Designers want to create a structure which optimizes each of these properties to yield the best ballistic properties.

2.5.1 Flexible ('soft') armor uses of filament composites

The most traditional way of applying filament lay-up composites to armor is in the arena of 'soft' body armor that encompasses the US NIJ threat levels I through IIIA. The present range of products made by this method include the previously mentioned Spectra Shield and Dyneema UD families, containing only extended chain, high-performance polyethylenes and the Goldflex products (Honeywell) that contain aramid fibers fixed in resin. Both of these product types retain a thinner profile than woven fabrics, and they are usually not fixed by stitching.

Resin-fixed PBO fiber structures have also been produced and marketed that exhibit very high ballistic performance. To date, there have been no documented uses of PIPD fibers in filament lay-up composites, but this is a certain logical evolution of that fiber.

2.5.2 Level III filament lay-up armors

One of the more astounding developments of the filament lay-up composite structure has been in rifle-resistant (NIJ Level III) armor. Both Spectra and Dyneema fibers have been successfully applied to this end so far. Studies from both US Army and Honeywell researchers were pursued in the early 1990s to define how best to back ceramic plates for rifle projectile defense. The studies reported,[12]

> Both woven fabric-reinforced laminates and angle-plied unidirectional fiber-reinforced laminates were found to exhibit sequential delamination, cut-out of a plug induced by through-the-thickness shear, and combined modes of shear and tensile failure of fibers as observed in the cases of glass and graphite fiber composites. At low areal density, both laminates demonstrated similar ballistic limits. However, as areal density increased, differences in ballistic limit became more apparent, with angle-plied composite laminates showing higher values. When subjected to the repeated impact of a constant striking velocity below the ballistic limit, a progressive growth of local delamination was observed until gross failure of composites occurred. The use of lower striking velocity of the projectile led to the increase in cumulative numbers of impacts for full penetration defining an impact fatigue lifetime profile. The results of impact testing indicated that Spectra fiber-reinforced composites with vinyl ester resin matrix have a higher ballistic limit and longer impact fatigue life at a given striking velocity than the polyurethane matrix composites. Less effective absorption of impact energy by flexible polyurethane matrix composites was attributed to much more restrained pattern of delamination growth. Correlated with the results of dynamic mechanical analysis, these trends indicated that the

stiffness of resin matrices plays an important role in controlling the ballistic impact resistance of Spectra fiber composites.

2.6   Historical uses of non-woven ballistic-resistant fabrics

The first instinct of the technology student or fiber engineer is to assume that non-woven ballistic-resistant armor is a relatively new idea, since the machine technology to produce it postdates that of weaving and knitting by a considerable time period. In truth, non-woven armor in the form of quilting has been used since at least the Middle Ages. Indeed, British historians have determined that Viking chain mail, reinforced and supplemented by quilted, fiber-filled underlays were likely the secret to its ability to withstand even spear attacks in battles.[13]

2.6.1   Test results from US Army Natick Soldier Center

The US Department of Defense has performed testing on ballistic-resistant non-wovens at its laboratories in Natick, Massachusetts and through other research facilities. The tests were designed to examine whether non-woven fabric could be used in military ballistic applications. The Natick studies found that a needlepunched structure could be produced at one-third the weight of woven fabrics for certain ranges of protection. These Army studies were inconclusive as to the extent of practicality that the use of non-wovens would bring to ballistic applications.[14]

2.6.2   Results from British researchers

The needlepunched structure has not been as thoroughly evaluated for geometrical and physical relationships as other fabric structures such as knits and woven fabrics. John Hearle[15] has offered the most complete explanation of the fabric which he describes in a geometric model of the needlepunched structure. This model shows the vertical structure consisting of tufts of fibers pulled through the web by felting needles. The horizontal structure consists of fibers following curved paths around the tufts. When looked at in a three-dimensional plane, individual fibers pass through both the horizontal and vertical sections.

2.6.3   Test results and developments from independent and commercial entities

Few commercial needlepunched non-wovens exist in the market yet. One reason for this is their greater bulk (volume) per unit area than their woven or filament composite lay-up competition. Many law enforcement and military personnel find thickness a less desirable trait than lighter weight even when the protection afforded by the non-woven is equal or better. Despite these limitations, a few companies such as DSM (Netherlands), National Nonwovens (Massachusetts, USA) and Plainsman Armor (Alabama, USA) are offering products of this nature in the marketplace.

DSM was the first commercial entity to have success in the marketplace with a 100% needled non-woven product that is known as Fraglight or FR10. This non-woven armor is composed entirely of DSM Dyneema staple fiber, and it has been used in fragment-resistant vests in European armies. DSM researchers found that the early versions of the product suffered from abrasion of fibers from the structure that deteriorated its ballistic performance over time. Further work is continuing with the Fraglight product to improve it now.

National Nonwovens has a standalone needled non-woven that has been certified for use in commercial airliners by the FAA. The Plainsman products have been successfully tested in this role as well, but are currently being developed more for modified body armor and vehicular armor use.

A hybrid armor of both needled non-woven and woven ground fabric has been jointly developed by Barrday (Canada) and TexTech Industries (USA). Further testing and marketing of this product is presently underway.

2.7   Methodologies for use of non-woven ballistic-resistant fabrics

As stated in the previous section, needled non-woven armors may be applied in standalone or in supplementation configurations. Regardless of the intended final product, careful consideration of the construction methodologies for individual components must be made and from these, rational decisions about the architecture and composition follow.

2.7.1 Single fiber components

The most common and natural scheme for assembly of ballistic-resistant fibers into a non-woven structure is a uniform assembly of the same fiber types. Almost all present, commercial, ballistic-resistant fabrics are made of the same fiber types, thicknesses and lengths. This scheme is easiest for a manufacturing facility because the fiber inputs are uniform and predictable, and minimal blending steps are required.

According to one producer of such fabrics, these structures can be produced to sufficiently rigorous standards to qualify for FAA flight deck protection against the standard test projectiles of NIJ Level IIIA.[16]

Such performance qualifications show that 100% needled non-wovens of uniform fiber types have great promise in a variety of ballistic-resistant applications.

2.7.2 Multiple layering of various single fibers

Layering of various kinds of non-woven fabrics and/or conventional fabrics was proposed as early as 1992 by a team from Allied Signal, the original owners of Spectra Fiber technology.[17] Although the scheme has been variously tested by military organizations, research institutions and universities, it is presently only applied commercially as combinations of woven and filament lay-up composites (shield-type fabrics). The application of needled non-wovens of individual fiber types in individual layers or combinations thereof has not been commercially applied.

2.7.3 Blended fiber constructions

Tests conducted by Auburn University indicated that combinations of aramid and HPPE/ECPE fibers in non-woven blends produced higher than anticipated performance beyond those of the advantages of both fiber types. Energy absorption properties 30% greater than in unblended structures were observed in initial tests of the material (Fig. 2.19). The combination of thermoplastic and non-thermoplastic fibers in the structure allowed an energy dissipation mechanism by phase change that boosted the fabric areal weight performance.

The original tests to develop a blended non-woven ballistic fabric, a 50% HPPE/ECPE and 50% aramid indicated that the new fabric thickness was significantly less than that of 100% aramid fiber blends. The observed effect was attributed to fiber denier differences between the aramid and HPPE/ECPE fibers. The HPPE/ECPE fibers were 5.5 dpf; the aramid fibers were 1.5 dpf. As a result of their higher denier, HPPE/ECPE fibers present in the blend afforded more voids in the blended needlepunched samples compared with the 100% aramid samples.

Energy absorption in ECPE, aramid blends

◆ Radiated strain energy
— transferred by aramids and ECPE outside impact
◆ Fibrillation of aramids
◆ Phase change induced in ECPE

2.19 Results of aramid and ECPE fabric ballistic impact.

Table 2.3 Comparative averages of fragment (FSP) testing on flexible armor system

Material	Number of plies	Areal density kg/m^2 (psf)	V_{50} m/s (ft/s)
GF	25	5.81 (1.19)	586 (1924)
AF + GF	2 + 23	5.81 (1.19)	575 (1887)
GF + AF	23 + 2	5.81 (1.19)	593 (1944)
AF	15	3.61 (0.74)	573 (1880)

Key: GF = Aramid filament lay-up composite; AF = Blended non-woven.

The HPPE/ECPE fibers/aramid blend had better ballistic resistance than 100% aramid blends in the tests. Ballistic resistance was also enhanced with increases from 4 layers to 8 layers. Web layers had less effect in the HPPE/ECPE fibers/aramid blends than in the 100% aramid samples. As the number of layers was increased, the differences between the blended conditions and the 100% aramid became less, but they retained significance. Variation in density showed a similar response of V_{50} ballistic resistance with varying fabric density for the different fiber-type conditions.

Further testing of the fragmentation stopping capability of blended non-woven fabrics continued between 1997 and 2001. Among findings during this development, it was clear that significant advantage exists where HPPE/ECPE fibers are 5.5 denier or finer. Disadvantage was observed when fiber blends with PBO present were tested because of the very low frictional characteristics of these fibers.[7]

2.7.4 Fragment protection

In 2002, blended, non-woven, needlepunched, ballistic-resistant fabrics were tested in 2002 at both Honeywell Performance Fibers Laboratory in Petersburg, Virginia and the US Army Aberdeen Proving Grounds, against woven aramid fabrics and against woven PBO fabric to compare performances in defeating explosion fragments. In those military specifications tests, flexible armor was tested against the most common specified fragment threat (MIL-STR-662F). Results of fragment testing at Honeywell are shown in Table 2.3.

2.7.5 Tests by US Army

Evaluation of the blended non-woven was conducted in 2002 as a part of the development of a fragment-resistant cover for the Army's LOSAT KEM trailer.

Fragment armor improvements with non-woven technology

◆ Results from US Army Aberdeen Proving Grounds test
 — .22 cal. 1.10 gram, fragment simulating projectile, steel
◆ Parameters
 — Weight < 3.42 kg/m^2
 — Projectile speed > 425 m/s (1400 fps)
◆ Non-woven materials were superior to woven aramid and woven PBO
◆ Historical development of non-woven armor
 — Original Kevlar 29 = 389 m/s
 — Original (1991) blend yielded 434 m/s (HPPE, 2nd quality and Kevlar 29)

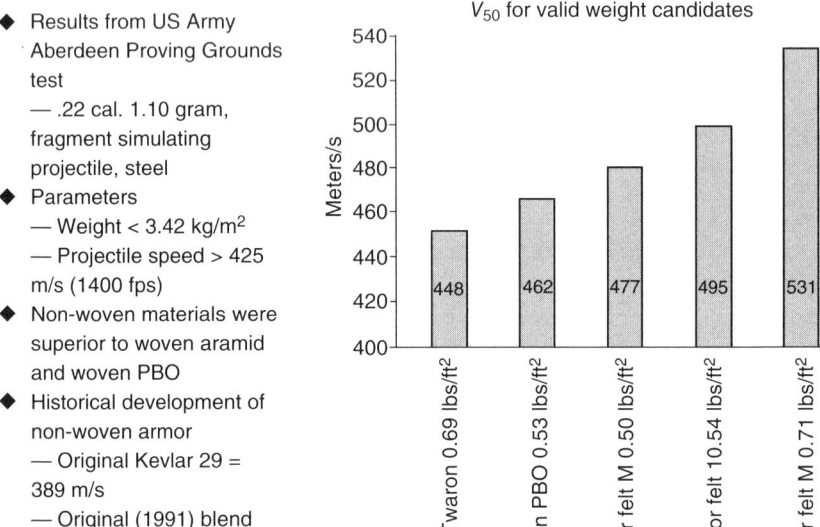

2.20 Performance of blended non-woven in fragment defeat.

In the test, at Aberdeen Proving Grounds, the parameters as specified by the US Army were weight of 0.75 pounds/square foot or less and projectile speed of 425 meters/sec (1400 feet/second) or more. The test results determined conclusively that blended non-woven outperformed woven aramid and woven PBO by a large difference (Fig. 2.20).

2.7.6 Combinations of non-wovens and conventional materials

A significant factor which has contributed to soft armor advances is the hybrid concept of combining more than one ballistic material in a single armor system. This technique allows armor design engineers to utilize the full potential of various ballistic materials.

Combinations of conventional materials and/or shield-based products with ArmorFelt have shown significant advantages when used against rated soft body armor threats. Testing of these systems using a modified NIJ 0101.04 Level IIIA Standard, .44 Magnum is shown in Tables 2.4, 2.5 and 2.6.

Table 2.4 Level IIIA baseline test results aramid filament lay-up only

Sample	Material	Bullet type	Speed m/s (ft/s)	Penetration	Backface deformation (mm)
1	24 Aramid filament lay-up composite	.44 mag JHP	433 (1422)	Partial	47
2	5.57 kg/m^2 (1.14 psf)	.44 mag JHP	438 (1438)	Partial	42
3		.44 mag JHP	442 (1450)	Partial	41
4		.44 mag JHP	438 (1438)	Partial	42
5		.44 mag JHP	442 (1449)	Partial	47
6		.44 mag JHP	435 (1427)	Partial	49
Averages			438 (1437)		44

Table 2.5 Level IIIA test results aramid filament lay-up + 4 ply blended non-woven

Sample	Material	Bullet type	Speed m/s (ft/s)	Penetration	Backface deformation (mm)
1	19 Aramid filament lay-up	.44 mag JHP	440 (1442)	Partial	39
2	1 felt 4 ply	.44 mag JHP	441 (1446)	Partial	38
3	5.57 kg/m^2	.44 mag JHP	440 (1443)	Partial	37
4	(1.14 psf)	.44 mag JHP	441 (1448)	Partial	43
5		.44 mag JHP	443 (1452)	Partial	35
6		.44 mag JHP	445 (1461)	Partial	42
Averages			439 (1440)		38

2.8 Future directions for non-woven fabric applications

The use of high-strength polymer materials created advances in armor protection far above those anticipated just 35 years ago. Further improvements may be anticipated by the advent of new materials and nanoscale technologies that will permit even better armor performances against very high-level ballistic threats.

Improvements that utilize the strongest characteristics of each fiber assembly method will yield the optimum ballistic protection device instead of simple reliance on standard and unitary assembly techniques.[2]

Table 2.6 Level IIIA test results aramid filament lay-up + 3 ply blended non-woven

Sample	Material	Bullet type	Speed m/s (ft/s)	Penetration	Backface deformation (mm)
1	18 Aramid filament lay-up	.44 mag JHP	427 (1402)	Partial	40
2	1 felt 3 ply	.44 mag JHP	430 (1411)	Partial	39
3	5.42 kg/m²	.44 mag JHP	430 (1410)	Partial	39
4	(1.11 psf)	.44 mag JHP	437 (1435)	Partial	40
5		.44 mag JHP	440 (1444)	Partial	38
6		.44 mag JHP	431 (1413)	Partial	37
Averages			433 (1419)		38

2.9   References

1 WARDER, B., 'History of Armor and Weapons Relevant to Jamestown', National Park Service, January 1995, http://www.nps.gov/colo/Jthanout/HisArmur.html

2 THOMAS, H.L., 'Armor and Materials for Combat Threat and Damage Protection', *SAMPE 2005 Conference and Exhibition*, Long Beach, CA, May 4, 2005.

3 'U.S. Body Armor (Flak Jackets) in World War II', http://www.olive-drab.com/od_soldiers_gear_body_armor_wwii.php

4 'Body Armor Development after World War II', http://www.olive-drab.com/od_soldiers_gear_body_armor_korea.php

5 IPSON, T.W. and WITTROCK, E.P., 'Response of Nonwoven Synthetic Fiber Textiles To Ballistic Impact', *Technical Report No. 67-8-CM*. US Army Natick Laboratories, Natick, MA, July 1966.

6 *The Nonwovens Handbook*, INDA Association of the Nonwoven Fabrics Industry, New York, NY, USA, 1988.

7 THOMAS, H.L., 'Needle-Punched Non-woven Fabric for Fragmentation Protection', *14th International Conference on Composite Materials*, Society of Manufacturing Engineers, July 14–18, 2003.

8 KRČMA, R., *Manual of Nonwovens*, Textile Trade Press, W.R.C. Smith Publishing Co., Atlanta, USA, 1971.

9 TOTORA, P.G., *Understanding Textiles*, Macmillan Publishing Co., New York, NY, USA, 1992.

10 MARVIN, J.H., *Textile Processing*, Vol. I, State Dept of Education, Office of Vocational Education, Columbia, SC, USA, 1973.

11 THOMAS, H.L. and THOMPSON, G.J., 'Characteristics and Performance of Needlepunched Flexible Ballistic Personal Protection Fabric Constructed from High Performance Fibers', *4th International Techtextil Symposium*, Frankfurt, Germany, June 1992.

12 LEE, B.L., SONG, J.W. and WARD, J.E., 'Failure of Spectra Polyethylene Fiber-reinforced Composites under Ballistic Impact Loading', *Journal of Composite Materials*, 28(13), 1202–1226, 1994.

13 LENT, C., Producer/Director, 'Secrets of the Viking Warriors', National Geographic Channel, Darlow Smithson Productions.

14 LAIBLE, R.C., *Methods and Phenomena, Ballistic Materials and Penetration Mechanics*, Elsevier Scientific Publishing Company, Inc., Amsterdam, 1980.

15 HEARLE, J.W.S. and PURDY, A.T., 'Report on Energy Absorption by Nonwoven Fabrics', *Contract No. DAJA37-1-C-0554*. European Research Office, United States Army, London, November 1971.

16 National Nonwovens, Performance Solutions E-News, Spring 2002, http://www.nationalnonwovens.com/enews/performance1.htm

17 CORDOVA, D.S. and KIRKLAND, K.M., Armor Systems, *US Patent 5,343,796*, Sept. 6, 1994.

3

Mechanical failure criteria for textiles and textile damage resistance

N. PAN, University of California, USA

3.1 Introduction: Material resistance, strength and failure

Adequate strength and durability are the pre-requirements for any engineering materials, simply because they have to be strong enough to function. In that sense, these are the principal attributes when examining the performance of fabrics or items made of the fabrics. In the latter case, the durability of the assembly adds another dimension to the problem. As such, it is not surprising that the industrial standards for quality assurance on fabrics and clothing overwhelmingly focus on strength and durability related issues. Industrial standards dealing with strength and durability were initially concerned with solid engineering materials. This is understandable, for failure of those materials often leads to immediate disastrous consequences: collapsing of buildings, bridges, ships and airplanes and associated human casualties. Therefore, the standard tests and the theories behind them, and the mechanisms of failure all have been developed with those materials in mind. Today's textile-related standard tests on strength and durability clearly bear resemblance to those preceding ones for solid materials. In that sense, a brief yet comprehensive review of all the related scientific knowledge on materials strength, failure and durability appears necessary here.

Some basics

The maximum load a material is able to carry without causing failure obviously depends on many factors including the major ones such as:

(i)   Material type;
(ii)   Material dimensions;
(iii) Nature of the load;

50

(iv) Ambient environmental conditions;
(v)   Testing schemes . . .

Among these factors, those in (i) correspond to specific load types, those in (iii) are the truly intrinsic variables, and the rest are subject to the user's choice and should therefore be standardized so that only the genuine material attributes are revealed.

3.2   Material strengths

To reflect the influence due to the nature of the load exerted on the materials, the load is classified as tensile, shear, bend, tear, and puncture/burst for sheet-like materials such as fabrics. So we will first take the tensile load (stretch) example.

3.2.1 Simple tensile

If a given material is stretched to break, we will call that extension load level P_m the *breaking load* of the material (Fig. 3.1(a)). Obviously, for the same type of materials, this P_m value will depend on how thick the material cross-section A_o is. Thus we have to define the ultimate strength $\sigma_u = P_m/A_o$. Likewise we define the breaking strain corresponding to σ_u as the ultimate breaking strain $\varepsilon_u = \Delta l_u/l_o$, i.e. the breaking elongation Δl_u per unit original

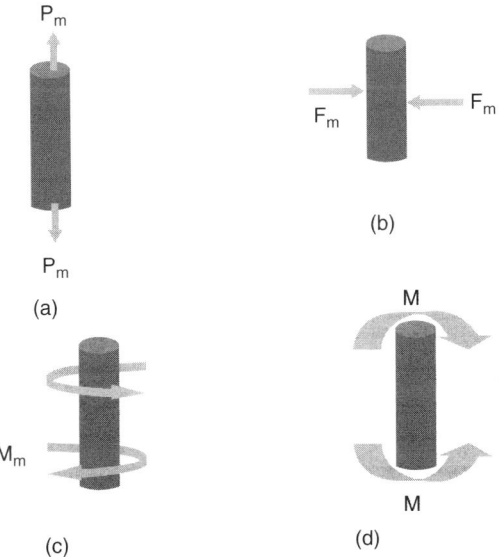

3.1 Various simple deformations, (a) uniaxial extension, (b) simple shearing, (c) axial torsion, (d) simple bending.

length l_o. Thus the pair $\{\sigma_u, \varepsilon_u\}$ provide the strength indicator for the material under tensile load. The case of axial compression is just a simple matter of changing the sign of the load assuming the material behaves the same for the same load but different signs. The breaking strain has no unit whereas the breaking strength is expressed as force per unit of cross-section area (N/m^2). The SI unit of stress is the Pascal, where $1\ Pa = 1\ N/m^2$. In Imperial units, the unit of stress is given as l bf/in^2 or pounds-force per square inch, often abbreviated as 'psi' where $1\ MPa = 145$ psi.

3.2.2   Simple shear

Shear force causes angular deformation as shown in Fig. 3.1(b). Again to eliminate the sample size influence, we define the ultimate shear strength $\tau_u = F_m/A_o$ and ultimate shear strain $\gamma_u = \Delta r_u/r$ where r and Δr_u are the radius and its displacement. They share the same units as their tensile counterparts.

3.2.3   Simple torsion

Torsional deformation actually belongs to the simple shear case, except that the shear stress is applied in the form of a torque as seen in Fig. 3.1(c). Consequently torsion cannot be treated as a new load type.

3.2.4   Simple bending

Bending is a little more complex, as seen in Fig. 3.1(d). As is well known when bending a piece of rod, the stress on the cross-section of the rod is not the same; one side is under tensile and the other side compressive load, separated by a neutral line across the cross-section. The exact location of the neutral line depends on both the shape of the cross-section and the load situation. The standard treatment of beam bending provides the following results:

The maximum stress occurs at the surface of the beam farthest from the neutral surface (axis) and is:

$$\sigma_{max} = \frac{Mc}{I} \qquad [3.1]$$

where M is the bending moment, c the distance from the neutral axis to outer surface where max stress occurs, and I the moment of inertia. So the failure of a beam under bending is caused by the tensile or compressive stress and the material will break as long as $|\sigma_{max}| \geq |\sigma_u|$ where σ_u is the corresponding strength in tension or compression. That is, bending is not a truly independent load type either and therefore does not have its own failure criterion in terms of bending moment.

3.2.5 Simple tear

Although tearing is a unique failure mode for sheet materials, it is not a new load type and hence no new breaking criterion is needed. Figure 3.2 illustrates a tearing process of a fabric. The failure process initiates by breakage of a few yarns (or fibers depending on the fabric composition), propagates by breaking yarns in the way, and completes when all the yarns in the path fail practically due to extension. Increasing the yarn tensile strength or allowing more yarn mobility so that they can retreat and group with more yarns will effectively enhance the material tear strength. However, it is the internal tensions that break the yarns.

For non-woven or paper sheets, it is the failure of either the bonding (reinforcing) points or the fibers. Bonding points break mainly due to shear, and fibers due to tension.

In either case, it is clear that tear failure is largely due to tensile breakage of the yarns and tear load is not itself an independent loading type.

To summarize, since fibers are best at carrying tension, when a fibrous material starts to break, at the micro-level fibers break almost exclusively

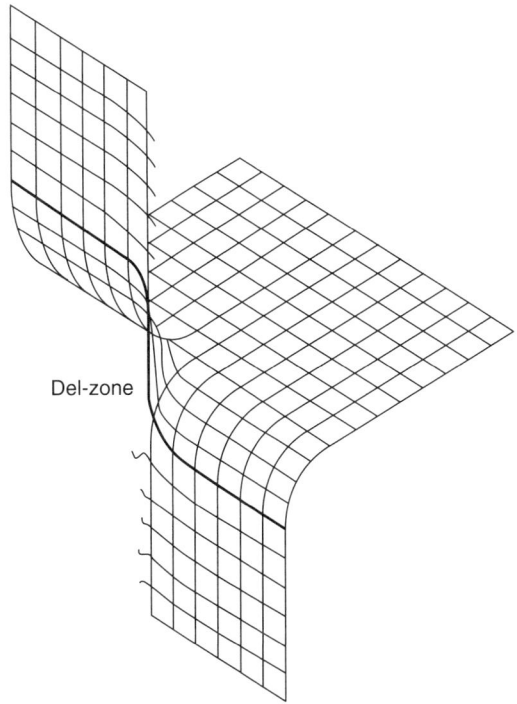

Del-zone

3.2 Fabric tearing deformation.

due to extension (and in much rarer cases due to shear), regardless of the nature of the macro deformation. In summary, there are only two types of failure, failure by tension (or compression when the stress is negative) and by shear. Bending is just a combination of both tension and compression caused by the bending moment; torsional deformation is a torque caused shear, and tear failure is a reflection of tensile breakage of yarns. Knowing all these greatly simplifies our discussion on fabric failure criteria.

3.3   The peculiarities of textile mechanics

As the definitions used above are all borrowed from the case of continuum, several considerations special to textile fibrous structures have to be noted.

3.3.1 The discrete nature of textiles

Because of the porous and soft structure with hairy surfaces, it is difficult to measure the fabric dimensions in order to calculate the stress in the fabric. A much more convenient way is to calculate the fabric stress as force/yarn or its strength in force/tex where tex is the thickness of the yarn expressed in the tex system. For the same reason, some of the analytical techniques in continuum mechanics become difficult or impossible to conduct. For instance, the vector and tensor tools, the internal force and stress resolution, and derivation of the principal eigen-stress components are unlikely to be applicable to fabrics.

3.3.2 The large deformation of textiles

Compared with other engineering materials, the scale of deformation in textiles is incredibly high; the bending and shearing deformation during a fabric draping highlight this unique feature of textiles, which should be astonishing for a civil or mechanical engineer. Along with this large deformation, the issues of non-linearity, the inter-yarn friction and true internal stress accounting for the cross-section change become significant.

3.3.3 Non-affinities between the macro- and micro-behaviors

For fibrous materials, the behaviors at the micro- and macro-levels often are of different natures. For instance, when sitting on a thick cushion filled with fibers, one is compressing the cushion; but a closer examination will reveal that most, if not all, the fibers are actually experiencing bending deformation (Carnaby and Pan, 1989; Neckar, 1997; van Wyk, 1946). Such a weak connection between, or even independence of, the properties of the system and its constituents renders a unique challenge for any attempt in formulating from the microstructural analysis to the macroscopic performance, a premise for any product design and application since fibers, in most cases, fail in tension (except in the case of cutting where fibers break because of shear).

3.3.4 Bi-modular nature of the fibrous materials

Anisotropy is responsible for many of the challenges in dealing with fibrous materials. However, even in the same direction, the material behaves differently depending on the sign of the force. In other words, for fibrous materials, the Young's modulus, as well as the entire stress–strain relationship, is quite different in tension versus in compression. Figure 3.3 depicts a typical example of a fabric under tension and compression in the longitudinal direction. Such so-called 'bi-modular' behavior is also prevalent in biomaterials. The pioneering work in this area was done mainly by Ambartsumyan and his collaborators and they further expanded the problem to two- and three-dimensional cases (Ambartsumyan, 1965, 1969; Ambartsumyan and Khachatryan, 1966). The problem has picked up new momentum as interest in biomaterials has increased.

Two issues are worth noting. First, the current approach treating materials in engineering as inherently identical in both compression and tension has to be re-examined. Also, any proposed model still needs to be evaluated to satisfy the following criteria: (i) the compliances and the moduli (stiffness) must be symmetric in any coordinate systems in order for the strain energy to be positive, and therefore a potential function exists; (ii) the values of the compliances are restricted in relation to one another such that the compliance matrix is definitely positive; (iii) the compliance matrix must be transformable between coordinate systems, i.e. it has to be a tensor (Bruno *et al.*, 1993; Eltahan *et al.*, 1989; Sacco and Reddy, 1992).

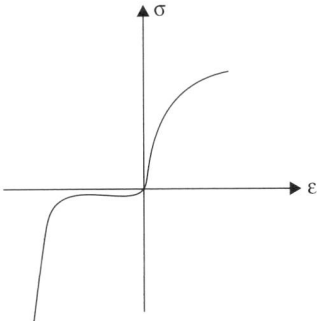

3.3 Bi-modular behavior under tension and compression.

3.4   Failure criteria for fabrics

Since fibers are best in carrying tension, when a fibrous material starts to break, at the micro-level, fibers break almost exclusively due to extension, regardless of the nature of the macro deformation as discussed above. The nature of the materials allows fibers to move and retreat from the loading. All of these lead to several scientific challenges including analyses of wrinkling or crumpling of fibrous sheets and of very peculiar fracture behaviors and failure criteria, as illustrated in Fig. 3.4. For an isotropic material, its strength is identical, regardless of direction, and can be represented by a circle (due to symmetry, only a quarter is drawn); whereas the strength direction relationship for an ordinary anisotropic solid can be illustrated by an ellipse. However, this relationship for a woven fabric is much more complex, because of the different degrees of internal yarn re-orientation and movement when stretched in different directions of the fabric. Finally, this yarn re-orientation to self-reinforce the resistance in the loading direction again reveals the adaptive nature of fibrous materials.

3.4.1   Failure criteria for different materials

Under a simple tensile test, the failure of the sample occurs when the stress caused by the actual load reaches the stress limit (the strength) of the sample. Correlation of the actual stress with the maximum stress (strength) is straightforward in this case because they are both uniaxial. But how can

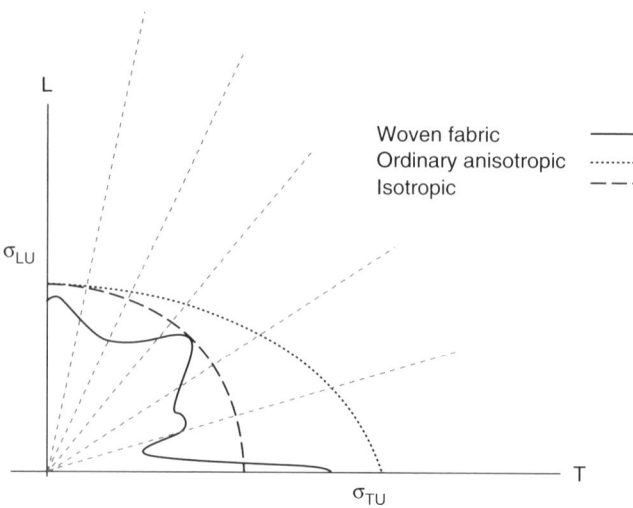

3.4 Various failure behaviors, where σ_{LU} is the strength in the longitudinal (L) direction, and σ_{TU} the strength in the transverse (T) direction.

we correlate the tri-axial stress state in a sample (whose material strength(s) is measured in uniaxial tests) to assess failure tendency? Unfortunately, there is, at present, no fundamental rationale for any such correlation.

We therefore postulate some attribute of the stress state as a descriptor of that state (e.g. an attribute such as the *maximum stress* or the *specific energy*) and then compare the values of that attribute for the given component tri-axial state on the one hand and the uniaxial test state on the other. This postulate is the *failure theory* based upon the particular attribute selected; it is a useful theory only if its predictions are confirmed by experiment.

Currently, no universal attribute has been identified which enables prediction of failure of both ductile and brittle materials to an acceptable degree of accuracy because of the complexities in the material failure process. This remains true for any of the currently employed failure evaluation tools including the failure criteria based on the Mohr's Circles; the Ductile Maximum Shear Stress Criterion also known as Tresca's or Guest's Criterion; the von Mises Criterion, also called the Maximum Distortion Energy Criterion, octahedral shear stress theory, or Maxwell–Huber–Hencky–von Mises theory; the Brittle Maximum Normal Stress Criterion, also termed the Normal Stress, Coulomb, or Rankine Criterion; or Mohr's theory, also known as the Coulomb–Mohr Criterion or Internal-friction Theory. Consequently, given the usefulness of such failure judgment rules and the non-conventional behaviors of the fabrics, we will develop specific failure criteria by focusing on the unique nature of the fabric material and then validate the theories by experiments so as to provide such useful rules for fabric design.

3.4.2 Establishing the failure criteria for textile fabrics

Introduction

Woven fabrics are well known for their property-direction dependence or property anisotropy. An experienced tailor would understand this well enough to choose different 'grains' for different fabric pieces on a garment so that a properly assembled dress can be made. However, for structural applications, this anisotropy will present a problem of irregularities in terms of performance or load-carrying capacity.

Woven fabrics are not only highly anisotropic, but dimensionally changeable; also, very susceptible to external loading and to its historical conditions. The important fabric properties critical to structural applications include tensile strengths, in-plane shear strengths, and normal compressive (in thickness direction) strength, as well as in-plane compressive strength, better known as the buckling strength.

Prediction of fabric strengths has its significance, both theoretically and practically, because, except for the uniaxial tensile strength, experimental determination of all other strengths is tedious with no convenient test methods and instruments available, and costs a great deal to perform, especially for sheet-form materials, whose flexural and torsional rigidity are very low. This generally requires elaborate devices to measure the shear and buckling strengths.

Kilby (1963) was probably the first researcher to deal with the mechanistic anisotropy of a woven fabric. He derived the so-called generalized modulus of a fabric, expressing the fabric tensile modulus in relation to the test direction. In the present study, however, we will focus on the anisotropy of the fabric strength. Following a paper (Pan, 1996) where we proposed a more realistic approach to predict the fabric tensile strengths at the principal directions under uniaxial and biaxial extension, we present in this study an attempt to investigate the direction-dependence or the anisotropy of the tensile strength of a woven fabric using a technique described by Cheng and Tan (1989) based on the experimental results.

Furthermore, by means of the Hill-type failure criterion (Hill, 1948; Theocaris and Philippidis, 1989; Tsai and Wu, 1971; Wu and Stachurski, 1984), widely applied in studying fiber-reinforced composites (Azzi and Tsai, 1965; Hoffman, 1967; Pipes and Cole, 1973), wood materials (Liu, 1984; Norris, 1950) and paper and geotextiles products (Minster, 1994; Pouyet *et al.*, 1990; Rowlands *et al.*, 1985), we will attempt to predict the shear strength of woven fabrics using the measured tensile strengths.

Prediction of the tensile strength anisotropy of woven fabrics

One approach of strength prediction is to employ the so-called failure criterion to derive other strength terms based on the given values of the strengths tested in a few particular directions. Of the various failure criteria for anisotropic materials, only three (Wu and Stachurski, 1984) have received wide attention; namely those of Hill (1948), Hoffman (1967) and Tsai and Wu (1971). Both Hill's and Hoffman's theories are limited only to ortho-tropic materials with plastic incompressibility. Tsai–Wu theory in this sense has a wider applicability. The basic assumption for the Tsai–Wu theory is that there exists a failure surface in the stress space which can be expressed in terms of a stress tensor polynomial function. In general, however, to apply the function, one has to know the compressive and shear strengths of the material besides its tensile ones.

Cheng and Tan (1989) have proposed an alternative technique by using a harmonic cosine series to represent the off-axial tensile strength of an anisotropic polymeric sheet or plate in any direction. That is

$$X_\phi = (\Sigma C_n \cos n_\phi) \tag{3.2}$$

where $n = 0, 2, 4, \ldots$, X_ϕ is the tensile strength at direction ϕ relative to the longitudinal direction L. C_n are the factors to be determined using the given or experimentally obtained tensile strength values in a few particular directions where tests are easy to carry out. Higher accuracy of the prediction using Equation [3.2] can be achieved by employing more pre-tested tensile strength values so that more factors C_n can be derived. The usefulness of Equation [3.2] is that, once C_n are determined, plotting of $X_\phi \sim \phi$ actually gives the fabric curves as in Fig. 3.4, illustrating the fabric strength anisotropy. The details of the examples using this technique can be found in Pan (1996).

From the curves in Fig. 3.4, it is clear that fabrics are vastly different from ordinary anisotropic materials whose failure curves are very close to an ellipse. The fabric failure loci are irregularly undulated due certainly to the fact that fabric structure is assembled by discrete yarns interlaced in two orthogonal directions, and the possible yarn–yarn relative movements and interactions at the crossing points are likely to be responsible for this irregularity of the fabric failure curve. Note that, for such an irregular shape, it is perhaps more advantageous to approximate the curve using a harmonic expression such as Equation [3.2], than a polynomial function of a regular failure criterion.

It is clear from Fig. 3.4 that maximum fabric strength occurs at either $\phi = 0°$ or $\phi = 90°$, i.e. along the longitudinal (L) or transverse directions (T) (warp or weft), for that is where the yarns have the best orientation to share and resist external loading. Local extremes of the fabric strength are due to the yarn re-orientation to better positions to defend. However, since they cannot achieve the same alignment level as in the L and T directions, the local extreme can never exceed the greater value of the two strengths σ_{Lu} or σ_{Tu}. However, when the difference between σ_{Lu} and σ_{Tu} is large enough, the strength in the bias direction can exceed the lower value of the two major strengths. Also, for most woven fabrics, the initial tensile modulus is at its minimum value in the direction $\phi = 45°$, that is, fabric is most stretchable in this direction; this is known to be true for most woven fabrics. The same cannot be said about the strength at $\phi = 45°$, for the strength depends on how yarns can better re-orient in that direction to resist the load.

It has to be pointed out that the criterion used above is based on the strength, or the maximum or breaking stress. For some applications, garments can become useless even though unbroken. An extreme example is for body armor: by using stretchable fibers, the strength of the fabric can reach a high value if the allowable strain is large enough. However, too much stretch allows hazardous sharp objects to injure the wearer even though the armor may still remain in good shape.

Prediction of the shearing strength based on the uniaxial tensile strengths of woven fabrics

As is well known, instrumental measurement with high accuracy of shear strength for anisotropic sheet materials such as woven fabrics is difficult and costly. Theoretical prediction of this strength based on the experimental data of uniaxial tensile testing thus becomes a very attractive alternative. One such approach is to utilize the Tsai–Wu theory on the material failure criterion.

$$f(\sigma) = F_i\sigma_i + F_{ij}\sigma_i\sigma_j = 1 \qquad [3.3]$$

where $i, j = 1, 2, 3$ and 6; F_i and F_{ij} are the strength tensors of the second and fourth rank, respectively. When $f(\sigma) < 1$, there will be no failure, whereas when $f(\sigma) \geq 1$, the material fails. The failure surface of Equation [3.3] is actually an ellipsoid when the following restrictions are satisfied (Suhling *et al.*, 1985):

$$F_{11}F_{66} > 0 \qquad [3.4]$$

$$F_{22}F_{66} > 0 \qquad [3.5]$$

and

$$F_{11}F_{22} - F_{12}^2 \geq 0 \qquad [3.6]$$

For fixed values of shear stress σ_6, the equation scribes ellipses in the $\sigma_1 - \sigma_2$ plane. For orthotropic sheet-type materials, the analysis is restricted to a plane stress state, Equation [3.3] can then be reduced (Rowlands *et al.*, 1985) to

$$F_{11}\sigma_1^2 + F_{22}\sigma_2^2 + 2F_{12}\sigma_1\sigma_2 + F_{66}\sigma_6^2 = 1 \qquad [3.7]$$

Let us assume that such a failure criterion is also valid for materials like woven fabrics; we can thus use this failure criterion to estimate the fabric shear strength. If we choose the L ~ T coordinate system, we will have

$$\sigma_1 = \sigma_L, \; \sigma_2 = \sigma_T, \; \sigma_6 = \tau_{LT} = \tau_{TL}$$

To determine the four coefficients F_{ij} according to Tsai and Wu (1971), we need to know the uniaxial tensile and compressive strengths in the L and T directions, and the pure shear in-plane strength as well as a uniaxial tensile strength of the material in a bias direction.

However, as pointed out in Rowlands *et al.* (1985), Equation [3.3] implies equal uniaxial strengths in tension and compression for the material concerned. This limitation is the result of an assumption associated with the theory that a hydrostatic stress has no effect on material strength. Since the applicability of theories that predict equal tensile and compressive strengths are restricted only to certain materials, Norris (1950) avoided this problem by making F_{11} and F_{22} functions of stress, instead of being constants. He then divided the stress plane into four quadrants so that the unknown coefficients can be derived specifically for each quadrant with different stress characteristics. In our particular case, in order to focus on prediction of fabric shear strength without the involvement of fabric in-plane compressive or buckling behavior, we will consider only the first quadrant of the four where the stresses, $\sigma_L \geq 0$ and $\sigma_T \geq 0$, are both of a tensile nature.

Denoting the uniaxial tensile strengths of the fabric in both the L and T directions as X and Y, respectively, since the fabric strengths are on the strength surface defined in Equation [3.7], this implies in the first quadrant that

$$F_{11} = \frac{1}{X^2} \tag{3.8}$$

$$F_{22} = \frac{1}{Y^2} \tag{3.9}$$

The various Hill-type predictions differ from one another only by the manner in which the coefficient F_{12} is determined (Rowlands et al., 1985). We here choose Norris' result (Norris, 1950) as

$$F_{12} = \frac{1}{2XY} \tag{3.10}$$

It is self-evident that F_{12}, thus defined, obeys the constraint in Equation [3.6].

The additional equation to derive the coefficient F_{66} can be established according to Pouyet and colleagues (1990) by completing an off-axis test applying tensile stress equal to the strength U_ϕ in the bias direction ϕ. By expressing all the resulting principal tensile and shear stresses σ_L, σ_T and τ_{LT} in terms of U_ϕ and ϕ, Equation [3.7] can be expanded into

$$U_\phi[F_{11}\cos^4\phi F_{22}\sin^4\phi + (2F_{12} + F_{66})\sin^2\phi\cos^2\phi] = 1 \tag{3.11}$$

from which the coefficient F_{66} can be determined. Then, the fabric shear strength S can be readily evaluated (Norris, 1950) as

$$S = \frac{1}{\sqrt{F_{66}}} \tag{3.12}$$

Since S cannot be negative, this relation, combined with Equations [3.4] and [3.5], implies that the coefficients F_{11}, F_{22} and F_{66} must possess positive values as well. The calculated results for the five fabrics are provided in Table 3.1 including the predictions of the fabric shear strength S.

Recall that the values of X and Y can be predicted theoretically using the results in Pan (1996). If we can somehow predict the off-axial tensile strength U_ϕ as well, we can then derive the fabric shear strength S based on the yarn properties and the fabric structure without even relying on the uniaxial fabric tensile tests.

Table 3.1 Coefficients of the strength tensor of the fabrics

Fabric	1	2	3	4	5
$F_{11}, 10^{-4}$	5.18	26.43	14.03	23.59	22.31
$F_{22}, 10^{-4}$	7.59	24.65	13.10	37.87	24.83
$F_{12}, 10^{-4}$	−3.13	−12.76	−6.78	−14.94	−11.77
$F_{66}, 10^{-4}$	68.59	168.93	83.52	143.69	120.60
X $(9.8 \cdot N)$	43.95	19.45	26.70	20.59	21.17
Y $(9.8 \cdot N)$	36.30	20.14	27.63	16.25	20.07
S $(9.8 \cdot N)$	12.08	7.69	10.94	8.34	9.11

There have been several experimental attempts (Pan *et al.*, 1992) using the tested off-axis tensile properties at $\phi = 45°$ to estimate the fabric shear properties. Equations [3.11] and [3.12] indicate that, although the shear strength is related to the off-axis tensile strength, the relation is not single-valued. Our study (Pan, 1996) demonstrated that $\phi = 45°$ is still the optimal direction at which to carry out the off-axis tensile test to approximate the shear behavior of the fabric. Also, it was concluded from our study that the shear strength of woven fabrics is lower than its lowest tensile strength in any direction.

3.5   Other forms of failure for fabrics and garments

3.5.1   Yarn cut

Given the fact that most fabric failure is originated by yarn breakage due to cutting with a sharp object, a measure of yarn cut resistance is needed for the material model. Unfortunately, yarn cut resistance is not commonly measured, and few data exist (Shin *et al.*, 2003, 2006). SRI recently developed a test procedure for evaluating the cut resistance of yarns under tension–shear loading conditions (Shin *et al.*, 2006). The test presses a knife blade transversely at a constant rate against a yarn gripped at its ends, measures the load–deflection relation, and computes the energy required to cut through the yarn. They reported that the cut resistance of all materials depends strongly first on slice angle. Cut energy dropped sharply when the slice angle deviated from 90°, falling about 50 to 75% at 82.5°, and decreasing further but more gradually at lower angles. At a 45° slice angle, cut energies were from 3% to 10% of the 90° values.

Next, they reported that the cut resistance also depends strongly on blade sharpness. At a 90° slice angle, a blade with a 2 mm tip radius required 47 to 75% less energy to cut yarns than a 20 mm blade, and at a slice angle of 45°, only 17 to 35%. Furthermore, the yarn pre-tension reduces cut resistance, and for 90° slice angles, axial loads on the yarns of up to 3 lb reduced the cut resistance by 32 to 40%. However, no comparison of the cut resistance with tensile yarn strength was given in the study.

3.5.2 Fabric tear

Increased application of coated fabrics has been demanding a better understanding of the behavior of the material, which will, in turn, help to optimize material design and textiles structural configurations for coated fabrics under complex loading conditions. As a critical indicator of serviceability of a fabric, tearing strength is rigorously examined when estimating the useful life of the fabric, for fabric is most vulnerable under a tearing load. A tear slit can propagate even with very low force for every step along the way; only a few yarns (or, in the worst case, a single yarn if the fabric is tight enough) are in the way to resist the propagation. That is why for a fabric the tensile strength is always much greater than the tear strength. Although there are simple ways to measure such strength, theoretical prediction and modeling remain difficult, due to the many variables that contribute to the complicated mechanisms involved in the tearing process (Hamkins and Backer, 1980; Krook and Fox, 1945; Mukhopadhyay et al., 2006; Scelzo et al., 1994a, 1994b; Taylor, 1959; Teixeira et al., 1955; Teutelink et al., 2003; Witkowska and Frydrych, 2004, 2005; Zhong et al., 2004). Examples include the development of a del-zone in a tongue tear test (see Fig. 3.2). The del-zone is a delta-shaped opening composed of the stretched part of the fabric that bridges the gap between the two tongues, and it serves to sustain the tearing load at the crack front, to prevent the remarkable yarn movement, slippage and even jamming during fabric tearing (to list just a few).

These are problems that have attracted much attention in the past 50 years (see the citations listed in the previous paragraph). However, for a description of the tearing behavior with acceptable accuracy, no satisfactory model is presently available, while current research work so intended has been mainly experimental and semi-empirical (Hamkins and Backer, 1980; Scelzo et al., 1994b; Taylor, 1959). Nonetheless, these models can still be useful if properly formulated. For instance, the prediction of tearing strength by Taylor (1959) (in a stress based model) was expressed as

$$T_R = T_R(\mu, f, \theta, p, D, f_{sn})   \text{[3.13]}$$

where T_R = predicted tongue tear strength, μ = coefficient of (yarn-on-yarn) sliding friction, f = mean breaking strength value of the del yarns, θ = half of the arc of contact (or wrap) angle (radians), p = inter-yarn spacing, D = sum of the warp and filling yarn diameters, f_{sn} = sliding force past n cross-yarns. As can be seen from the equation, some of the parameters are unlikely to be independent. Therefore a dimensional analysis could combine them into fewer yet independent variables to enable development of more useful and robust models.

Zhong *et al.* (2004) employed a stochastic approach, using the Ising model combined with Monte Carlo simulation, to study the phenomenon of tongue tear failure in coated fabrics. In this approach the complicated mechanisms involved can be realistically simulated with a relatively simple algorithm. The important factors, especially the effects of the interphase between the coating and fabric, and the stretched part of the material at the crack front (the del-zone) can be represented by corresponding coefficients in the Hamiltonian expression of the system. The minimization of the system Hamiltonian yields the most likely new steps for crack propagation, while the Monte Carlo method is used to select the one that will actually occur, reflecting the stochastic nature in the behavior of real testing. However, this model, like many others, needs to be calibrated based on actual testing data for quantitative and accurate predictions.

Note: One thing for sure is that how movable the yarns are in a fabric is *the* decisive factor. High yarn mobility allows yarns to retreat from being broken, and thus jammed yarns will then collectively resist the tear force. This explains the tear strength of a piece of cheesecloth being much higher than other thick and stiff fabrics.

3.5.3 Dynamic failures

Textiles are, in general, made from polymeric fibers, and thus can be treated as polymer sheets. Because of the complexity of the macromolecular morphology, polymers exhibit viscoelasticity under loading. In other words, the mechanical behaviors of textile fabrics are time or strain rate dependent. At different loading rates, the same fabric can behave in vastly distinctive ways. To predict fabric behaviors in high deformation speed based on the static data is usually unreliable, and even fatal in extremely high rate cases, such as ballistic processes.

Stab resistance

One important application for textile fabrics is acting as body armor, stabbing assaults being a likely threat (Chadwick *et al.*, 1999; Decker *et al.*, 2007; Jones *et al.*, 1994; Mahfuz *et al.*, 2004; Walker *et al.*, 2004). For military applications, the increasing relevance of close-quarters, urban conflict necessitates the development of armor systems with stab-resistant capabilities. Stab threats encountered by soldiers in the field include direct attacks from knives and sharpened instruments, as well as physical contact with debris, broken glass, and razor wire.

Stab threats can be classified into two categories (Decker *et al.*, 2007): puncture and cut. Puncture refers to penetration by instruments with sharp tips but no cutting edge, such as ice picks or awls. These threats are of primary concern to correctional officers, since sharply-pointed objects are relatively easy to improvise. Cut refers to contact with knives with a continuous cutting edge. Knife threats are generally more difficult to stop than puncture, since the long cutting edge presents a continuous source of damage initiation during the stab event.

Bullet penetration

Numerous studies have been conducted on the ballistic impact of high-strength fabric structures (Bazhenov, 1997; Billon and Robinson, 2001; Briscoe and Motamedi, 1992; Cheeseman and Bogetti, 2003; Cunniff, 1992, 1996; Pargalanda and Hernandezolivares, 1995; Roylance *et al.*, 1973; Shim *et al.*, 1995, 2001; Starratt *et al.*, 2000; Tan *et al.*, 1997, 2003; Tan and Khoo, 2005; Wilde *et al.*, 1973).

Cunniff (1992) states that the energy absorption characteristics of fabric systems under ballistic impact are influenced by a number of factors including fiber properties, weave style, the number of fabric layers, areal density, projectile parameters, and impact parameters, and later on develops a semi-empirical model to study the ballistic process (Cunniff, 1996). Additionally, Bazhenov (1997), Briscoe and Motamedi (1992), and Tan *et al.* (1997, 2003, 2005) have shown, through experiments, that interfacial friction within ballistic impact systems is also an important factor that affects fabric energy absorption capacity.

Among the parameters involved, the influences of the fabric frictional behavior seem to be most complex. Work using finite element analysis by Duan *et al.*, (2005) revealed that the friction contributed to delaying fabric failure and increasing impact load. The delay of fabric failure and increase of impact load allowed the fabric to absorb more energy, as shown in Fig. 3.5. Results in the figure from the modeling effort also indicate that the fabric boundary condition is also an important factor that influenced the effect of friction. The fabric more effectively reduced the projectile residual velocity when only two edges were clamped so that the fabric had more freedom to 'give in'.

3.6   Fabric and garment failure reduction

Fabrics are most vulnerable under tear loading so the most effective way to increase fabric durability is not choosing a stronger fiber or yarn, but making the fabric more tear resistant. As discussed above, improving the yarn mobility is the most effective way for that.

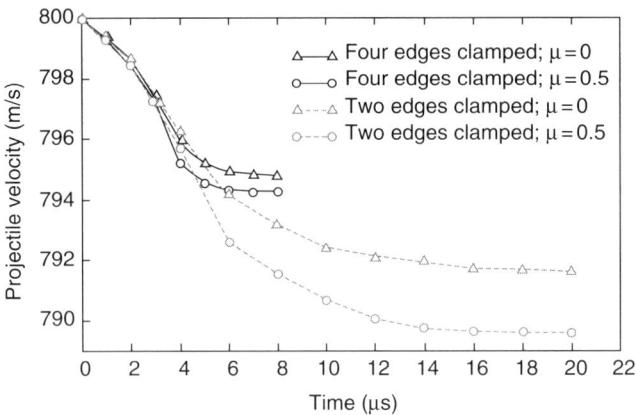

3.5 Time history of the projectile velocity for the four cases with different boundary and friction conditions (Duan *et al.*, 2005).

It seems logical to expect that stronger fiber or yarn will improve the fabric resistance to load. However, since fibers and yarns are best for carrying tensile loads, how much the strength of a stronger fiber or yarn can be translated into the strength of the fabric is a very complex matter (Pan, 1996). In other words, stronger fibers or yarns do not necessarily lead to a stronger fabric, for the way fibers and yarns are assembled in forming the fabric, and thus the interactions among them, is just as important. More importantly, a fiber stronger in tensile strength can reinforce the tensile strength of the fabric only under the same testing conditions. Given that fibers are tested only in extension at a pseudo-static strain rate and under standard atmospheric conditions according to all the industry standards, extra caution must be taken at the design stage when trying to predict the fabric performance under extreme conditions.

By the same token, the durability of a garment is determined by both the fabric properties and the garment assembling techniques. A garment becomes non-functional if the seams fail, regardless whether the fabric pieces are still intact or not.

In improving the resistance to stabbing and bullets, the biggest constraint is the tolerance of the armor wearer. Here the obvious challenge lies in a proper balance among the competing factors such as weight, breathability/comfort and resistance of the armor, especially the need to maintain the comfort level of the garment within the tolerable range. Materials light in weight yet high in toughness are readily available, but nothing has shown serious potential to replace polymeric fibers in the near future as the textile material for the next generation.

3.7   References

AMBARTSUMYAN, S. A. (1965). The axisymmetric problem of circular cylinder shell made of materials with different stiffnesses in tension and compression. *Izv. Akad. Nauk. SSSR Mekh.*, *4*, 77–85.

AMBARTSUMYAN, S. A. (1969). Basic equations and relations of the different modulus theory of elasticity of an anisotropic body. *Mechanics of Solids*, *4*, 48–56.

AMBARTSUMYAN, S. A. & KHACHATRYAN, A. A. (1966). The basic equations of the theory of elasticity for materials with different stiffnesses in tension and compression. *Mechanics of Solids*, *1*, 29–34.

AZZI, V. D., & TSAI, S. W. (1965). Anisotropic strength of composites. *Experimental Mechanics*, *5*, 283.

BAZHENOV, S. (1997). Dissipation of energy by bulletproof aramid fabric. *Journal of Materials Science*, *32*(15), 4167–4173.

BILLON, H. H., & ROBINSON, D. J. (2001). Models for the ballistic impact of fabric armour. *International Journal of Impact Engineering*, *25*(4), 411–422.

BRISCOE, B. J., & MOTAMEDI, F. (1992). The Ballistic Impact Characteristics of Aramid Fabrics – The Influence of Interface Friction. *Wear*, *158*(1–2), 229–247.

BRUNO, D., LATO, S., & ZINNO, R. (1993). Nonlinear Analysis of Doubly Curved Composite Shells of Bimodular Material. *Composites Engineering*, *3*(5), 419–435.

CARNABY, G. A., & PAN, N. (1989). Theory of the Compression Hysteresis of Fibrous Assemblies. *Textile Research Journal*, *59*(5), 275–284.

CHADWICK, E. K. J., NICOL, A. C., LANE, J. V., & GRAY, T. G. F. (1999). Biomechanics of knife stab attacks. *Forensic Science International*, *105*(1), 35–44.

CHEESEMAN, B. A., & BOGETTI, T. A. (2003). Ballistic impact into fabric and compliant composite laminates. *Composite Structures*, *61*(1–2), 161–173.

CHENG, S., & TAN, S. C. (1989). Failure criteria for fibrous anisotropic materials. In *Mechanics of Cellulosic and Polymeric Materials, AMD-V. 99, MD-V. 13, Proc. of the third joint ASCE/ASME Mechanics Conference* (p. 87).

CUNNIFF, P. M. (1992). An Analysis of the System Effects in Woven Fabrics under Ballistic Impact. *Textile Research Journal*, *62*(9), 495–509.

CUNNIFF, P. M. (1996). A semiempirical model for the ballistic impact performance of textile-based personnel armor. *Textile Research Journal*, *66*(1), 45–59.

DECKER, M. J., HALBACH, C. J., NAM, C. H., WAGNER, N. J., & WETZEL, E. D. (2007). Stab resistance of shear thickening fluid (STF)-treated fabrics. *Composites Science and Technology*, *67*(3–4), 565–578.

DUAN, Y., KEEFE, M., BOGETTI, T. A., & CHEESEMAN, B. A. (2005). Modeling the role of friction during ballistic impact of a high-strength plain-weave fabric. *Composite Structures*, *68*(3), 331–337.

ELTAHAN, W. W., STAAB, G. H., ADVANI, S. H., & LEE, J. K. (1989). Structural-Analysis of Bimodular Materials. *Journal of Engineering Mechanics–ASCE*, *115*(5), 963–981.

HAMKINS, C. P., & BACKER, S. (1980). On the Mechanisms of Tearing in Woven Fabrics. *Textile Research Journal*, *50*(5), 323–327.

HILL, R. (1948). A Theory of the Yielding and Plastic Flow of Anisotropic Metals. *Proceedings of the Royal Society of London, Series A–Mathematical and Physical Sciences*, *193*(1033), 281–297.

HOFFMAN, O. (1967). The brittle strength of orthotropic materials. *J. Composite Materials*, *1*, 200.

JONES, S., NOKES, L., & LEADBEATTER, S. (1994). The Mechanics of Stab Wounding. *Forensic Science International*, *67*(1), 59–63.

KILBY, W. F. (1963). Plannar stress–strain relationship in woven fabrics. *J. Textile Inst.*, *54*, T9.

KROOK, C. M., & FOX, K. R. (1945). Study of the Tongue-Tear Test. *Textile Research Journal*, *15*(11), 389–396.

LIU, J. Y. (1984). Evaluation of the Tensor Polynomial Strength Theory for Wood. *Journal of Composite Materials*, *18*(3), 216–226.

MAHFUZ, H., MAJUMDAR, P., SAHA, M., SHAMERY, F., & JEELANI, S. (2004). Integral manufacturing of composite skin–stringer assemblies and their stability analyses. *Applied Composite Materials*, *11*(3), 155–171.

MINSTER, J. (1994). Failure Criteria for 2-Dimensional Orthotropic Fibrous Composites of Low Bending Stiffness. *Geotextiles and Geomembranes*, *13*(2), 119–126.

MUKHOPADHYAY, A., GHOSH, S., & BHAUMIK, S. (2006). Tearing and tensile strength behaviour of military khaki fabrics from grey to finished process. *International Journal of Clothing Science and Technology*, *18*(3–4), 247–264.

NECKAR, B. (1997). Compression and packing density of fibrous assemblies. *Textile Research Journal*, *67*(2), 123–130.

NORRIS, C. B. (1950). *Strength of orthotropic materials subjected to combined stress.* Madison, WI USDA Forest Service, Forest Products Laboratory.

PAN, N. (1996). Analysis of woven fabric strengths: Prediction of fabric strength under uniaxial and biaxial extensions. *Composites Science and Technology*, *56*(3), 311–327.

PAN, N., ZERONIAN, H., & RYU, H. S. (1992). An alternative approach to the objective measurement of fabrics. *Textile Research Journal*, *62*, 33.

PARGALANDA, B., & HERNANDEZOLIVARES, F. (1995). An Analytical Model to Predict Impact Behavior of Soft Armors. *International Journal of Impact Engineering*, *16*(3), 455–466.

PIPES, R. B., & COLE, B. W. (1973). Off-axis Strength Test for Anisotropic Materials. *Journal of Composite Materials*, *7*(Apr), 246–256.

POUYET, J., HUCHON, R., & VIDAL, F. (1990). Predicting compressive and shear strengths of polymer or paper sheets from off-axial tensile tests. In *Mechanics of Wood and Paper Materials* (Vol. ASME AMD-V. 112, MD-V. 23, p. 99).

ROWLANDS, R. E., GUNDERSON, D. F., SUHLING, J. C., & JOHNSON, M. W. (1985). Biaxial Strength of Paperboard Predicted by Hill-type Theories. *Journal of Strain Analysis for Engineering Design*, *20*(2), 121–127.

ROYLANCE, D., WILDE, A., & TOCCI, G. (1973). Ballistic Impact of Textile Structures. *Textile Research Journal*, *43*(1), 34–41.

SACCO, E., & REDDY, J. N. (1992). A Constitutive Model for Bimodular Materials with an Application to Plate Bending. *Journal of Applied Mechanics–Transactions of the ASME*, *59*(1), 220–221.

SCELZO, W. A., BACKER, S., & BOYCE, M. C. (1994a). Mechanistic Role of Yarn and Fabric Structure in Determining Tear Resistance of Woven Cloth.1. Understanding Tongue Tear. *Textile Research Journal*, *64*(5), 291–304.

SCELZO, W. A., BACKER, S., & BOYCE, M. C. (1994b). Mechanistic Role of Yarn and Fabric Structure in Determining Tear Resistance of Woven Cloth.2. Modeling Tongue Tear. *Textile Research Journal*, *64*(6), 321–329.

SHIM, V. P. W., LIM, C. T., & FOO, K. J. (2001). Dynamic mechanical properties of fabric armour. *International Journal of Impact Engineering*, *25*(1), 1–15.

SHIM, V. P. W., TAN, V. B. C., & TAY, T. E. (1995). Modeling Deformation and Damage Characteristics of Woven Fabric under Small Projectile Impact. *International Journal of Impact Engineering*, *16*(4), 585–605.

SHIN, H. S., ERLICH, D. C., & SHOCKEY, D. A. (2003). Test for measuring cut resistance of yarns. *Journal of Materials Science*, *38*(17), 3603–3610.

SHIN, H. S., ERLICH, D. C., SIMONS, J. W., & SHOCKEY, D. A. (2006). Cut resistance of high-strength yarns. *Textile Research Journal*, *76*(8), 607–613.

STARRATT, D., SANDERS, T., CEPUS, E., POURSARTIP, A., & VAZIRI, R. (2000). An efficient method for continuous measurement of projectile motion in ballistic impact experiments. *International Journal of Impact Engineering*, *24*(2), 155–170.

SUHLING, J. C., ROWLANDS, R. E., JOHNSON, M. W., & GUNDERSON, D. E. (1985). Tensorial Strength Analysis of Paperboard. *Experimental Mechanics*, *25*(1), 75–84.

TAN, P., TONG, L., & STEVEN, G. P. (1997). Modelling for predicting the mechanical properties of textile composites – A review. *Composites, Part A–Applied Science and Manufacturing*, *28*(11), 903–922.

TAN, V. B. C., & KHOO, K. J. L. (2005). Perforation of flexible laminates by projectiles of different geometry. *International Journal of Impact Engineering*, *31*(7), 793–810.

TAN, V. B. C., LIM, C. T., & CHEONG, C. H. (2003). Perforation of high-strength fabric by projectiles of different geometry. *International Journal of Impact Engineering*, *28*(2), 207–222.

TAYLOR, H. M. (1959). Tensile and Tearing Strength of Cotton Cloths. *J. Textile Inst.* (50), T161–T188.

TEIXEIRA, N. A., PLATT, M. M., & HAMBURGER, W. J. (1955). Mechanics of Elastic Performance of Textile Materials, Part XII: Relation of Certain Geometric Factors to the Tear Strength of Woven Fabrics. *Textile Res. J.*, *25*, 838–861.

TEUTELINK, A., VAN DER LAAN, M. J., MILNER, R., & BLANKENSTEIJN, J. D. (2003). Fabric tears as a new cause of type III endoleak with Ancure endograft. *Journal of Vascular Surgery*, *38*(4), 843–846.

THEOCARIS, P. S., & PHILIPPIDIS, T. P. (1989). Extremum Properties of the Failure Function in Initially Anisotropic Elastic Solids. *International Journal of Fracture*, *41*(1), R9–R13.

TSAI, S. W., & WU, E. M. (1971). General Theory of Strength for Anisotropic Materials. *Journal of Composite Materials*, *5*(Jan), 58.

VAN WYK, C. M. (1946). Note on the compressibility of wool. *Journal of Textile Institute*, *37*, 282.

WALKER, C. A., GRAY, T. G. F., NICOL, A. C., & CHADWICK, E. K. J. (2004). Evaluation of test regimes for stab-resistant body armour. *Proceedings of the Institution of Mechanical Engineers, Part L – Journal of Materials – Design and Applications*, *218*(L4), 355–361.

WILDE, A. F., ROYLANCE, D. K., & ROGERS, J. M. (1973). Photographic Investigation of High-speed Missile Impact Upon Nylon Fabric .1. Energy Absorption and Cone Radial Velocity in Fabric. *Textile Research Journal*, *43*(12), 753–761.

WITKOWSKA, B., & FRYDRYCH, I. (2004). A comparative analysis of tear strength methods. *Fibres & Textiles in Eastern Europe*, *12*(2), 42–47.

WITKOWSKA, B., & FRYDRYCH, I. (2005). Protective clothing – test methods and criteria of tear resistance assessment. *International Journal of Clothing Science and Technology*, *17*(3–4), 242–252.

WU, R. Y., & STACHURSKI, Z. (1984). Evaluation of the Normal Stress Interaction Parameter in the Tensor Polynomial Strength Theory for Anisotropic Materials. *Journal of Composite Materials*, *18*(5), 456–463.

ZHONG, W., PAN, N., & LUKAS, D. (2004). Stochastic modelling of tear behaviour of coated fabrics. *Modelling and Simulation in Materials Science and Engineering*, *12*(2), 293–309.

4

The sensory properties and comfort of military fabrics and clothing

A. V. CARDELLO, US Army Natick Soldier Research, Development and Engineering Center, USA

4.1   Introduction

4.1.1 The role of comfort in military clothing

Throughout the ages, fighting men have worn protective clothing or armor, and since the Middle Ages military forces have adopted standards of military clothing that we refer to as 'uniforms'. The primary purposes of military clothing have always been protection, functionality, and identification – protection from projectiles, explosions, fire, extreme environments, chemical and biological toxins, or radiation; functionality to aid in the performance of military tasks quickly, effectively and with a minimum of energy expenditure; and identification of friend and foe. In general, comfort has taken a secondary role to these other factors.

In a paper on protective clothing entitled 'Comfort or protection: the clothing dilemma', Slater (1996) captured the difficult trade-offs that must be made between providing protection *vs* comfort in clothing. While citing the obvious need for protection, Slater (1996) makes the point that 'human beings cannot function satisfactorily if they are not completely comfortable'. Thus, protective fabrics with low moisture permeability can create heat stress and profuse sweating in the wearer, impeding visual, cognitive, and physical performance. Abrasive materials can cause chafing of the skin and accompanying discomfort that can interfere with attention and performance. In addition, psychological factors related to attitudes and beliefs toward the garment and its ability to protect the wearer can create psychological discomfort that interferes with motivation and willingness to perform high-risk assignments.

During the past several years, there has been a growing realization that effective military clothing design requires greater consideration of comfort factors. The United States Department of Defense procures over 1.1 billion dollars of clothing and individual equipment each year. In order to make the most effective use of these expenditures, all clothing designed for military use has the multi-purpose goal of protecting the soldier and enabling him/her to function effectively, while at the same time maintaining his/her comfort within a range that minimizes physical, cognitive, or other performance decrements on the battlefield.

4.1.2 Defining comfort and its components

The word 'comfort' has a variety of meanings as it relates to clothing and to the wearer. Foremost among these for military clothing has been the notion of 'thermal comfort', i.e. the comfort or discomfort associated with how hot or cold the individual feels. Thermal comfort is closely associated with changes in physiological variables, such as skin and core temperature, and is a function of environmental variables, e.g. temperature, humidity, and wind speed; the activity level of the individual; and clothing properties, such as the fabric's insulation value and water vapor permeability. Due to its close association with changes in physiologically measured variables, thermal comfort has often been quantified using physiological parameters. However, thermal 'comfort' is a *psychological* concept. The word comfort refers to how the individual 'feels'. Under the same environmental conditions and with the same clothing, one individual may feel 'hot' and the other may feel 'cool'. Similarly, identical skin and core temperatures in two different individuals do not mean that they will *feel* equally hot or cold. Furthermore, two people who feel equally hot or cool from a perceptual standpoint, may not be equally comfortable or uncomfortable. The thermal comfort of an individual is a relative concept that can only be assessed through subjective assessments made by the individual.

Another interpretation of comfort derives from the tactile sensations that result from fabrics in contact with the skin. For example, a military garment may feel smooth or rough against the skin. Depending upon the degree of smoothness or roughness, that sensation might be characterized as comfortable or uncomfortable. Unfinished wool garments worn by early troops were notorious for their coarse feel and associated discomfort. Other physical and tactile aspects of fabrics, such as stiffness, thickness, fuzziness, or thermal 'feel' can also impact tactile comfort. The importance of the skin feel characteristics of fabrics used in clothing and their role in garment comfort can easily be seen by walking through a department store and observing shoppers as they feel garments by rubbing the fabric between their fingers, passing their palm over the surface of the garment or even brushing the fabric against their face. Since many items of military clothing are worn on a daily basis in routine, non-combat situations, e.g. in garrison, the tactile comfort or discomfort of the clothing in these situations is likely to be an equally important factor to its overall comfort and performance as either its protective or insulative properties.

A third component of comfort is that which arises from the fit of the garment. A poorly fitting garment, especially if too small, will produce discomfort and impede mobility and performance. If too large, the garment may also impede mobility and performance, although the impact on comfort may not be as great. The fit of the garment can also influence psychosocial perceptions of the self through personal or cultural preferences regarding fit and fashion–size trends. Although the latter factors may play less of a role in military clothing, the protective element of military clothing can influence other aspects of the psychological comfort with these garments.

Soldier attitudes and beliefs regarding the efficacy of the protective aspects of military clothing can significantly impact the 'psychological comfort' with clothing. If a military garment is designed to protect the soldier against chemical or biological threats, but the soldier does not have confidence in the garment to do that, he or she may experience anxiety and a state of psychological discomfort. Slater (1996) has discussed the concept of psychological discomfort within the context of trade-offs between clothing comfort and protection. He cites the example of protective vests that are designed to be thin and lightweight for comfort, but that leave the wearer with the perception that the vest is too thin and lightweight to be of sufficient protective value, or the situation in which a vest, though protective of the torso, still exposes the wearer's head and limbs. These psychological sources of discomfort can influence the wearer when he/she must make decisions regarding exposure to chemical, biological, ballistic or other battlefield threats.

Finally, whether we are discussing thermal comfort, sensory skin-feel comfort, comfort due to fit, or the psychological comfort of clothing, each of these can have considerable impact on the individual's physical and cognitive performance and, in turn, on mission performance. For this reason, comfort must be seen as an essential element in all areas of military clothing design.

4.1.3 Chapter goals

As suggested by this Introduction, the problem of military clothing comfort is a complex one that involves a wide range of physiological, sensory, social, cultural and psychological factors. The focus of the present chapter will be on the *sensory* and *psychological* factors that influence the comfort or discomfort of military fabrics and clothing. From this perspective, we will examine the influence of fabric sensory characteristics on skin feel comfort, the ability to predict comfort from the sensory properties of fabrics, the relationship of sensory fabric comfort to wear comfort, the influence of attitudes and beliefs on perceptions of clothing comfort, and the influence of comfort factors on cognitive performance of the wearer. This chapter will provide a psychological perspective on military clothing comfort and will focus on contemporary methods and approaches to quantifying the sensory and psychological dimensions of comfort. Finally, the chapter will present a variety of empirical applications of these methods to problems of military clothing.

4.2   The sensory and perceptual properties of fabrics and clothing

4.2.1 Sensory experience, cognition and affect

The *sensory* elements of fabrics and clothing are the discrete sensations that arise from stimulation of human sensory receptors. The sensations that can arise from clothing include those related to touch (somesthesis), e.g. the perception of fabric roughness; limb position (kinesthesis), e.g. the perception of restricted range of arm positions; vision, e.g. the color and appearance of the garment; audition, e.g. the perception of sound being emitted upon motion or when the fabric brushes together; and olfaction, e.g. the smell of clothing worn for prolonged periods without laundering.

All sensory experiences, regardless of the sensory system through which they arise, consist of the two distinct psychological dimensions of *quality* and *magnitude*. Quality refers to the *qualitative* nature of the sensation, i.e. it is a *soft* garment *vs* a *stiff* one, it is a *beige* color *vs* a *blue* one, it makes a *high pitched* noise *vs* it makes a *low pitched* rustle. Magnitude, on the other hand, refers to the *intensity* of the sensation. Thus, the garment is *very* stiff or only *slightly* so, the fabric is *moderately* rough *vs extremely* rough, or the garment makes a *loud* noise during movement or a *barely perceptible* one.

Multiple sensations arising from a garment will combine with elements of past experiences, memory, attitudes and beliefs to form an overall perception of the garment. These non-sensory contributors to perception are referred to as *cognitive* elements, and their impact can be to alter perceptions of garment feel, comfort and perceived efficacy. Thus, understanding the role of cognitive variables in clothing can improve knowledge of how the comfort or discomfort of clothing is influenced by factors unrelated to the garment or garment fabric itself.

Comfort, unlike sensory experience or cognition, is an emotional experience that arises as a result of the combined effects of sensory and cognitive elements. Comfort and discomfort are *affective* emotions that relate to the pleasantness or unpleasantness that is experienced when the fabric or garment is perceived. Thus, the three distinct psychological elements of the human experience of fabrics and garments that we will be discussing in this chapter are *sensory* experiences, *cognitive* influences and *affective* emotions.

4.2.2 The sensory or 'handle' properties of fabrics and clothing

Early approaches to handfeel analysis

The sensory system by which we experience sensations from the skin is known as 'somesthesis'. In spite of the seemingly large range of human experiences that are produced when we hold or feel objects, the sensory qualities of skin sensations are limited. They include light pressure (touch), deep pressure, vibration, pain, cold, and heat. For each of these sensory 'qualities' there is a set of sensory receptors in the skin that mediate that sensation. However, when fabrics or garments are felt in the hand, or otherwise come into contact with the skin and body, there is a complex interaction of these *somesthetic* sensations with *kinesthetic* sensations that are produced by receptors in the joints of the fingers, wrists, arms, legs, ankles, toes and elsewhere that respond to body and limb movement and position. The combination of these tactile and kinesthetic sensations is what produces the complex set of perceptions that we experience when holding objects or fabrics in our hand or when wearing clothing and garments.

In textile applications, the 'feel' of fabrics, which is typically done with the human hand, has come to be called its 'hand' or 'handle'. This term generally refers to the range of sensory and perceptual experiences that are encountered when fabrics or garments are felt, handled, or otherwise manipulated by humans. Thus, the softness, stiffness, roughness, etc. of a fabric or garment are all considered properties of its 'hand' or 'handfeel'. In the past, experts were utilized to make these judgments. In some arenas, the term 'handle' has been used to describe the mechanical forces measured by instruments when they come into contact with or are used to manipulate fabrics. However, this usage of the term is misleading, both semantically, because these instrumental methods do not involve the use of a hand, and logically, because they are merely secondary parameters related to measurable physical forces of the fabrics that only derive their validity through established associations with sensory measures obtained using humans.

Early investigators devised a variety of methods and terminology to describe the subjective responses to fabrics and clothing, including its handfeel (e.g. Binns, 1926; Pierce, 1930; Winslow *et al.*, 1937). However, much of this research failed to use systematic approaches for defining terminology, measuring operational constructs or defining who *is* or *should be* qualified to make handfeel judgments (see Brand, 1964; Winakor *et al.*, 1980; Yick *et al.*, 1995). It was not until the 1970s that better theoretical and methodological approaches for describing and measuring the sensory and perceptual attributes of fabrics and clothing were developed (Fourt and Hollies, 1970; Rohles, 1971; Slater, 1977, 1985; Pontrelli, 1977; see Branson and Sweeney (1991) for an historical summary of this work).

A contemporary approach to quantifying handfeel

In a review of the area of handfeel analysis, Civille and Dus (1990) concluded that existing methods for assessing the sensory properties of fabrics had numerous problems, including the failure to identify primary tactile characteristics, lack of standardized methods, improper approaches to scaling, lack of specification of subject/panelist training, and failure to use proper test controls. As a consequence, Civille and Dus (1990) developed the Handfeel Spectrum Descriptive Analysis (HSDA) method for the evaluation of woven and non-woven fabrics, patterning it after highly successful, descriptive methods that have been used to assess the sensory characteristics of foods, beverages, perfumes and skin care products. This new methodology standardized handfeel terminology and operationally defined its methods of analysis. It employs a wide range of tactile and sound attributes, as shown and defined in Table 4.1. Although any one of a variety of psychophysical scaling methods can be used to rate the magnitude or strength of each attribute, Civille and Dus (1990) proposed the use of a 15-point intensity scale with physical fabric standards serving as references along the attribute scales. The attribute terms and protocols for the HSDA method have been approved by the Other Senses Task Group (E18.02.06.03) of ASTM Committee E-18, and a number of researchers have now adopted this method for the descriptive analysis of textiles (Robinson *et al.*, 1994, 1997; Cardello *et al.*, 2003). Other approaches to handfeel evaluations have also been proposed during the past 15 years (e.g. Jacobsen *et al.*, 1992 and Philippe *et al.*, 2004), but the HSDA method is the most well standardized and validated.

4.3   The comfort properties of fabrics and clothing

4.3.1   Comfort as a theoretical and measurement construct

As noted in Section 4.2.1, comfort is an emotional experience that results from a variety of factors related to the individual, his/her clothing, the environment, cognitive and psychological influences, and past learning and experience. A number of methods for quantifying the comfort (or discomfort) of fabrics and clothing have been developed over the years. One of the first widely accepted methods was developed by Gagge *et al.* (1967). This method employed a simple 4-point category scale that ranged from

Table 4.1 The 17 handfeel attributes of the Handfeel Spectrum Descriptive Sensory method and their definitions

Attributes	Definitions
Grainy	The amount of small, round particles in the surface of the sample.
Gritty	The amount of small, abrasive, picky particles in the surface of the sample.
Fuzziness	The amount of pile, fiber, fuzz on the surface of the sample.
Thickness	The perceived distance between the thumb and index finger (when the sample is placed between the two).
Tensile stretch	The degree to which the sample stretches from its original shape.
Hand friction	The force required to move the palm of the hand across the surface of the sample.
Fabric–fabric friction	The force required to move the fabric over itself.
Depression depth	The amount that the sample depresses when downward force is applied.
Springiness	The rate at which the sample returns to its original position after the downward force is released.
Force to gather	The amount of force required to compress the gathered sample into the palm.
Stiffness	The degree to which the sample feels pointed, ridged and cracked; not pliable.
Force to compress	The amount of force required to compress the gathered sample into the palm.
Fullness/volume	The amount of material felt in the hand.
Compression resilience intensity	The perceived force with which the sample exerts resistive pressure against the cupped hands.
Compression resilience rate	The rate at which the sample returns to its original shape or the rate at which the sample opens after compression.
Noise intensity	The loudness of the noise.
Noise pitch	The pitch (frequency) of the noise.

'comfortable' through 'slightly uncomfortable', 'uncomfortable' and 'very uncomfortable'. Unfortunately, this scale suffered from a limited number of scale points, which limited its sensitivity, and from imbalance in the scale, i.e. there were three levels of discomfort but only one level of comfort. Other category scales have been developed to quantify 'thermal comfort', such as the McGinnis Thermal Scale (Hollies *et al.*, 1979) which requires individuals to rate their subjective experience on a 13-point scale that ranges from 'I am so cold, I am helpless' to 'I am so hot, I am sick and nauseated'. Although this scale employs a sufficient number of scale points, the labels on the scale use a mix of sensory, affective and behaviorally oriented terminology, which confound multiple dimensions of comfort experience and behavior.

In most areas of psychological measurement, category scales have given way to better and more sophisticated psychophysical methods. The reasons for this are (i) that the points on category scales have been shown to be unequal in their subjective intervals (Stevens and Galanter, 1957), (ii) that subjects tend not to use the end categories (Stevens and Galanter, 1957; Guilford and Dingman, 1955), thus reducing the effective length of the scale, and (iii) that bi-directional category scales that employ a 'neutral' or null category encourage subjects to be non-committal in their responses (Gridgeman, 1961).

4.3.2 A new method for scaling comfort

Recently, a scale for measuring comfort was developed at the US Army Natick Soldier RD&E Center (NSRDEC) using contemporary psychophysical scaling techniques (Cardello *et al.*, 2003). This Comfort Affective Labeled Magnitude (CALM) scale was modeled after earlier labeled magnitude scales developed by Borg (1982) for perceived exertion, by Green *et al.* (1993) for the magnitude of oral sensations, and by Schutz and Cardello (2001) for measuring liking/disliking. The scale was developed by having consumers rate the semantic meaning of 43 different words and phrases that can be used to describe comfort or discomfort. Each word or phrase was judged for the magnitude of comfort/discomfort that it expressed using the method of magnitude estimation (Stevens, 1957; Sweeney and Branson, 1990).

Table 4.2 shows the data obtained by this method. The data are the geometric means of the magnitude estimates assigned by subjects to index the semantic meaning of the 43 phrases of comfort or discomfort. Examination of the data reveals the non-equivalence of intervals between points on the Gagge *et al.* (1967) comfort sensation scale (bolded in Table 4.2). Note that the interval between the phrases 'uncomfortable' and 'very uncomfortable' on the Gagge *et al.* scale is 113 perceptual units, while the interval between the phrases 'uncomfortable' and 'slightly uncomfortable' on that scale is only 43 perceptual units.

Using the data in Table 4.2, the CALM scale shown in Fig. 4.1 was created. This comfort scale employs a line with the end-points labeled 'greatest imaginable discomfort' and 'greatest imaginable comfort' and with 'neither comfortable nor uncomfortable' located in the middle. The two end-point labels are critical to the psychophysical theory underlying the scale, because these labels enable the valid comparison of ratings among people who may

Table 4.2 Geometric mean magnitude estimates, standard errors and standard errors of the geometric means for the semantic meaning of 43 different comfort-related phrases as determined using a magnitude estimation procedure (from Cardello *et al.*, 2003). Bolded phrases are those used in Gagge *et al.*'s (1967) comfort sensation scale

Comfort/discomfort word phrases	Geom. mean mag. est.	Standard error	Standard error/G.M.
Greatest imaginable comfort	366.72	34.88	0.10
Greatest possible comfort	345.28	28.76	0.08
Exceptionally comfortable	280.20	16.03	0.06
Superior comfort	279.71	19.27	0.07
Intensely comfortable	268.44	19.82	0.07
Extremely comfortable	260.75	23.51	0.09
Highly comfortable	224.01	15.80	0.07
Very comfortable	203.99	13.96	0.07
Terribly comfortable	135.93	48.72	0.36
Moderately comfortable	130.18	10.51	0.08
Comfortable	109.22	10.81	0.10
Satisfactory comfort	86.11	11.68	0.14
Fairly comfortable	85.16	8.62	0.10
Average comfort	77.58	17.30	0.22
Acceptable comfort	72.17	8.85	0.12
Somewhat comfortable	59.98	9.07	0.15
Slightly comfortable	38.26	9.96	0.06
A little comfortable	28.77	7.82	0.27
Mediocre comfort	22.63	9.60	0.42
Barely comfortable	15.42	4.77	0.31
Neutral	0	0	N.A.
Neither comfortable nor uncomfortable	0	0	N.A.
Barely uncomfortable	−27.61	4.38	0.16
A little uncomfortable	−40.90	5.05	0.12
Slightly uncomfortable	−52.95	5.73	0.11
Somewhat uncomfortable	−71.56	6.74	0.09
Average discomfort	−76.64	13.55	0.18
Mediocre discomfort	−79.56	10.96	0.14
Uncomfortable	−96.34	8.21	0.09
Fairly uncomfortable	−99.38	10.07	0.10
Moderately uncomfortable	−145.63	7.23	0.05
Very uncomfortable	−209.86	11.00	0.05
Awfully uncomfortable	−228.96	10.71	0.05
Highly uncomfortable	−231.80	11.42	0.05
Terribly uncomforable	−257.78	14.51	0.06
Exceptionally uncomfortable	−272.76	12.41	0.05
Intensely uncomfortable	−274.34	18.28	0.07
Oppressively uncomfortable	−279.70	15.71	0.06
Horribly uncomfortable	−283.88	22.86	0.08
Extremely uncomfortable	−290.84	15.57	0.05
Unbearably uncomfortable	−298.44	21.79	0.07
Greatest possible discomfort	−345.82	24.29	0.07
Greatest imaginable discomfort	−350.67	35.85	0.10

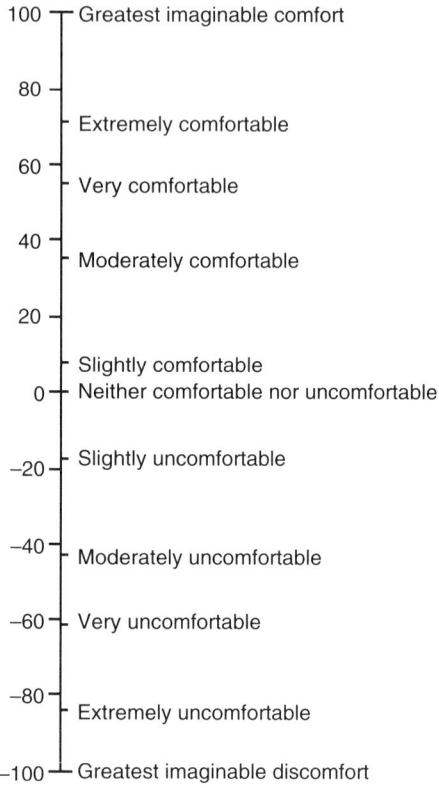

4.1 The Comfort Affective Labeled Magnitude (CALM) scale.

differ in their comfort perceptions. For example, when using a category scale, what one person calls 'moderately uncomfortable', another person might call 'very uncomfortable'. However, by placing everyone's ratings on a scale anchored with 'greatest imaginable (dis)comfort', the end-point anchors are effectively equalized for everyone. Other descriptive comfort labels are located along the line in accordance with their semantic meaning as quantified in Table 4.2. The validity, sensitivity and reliability of this scale for measuring the comfort of fabrics and clothing have been demonstrated now in a variety of studies (Cardello *et al.*, 2003, Bell *et al.*, 2003; 2005; Santee *et al.*, 2006).

The CALM scale shown in Fig. 4.1 has several advantages over other comfort scales. First, the scale is simple to use, merely requiring individuals to place a slash mark somewhere on the vertical line. Secondly, since the labels are located along the scale at points that represent the magnitude of their semantic meaning as determined by a ratio scaling procedure (magnitude estimation), the measured distances along the scale can be treated as ratio-level data. This property of the CALM scale makes it possible to describe a fabric as one-third, 50%, 3 times, etc. as comfortable (or uncomfortable) as another fabric. Third, the CALM scale labels of 'greatest imaginable liking/disliking' enable more extreme ratings than 'extremely comfortable (or uncomfortable)', allowing greater sensitivity to differences among very comfortable (or uncomfortable) fabrics/garments. This can be an important advantage, because in many evaluations the fabrics and/or garments have already been down-selected to be all relatively high in comfort. Lastly, the CALM scale can be used in both laboratory and wear trial evaluations to assess either skin contact or overall comfort.

4.4   Cognitive influences on fabrics and clothing

4.4.1   Measuring attitudes and beliefs about fabrics and clothing

The perception, and comfort, of fabrics and clothing is not simply a function of their physical properties or design features. Rather, their perception and comfort can also be influenced by a variety of cognitive factors, such as attitudes, beliefs and expectations about them. Such attitudes and beliefs may be formed through prior experiences with the fabric or garment or with fabrics/garments that are conceptually similar. Alternatively, they may be formed through information obtained about the garments or fabrics. Once formed, these attitudes and beliefs can outweigh the actual physiological, comfort, or other performance properties of the garment and can become the primary determinants of consumer behavior.

In order to assess consumer attitudes toward fabrics and clothing, a number of behavioral approaches have been used. For example, DeLong et al. (1986) analyzed the content of the words used by consumers to describe sweaters in order to identify the important factors underlying their concept of 'sweater'. Workman (1990) used Likert-type rating scales to determine consumer attitudes and stereotypes toward such fabric labels as 'cotton', 'polyester', and 'blended fabrics'. They then used the attitude ratings to assess how the fabric labels contributed to consumer preferences for jeans. Byrne et al. (1993) used the semantic differential method to study consumer attitudes toward silk, nylon and polyester for use in sport shirts and undershirts. These researchers found that the attitudes toward the different fabric names were distinct and reliable and that the intended end-uses greatly impacted the perceptions of the adequacy of the fabric. Likewise, Forsythe and Thomas (1989) examined preferences for natural, synthetic and blended fibers. They and other researchers have found consumers to have well-defined attitudes toward fibers and fabrics that are consistent across various demographic groups.

4.4.2 Item by use appropriateness scaling and conjoint analysis

Another approach to assessing attitudes and beliefs toward clothing is through the use of *item by use appropriateness scaling*. Using this approach, fabrics or garments are rated on a scale for how 'appropriate' they are for use in different situations. Schutz and Phillips (1976) used this technique to study women's attitudes toward a variety of fabrics. Their study produced important information about women's attitudes for clothing fabrics and the conceptual dimensions of fabrics that drive women's clothing choices.

Another research technique for assessing consumer attitudes about fabrics and clothing is *conjoint analysis*. This technique enables the clothing researcher to uncover the important factors underlying consumer attitudes by using multi-attribute choice alternatives utilizing a pre-determined experimental design (Green and Srinivasan, 1978a, b). In a conjoint analytic study, consumers are presented with a large set of conceptual fabric descriptions in a survey. Each fabric or garment description is composed of a set of independent statements on each of a set of variables believed important to the underlying attitudes and beliefs. By varying the attributes and their levels according to the statistically determined design, this method enables the researcher to work backwards from the choices/ratings to uncover the relative importance of each factor to the consumer's decision process, but without the need to directly ask their importance from the consumer. Conjoint analysis has been used in clothing research to study a number of factors important to standard commercial and protective garments (Crown and Brown, 1984; Wagner *et al.*, 1990; Eckman, 1997).

4.5   Handfeel and comfort evaluations of military fabrics

4.5.1 Sensory handfeel evaluations of military fabrics

At Natick Soldier RD&E Center, the Handfeel Spectrum Descriptive Analysis (HSDA) method (see Section 4.2.2) has been used to assess the handfeel properties of military fabrics used in US and other combat uniforms. The NSRDEC handfeel panel is composed of approximately 15 volunteer employees, both males and females, who were chosen on the basis of interest, availability, and successful completion of a screening test to establish minimum tactile sensitivity. All panelists were enrolled in a 6-month training program during which time they were trained in the HSDA methodology, participated in repeated practice sessions using the attribute definitions and rating scales, and received training in the application of these methods to military clothing fabrics. As noted, Table 4.1 lists the 17 sensory attributes employed in testing, and the operational techniques of evaluation can be found in Cardello *et al.* (2002) and Meiselman and Cardello (2002). In applications of this methodology to military and commercial products, the test–retest reliability coefficients have ranged from 0.89 to 0.95 for data obtained up to 6 months apart (Cardello *et al.*, 2003).

Figures 4.2 and 4.3 show sensory handfeel data obtained at NSRDEC using the HSDA method in a study of 13 fabrics used in US, British, Canadian and Australian military garments (Cardello *et al.*, 2003). Table 4.3 lists the fabric compositions. The 17 HSDA sensory handfeel attributes are shown along the bottom of the figures and the mean intensity ratings for each attribute are plotted along the ordinates.

As can be seen in both figures, the HSDA method provides informative sensory 'profiles' of the fabrics that enable ready comparison of the differences among fabrics. For example, in Fig. 4.2, it can be seen that, while the Army Aircrew and the Temperate Weather BDU fabrics (black circles/ squares) are relatively similar, the Army Hot Weather BDU fabric (gray circles) differs greatly from these two, having lower 'fuzziness', 'hand

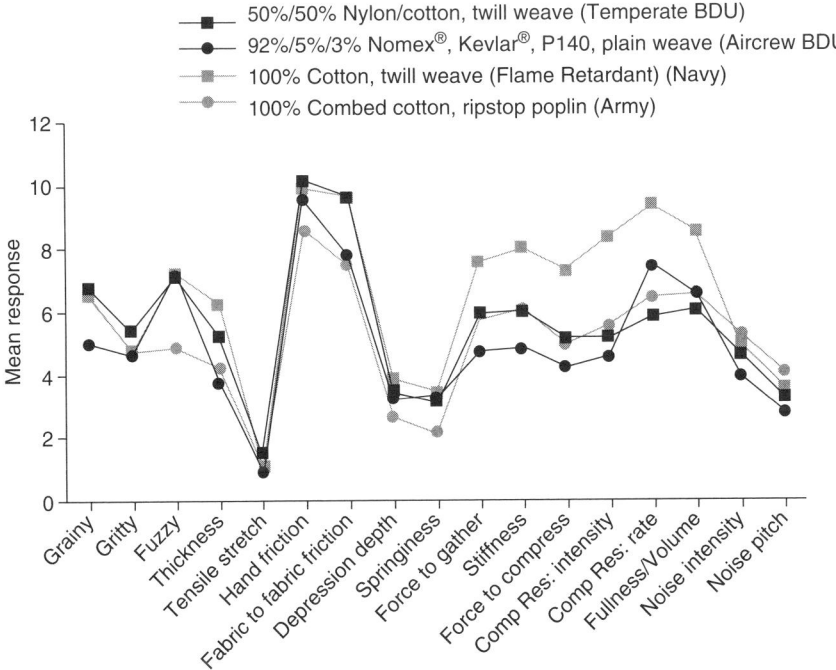

4.2 Mean panel ratings of handfeel attributes averaged over three replicates for eight different military clothing fabrics (from Cardello *et al.*, 2003).

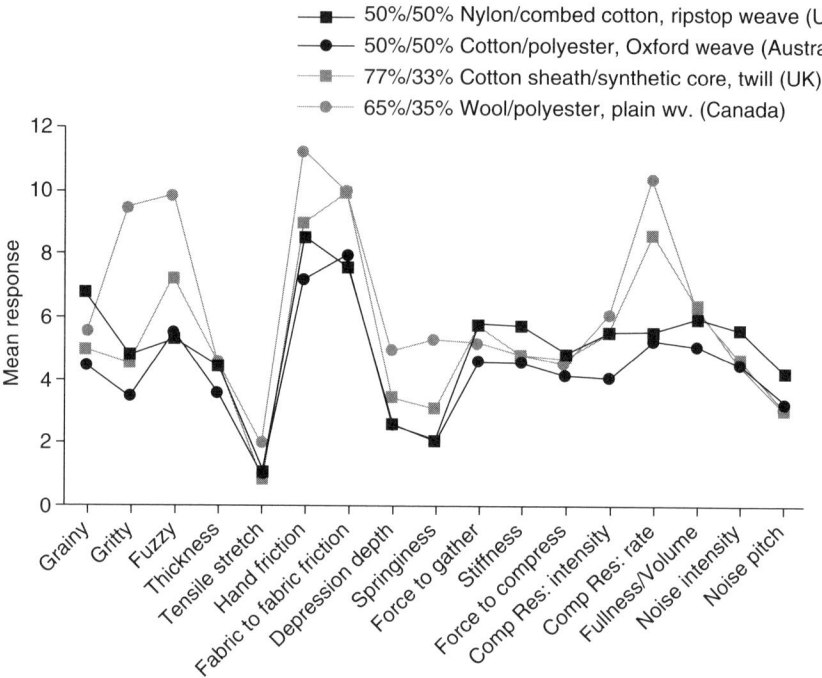

4.3 Mean panel ratings of handfeel attributes averaged over three replicates for four additional military clothing fabrics (from Cardello *et al.*, 2003).

friction', 'depression depth' and 'springiness'. The Navy fabric (gray squares) is unique in several of its handfeel characteristics, being 'thicker', and having greater 'force to gather', 'stiffness', 'compressive resilience' and 'fullness/ volume' than any of the other fabrics. The Army flame-resistant fabric is similar in terms of 'fuzziness', 'tensile stretch', 'hand friction', 'depression depth' and 'springiness', but is a thinner, smoother (less grainy) fabric, and has lower 'force to gather', 'stiffness', and 'compressive resistance' than the Navy material. Examining Fig. 4.3, it can be seen that the Canadian fabric rates very highly on the attributes of 'gritty', 'fuzzy', 'hand friction', 'depression depth', 'springiness' and 'compressive resilience', compared to all other fabrics, with the US hot weather fabric also scoring high on many of these same attributes. On the other hand, the Australian fabric has the lowest overall profile, scoring lowest on almost all sensory attributes.

Statistical analysis of the data for all 13 fabrics tested showed that each of the sensory attributes significantly discriminated among one or more of the tested fabrics, and several attributes enabled the fabrics to be differentiated into as many as five distinct subsets. The large differences among

Table 4.3 Fabrics used in the study of the handfeel and comfort properties of military fabrics (Cardello *et al.*, 2003). Fabrics designated with single-letter codes after them are those used in the comfort wear study (Santee *et al.*, 2006)

Fabric composition		Fabric code
50%/50% Nylon/combed cotton, ripstop poplin weave	(B)	10R*
50%/50% Nylon/polyester, Oxford weave (Australian)	(A)	11A*
50%/50% Nylon/cotton, twill weave		12T*
92%/5%/3% Nomex®, Kevlar®, P140, plain weave		13P*
100% Cotton, twill weave (Flame retardant treated)		14N*
77%/33% Cotton sheath/synthetic core, twill (U.K.)		15B*
100% Combed cotton, ripstop poplin (Former hot weather BDU)		16C*
65%/35% Wool/polyester, plain weave. (Canada–unlaundered)		17C*
65%/35% Wool/polyester, plain weave. (Canada–laundered)	(C)	18L
92%/5%/3% Nomex®, Kevlar®, P140, Oxford weave	(D)	19N
Carded cotton sheath/nylon core, plain weave (Canada)		20J
100% Pima cotton ripstop poplin (experimental)		124
50%/50% Nylon carded cotton ripstop poplin weave		176

*Fabrics for which Kawabata data were obtained.

fabrics seen in Fig. 4.2 and Fig. 4.3, combined with the demonstrated sensitivity and reliability of the HSDA method when applied to military fabrics (see Cardello *et al.*, 2003), has established this handfeel approach as a significant advance in sensory characterization of military fabrics. The method also offers a unique, standard protocol for use in inter-laboratory studies or for establishing functional, performance-based specifications for military and other clothing fabrics.

4.5.2 Comfort evaluations of military fabrics

Although the evaluation of the sensory handfeel attributes of fabrics requires trained individuals who have a common vocabulary and procedures for evaluating fabrics, the evaluation of comfort must be done by naïve consumers. The reasons for this are (i) that trained panelists, because of their greater knowledge of and experience with fabrics, are unlike naïve consumers and may be biased in their perceptions of comfort, and (ii) as noted previously, while two individuals may agree that a fabric possesses certain sensory properties, they may disagree as to whether the feel of that fabric is comfortable or uncomfortable. Thus, 'training' individuals in comfort assessments is antithetical to the subjective nature of comfort experience.

In a laboratory study, the same fabrics listed in Table 4.3 were evaluated for handfeel comfort by 40 consumers who had no formal training in textiles. The consumers were instructed that they could 'hold, touch, feel or squeeze the material in any manner'. Comfort judgments were made in individualized consumer testing booths using the CALM scale (Fig. 4.1).

Table 4.4 shows the mean comfort ratings for the 13 test fabrics. An ANOVA with *post-hoc* tests showed that the Australian fabric was the most comfortable and the Canadian fabric was the least comfortable. The other fabrics rated between these two. The large difference in comfort between the Australian and Canadian fabrics is consistent with the large difference in their handfeel profiles as seen in Fig. 4.2, where the Australian fabric has a very shallow handfeel profile versus the Canadian fabric, which has a peaked and jagged profile.

4.5.3 Relating the sensory and comfort properties of fabrics

One of the most important questions that textile technologists and clothing designers encounter is 'what are the fabric characteristics that predict fabric or clothing comfort?' Since comfort is a subjective judgment of the user that is greatly dependent upon the perceived sensory properties of the fabric, it would only make sense that sensory handfeel judgments of fabrics should predict fabric comfort. In order to statistically assess the predictive relationship between sensory handfeel and comfort ratings, the handfeel data obtained on the fabrics in Table 4.3 were correlated with the consumer comfort data in Table 4.4. In addition, the mean of the descriptive attribute intensity ratings across all handfeel attributes was calculated for each fabric to index the total magnitude or 'sensory salience' of the fabric.

Table 4.5 shows the Pearson product–moment correlation coefficients between comfort ratings, sensory handfeel attributes and mean fabric salience rating. As can be seen, several of the sensory attributes are significantly correlated with consumer comfort ratings and all 17 are *negatively* correlated with comfort. Thus, regardless of the nature of the handfeel attribute that was experienced, the greater the perception of the intensity

Table 4.4 Mean comfort ratings obtained for the 13 test fabrics utilized in the study by Cardello *et al.* (2003). The fabrics used in the Santee *et al.* (2006) wear trial are indicated by fabric code for comparison with the data in Figures 4.4–4.6

Fabric code (Cardello *et al.*, 2003)	Fabric code (Santee *et al.*, 2006)	Mean comfort rating
18L	C (Canadian)	-9.8^a
17C		-1.4^{ab}
176		2.4^{ab}
124		9.8^{bc}
20J		10.9^{bcd}
16C		22.0^{cde}
12T		23.6^{cde}
14N		24.2^{cde}
19N	D (Nomex®/Kevlar®)	28.5^{cdef}
10R	B (Light weight BDU)	28.9^{def}
13P		37.4^{ef}
15B		46.4^f
11A	A (Australian)	47.2^f

Means with different superscript letters are different at $P < 0.05$.

Table 4.5 Pearson product–moment correlation coefficients for the associations between each of the 17 HSDS handfeel attributes (and the mean overall attributes) with comfort, as obtained in the study by Cardello *et al.*, 2003)

Handfeel attribute	r with comfort
Grainy	−0.41
Gritty	−0.92**
Fuzzy	−0.60
Thickness	−0.32
Tensile stretch	−0.92**
Hand friction	−0.77*
Fabric to fabric friction	−0.36
Depression depth	−0.71*
Springiness	−0.72*
Force to gather	−0.17
Stiffness	−0.17
Force to compress	−0.17
Compression resilience/intensity	−0.42
Compression resilience/rate	−0.53
Fullness/volume	−0.17
Noise intensity	−0.25
Noise pitch	−0.03
Mean overall attributes (fabric salience)	−0.70*

* $P < 0.05$.
** $P < 0.01$.

of that attribute, the lower the perceived comfort. In keeping with the notion that the comfort of a fabric may well be dependent on a minimal tactile profile, the correlation of the mean intensity rating across all attributes (the handfeel salience) was negative ($r = -0.70$) and accounted for about 50% of the variance in the comfort responses.

4.5.4 Predicting comfort from instrumental data

Since comfort is a human experience based on sensory experiences of clothing and the environment, it is only logical that the sensory attributes of a fabric should better predict the perceived comfort of the fabric than any instrumental measure obtained on the fabric itself. Of course, instrumental measures are convenient and may be desirable for certain manufacturing applications. In order to determine the potential relationships among instrumental fabric measures, fabric hand and the comfort of military fabrics, Kawabata mechanical measures (Kawabata, 1980; Kawabata and Niwa, 1975) were obtained on the eight fabrics asterisked in Table 4.3 (see Cardello *et al.*, 2002, 2003). Table 4.6 shows Pearson product–moment correlation coefficients greater than 0.50 between the Kawabata hand parameters and the sensory handfeel ratings. The specific sensory attributes that correlate best with each primary hand expression demonstrate a good degree of conceptual agreement between the two methods. Both Kawabata 'stiffness' and 'anti-drape stiffness' are associated with the same six handfeel attributes, and both are correlated very highly with 'stiffness' ($r = 0.80$, .84). Likewise, the attributes that best correlate with Kawabata 'fullness/softness', which is defined as 'bulky', 'rich' and 'springy' sensations, are 'depression depth', 'fuzziness' and 'springiness'. Lastly, Kawabata smoothness, which is defined as 'limber' and 'soft' like 'cashmere fiber', is most highly associated with 'fuzziness' and 'fabric to fabric friction', two attributes that would be expected to be positively associated with softer pile fabrics.

Since the Kawabata hand values are based on predictive equations derived from men's winter suit fabrics, Principal Component Analysis (PCA) was used to analyze the Kawabata mechanical parameters and to reduce them to a smaller number for correlation with the sensory and comfort data. This same approach was also used to reduce the number of handfeel attributes. The reader is referred to Cardello *et al.* (2003) for a description of these statistical procedures. However, the results of these analyses for the Kawabata data identified five important instrumental factors related to: 'shear properties', 'bending properties', 'compression/friction', 'tensile properties' and 'surface roughness'. The PCA on the sensory data produced three major factors: 'surface texture/depth', 'volume' and 'noise'.

Table 4.6 All Pearson product–moment correlations greater than 0.50 for the associations between Kawabata hand values and sensory handfeel attributes as obtained in the study by Cardello *et al.* (2003)

Kawabata hand value	Handfeel attribute	r
Stiffness		
	Force to compress	0.83*
	Stiffness	0.80*
	Force to gather	0.79*
	Compression resilience:	
	intensity	0.71
	Thickness	0.68
	Fullness/volume	0.63
Anti-drape stiffness		
	Force to compress	0.87*
	Stiffness	0.84*
	Force to gather	0.80*
	Compression resilience:	
	intensity	0.73*
	Fullness/volume	0.71*
	Thickness	0.67
Fullness/softness		
	Springiness	0.87*
	Depression depth	0.85*
	Fuzzy	0.85*
	Hand friction	0.77*
	Gritty	0.76*
	Tensile stretch	0.67
Smoothness		
	Fuzzy	0.55
	Fabric to fabric friction	0.50
Crispness		
	No correlation	>0.50

* $P < 0.05$.

To predict the comfort of the test fabrics from instrumental measures, the five Kawabata factors were regressed against consumer ratings of the comfort of the fabrics, producing the regression equation:

$$\text{COMFORT} = 11.8 \text{ (shear)} - 3.1 \text{ (bending)} - 0.3 \text{ (compression/friction)} - 11.9 \text{ (tensile)} + 0.4 \text{ (surface roughness)} + 27.5 \quad (R^2_{\text{Adj}} = 0.60)$$

This contrasted with the regression of the sensory factor components against comfort ratings which produced the equation:

$$\text{COMFORT} = -15.6 \text{ (surface texture/depth)} - 1.07 \text{ (volume)} - 7.67 \text{ (noise)} + 27.5 \quad (R^2_{\text{Adj}} = 0.87)$$

which explained considerably more of the variability in comfort ratings. Lastly, combining both the sensory and Kawabata component scores into a stepwise multiple regression model to predict comfort resulted in the equation:

$$\text{COMFORT} = -16.3 \text{ (sensory surface texture/depth)} - 8.7 \text{ (sensory noise)} - 4.3 \text{ (Kawabata surface texture)} + 27.5 \qquad (R^2_{\text{Adj}} = 0.96)$$

suggesting that almost all of the variance in comfort could be accounted by a combination of the sensory and instrumental parameters of the fabrics.

The approach used in the above research to uncover sensory–instrumental–comfort relationships is valuable for understanding the complex factors that contribute to the perceived comfort of military fabrics and clothing. By reducing the large array of sensory and mechanical properties that can be measured on fabrics to a small number of independent components, it is possible to derive simple regression models to predict the perceived comfort of the fabrics.

4.5.5 Sensory and comfort analyses in wear trials

Although laboratory assessments of the sensory, instrumental and/or comfort properties of fabrics are important for product development, the ultimate test of the feel and comfort of a garment is to be found in a wear trial using controlled environmental or use conditions. It is only in such trials that the fabric feel characteristics can be realistically perceived on all parts of the body and under actual conditions of movement. Most wear trials are conducted in either field environments or in environmentally controlled 'chamber' studies. In a recent study (Santee *et al.*, 2006), it was possible to assess the sensory and comfort properties of garments fabricated from four of the fabrics that were used in the sensory and laboratory hand comfort studies reported here.

Wear trial study design

In the study by Santee *et al.* (2006), four combat uniforms (standard battle dress uniform: MIL-C-44048 Coats, Camouflage Pattern, Combat and MIL-T-44047 Trousers, Camouflage Pattern, Combat) were evaluated for their sensory, comfort, and physiological effects in two test environments. The garments were fabricated using four fabrics chosen from among those used in the laboratory evaluations of sensory and comfort properties. The four uniform fabrics are those labeled A, B, C, and D in Table 4.3. The four fabrics were chosen to represent a wide range of comfort, based on the consumer handfeel evaluations (see Table 4.4).

Nine soldiers wore the four garments in two different environmental conditions in a climatically controlled test chamber. The two conditions were a neutral condition of 20°C (68°F) and 50% RH, and a warm humid condition of 27°C (80.6°F) and 75% RH. Soldiers alternated walking on a treadmill at 1.34 m·s⁻¹ (3 mph) for 30 minutes, then sitting for a 10-minute period. The cycle was repeated four times. Prior to entering the test chamber, during walking, while seated in the chambers, and after exiting the chambers, soldiers rated their comfort, tactile and thermal sensations. Since wear trials involve naïve users of the garments, detailed sensory analysis of fabric skin feel sensations is not possible. Instead, terminology that is easily understood by consumers is employed. In the present study, the skin feel sensations of 'hot/cold', 'sweatiness', and 'skin wettedness', along with the fabric tactile sensations of 'softness', 'scratchiness', 'stickiness', 'stiffness', 'clamminess' and 'clinginess' were evaluated. In addition, physiological measures were obtained, including relative humidity (RH_{uc}) and skin temperature (T_{sk}) at three body locations, rectal temperature (T_{re}), and heart rate.

Selected results of the wear trial

With regard to tactile sensations, Fig. 4.4 shows the data for the tactile sensation of 'scratchiness'. For both environmental conditions and across all time periods, the Canadian fabric (C) felt significantly more scratchy than all the other fabric types. The other three fabrics were more similar to one another, with the Nomex® fabric (D) being perceived as scratchier than the hot weather BDU fabric (B), which was perceived as being marginally more scratchy than the Australian fabric (A). Figure 4.5 shows the data for perceptions of 'softness'. As expected, these data are inversely related to those for scratchiness. The softest feeling fabric was the Australian fabric (A), followed by the hot weather BDU fabric (B), especially during the first two hours of the study. The Nomex® fabric (D) and the Canadian fabric (C) had the lowest levels of softness.

Figure 4.6 shows the overall comfort ratings. ANOVAs showed that comfort was significantly lower in the warm humid condition than in the neutral condition and that there was a gradual decrease in comfort over time in the warm humid condition. In both test conditions, comfort was significantly lower for the Canadian fabric (C) than for all other garments. Although there were only small differences in comfort among the other fabrics tested, the Australian fabric had highest comfort ratings under almost all environmental and time conditions. In addition, ratings of comfort taken both before and at the completion of the test also showed significant effects of both the environmental condition and fabric type. In particular, pre-test ratings of both the 'liking of the feel' of the garment and 'comfort' showed significant differences by garment type (Liking of the Feel: $F = 31.7$,

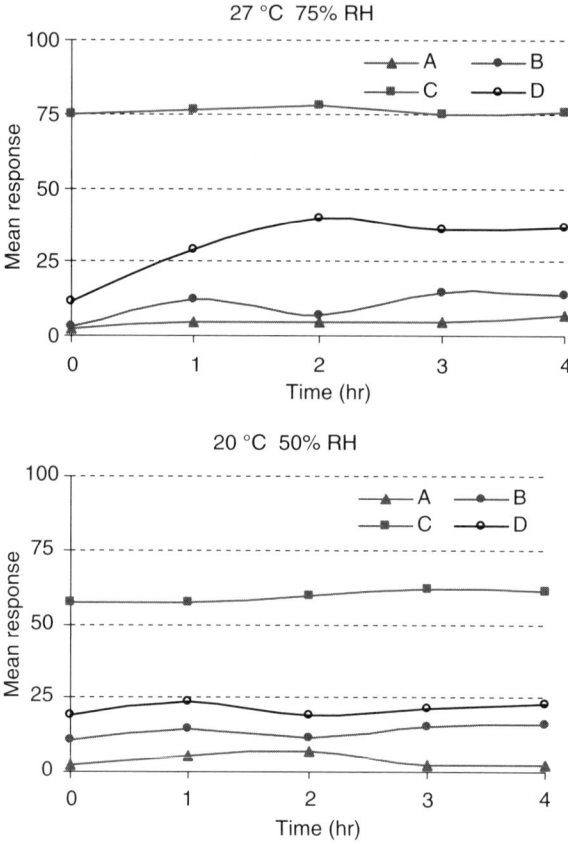

4.4 Mean ratings of perceived 'scratchiness' plotted over time for each of the four garments and two environmental conditions used in the wear trial (from Santee *et al.*, 2006).

df = 3,24, $P < 0.001$; Comfort: $F = 20.0$, df = 3,24, $P < 0.001$). Examination of the data showed that both the liking of feel and comfort of the garment fabricated with the Canadian material were significantly lower than for all the other garments and that the Canadian garment was the only one that received negative ratings. The garment fabricated from the Australian material was rated highest on these dimensions and slightly more positively than the Hot Weather BDU garment, and both were rated higher than the Nomex garment.

Thus, the fabric characterized by the handfeel panel as the most gritty and as having the most friction-related properties, and by the laboratory consumer panel as having the most uncomfortable hand, was the Canadian fabric (C) and this was the same fabric that was rated the scratchiest and

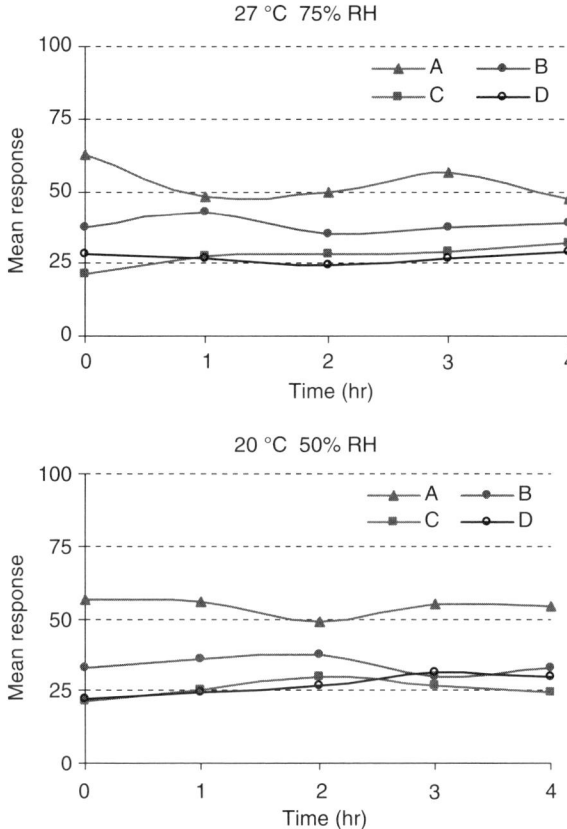

4.5 Mean ratings of perceived 'softness' plotted over time for each of the four garments and two environmental conditions used in the wear trial (from Santee *et al.*, 2006).

least comfortable in the wear trial. Similarly, the Australian (A) fabric, which was rated lowest on all handfeel dimensions by the sensory handfeel panel and highest in comfort by the laboratory consumer panel, was also rated as being the softest and most comfortable in the wear trial. These data establish convergent validity for the sensory methods used in these studies and suggest that laboratory handfeel and comfort analyses using swatches of fabric are good predictors of the perceived skin feel and comfort of the fabrics when manufactured into, and worn as, garments.

It should be noted here that the fabrics used in this study likely differed in moisture vapor transmission rate and in other *thermal* comfort related properties. However, the large differences in the skin feel sensations of the garments, combined with the observed pre-test differences among the

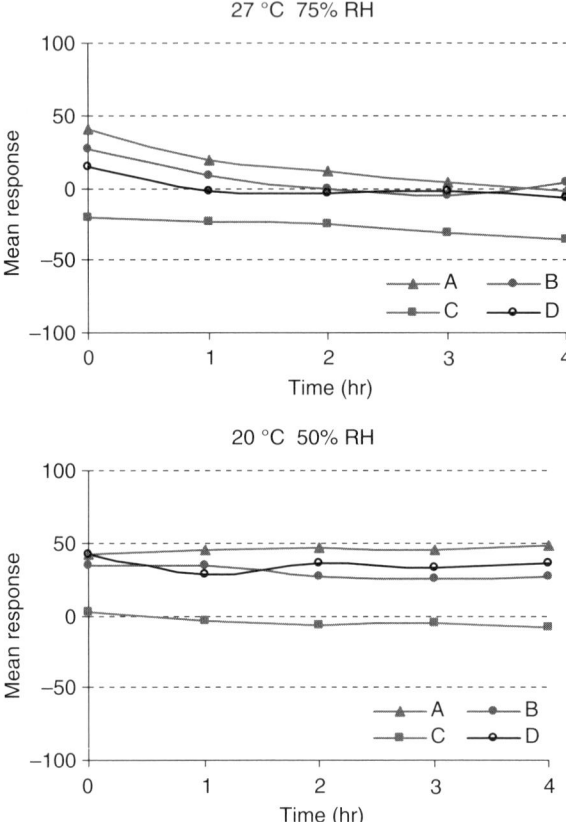

4.6 Mean ratings of comfort plotted over time for each of the four garments and two environmental conditions used in the wear trial (from Santee *et al.*, 2006).

garments for 'liking of feel' and 'comfort' suggest that the tactile character-istics of the fabrics were a primary contributor to the overall assessment of the comfort of the garments during the wear trial.

4.6   Cognitive influences on fabric and clothing perception

4.6.1   Item by use appropriateness scaling of fibers and fabrics

All of the above approaches to assessing military fabrics and garments are based on sensory or perceptual evaluations of physical fabrics and gar-ments. However, a complete understanding of how fabrics and garments are perceived by military users requires the analysis of attitudes and beliefs about the fabrics and/or garments. In a recent study (Schutz *et al.*, 2005), US soldiers' attitudes toward clothing fibers and fabrics used in military and commercial clothing were assessed using item by use appropriateness scaling (see Schutz and Phillips, 1976). Sixteen fiber and fabric names were examined, as well as the term 'ideal fabric'. Thirty possible characteristics and uses of the fibers/fabrics were also examined. Table 4.7 lists the fiber/ fabric names and the 30 'characteristics/uses'. Respondents rated each fiber/ fabric for its 'appropriateness' on each of the characteristics/uses, using a seven-point scale (1 = 'never appropriate'; 7 = 'always appropriate').

Table 4.7 Fiber and fabric names and the use characteristics utilized in the appropriateness study by Schutz *et al.* (2005)

Fibers and fabrics	Uses/characteristics
	To wear for a long time
Cotton	breathable
Kevlar®	heavy
Polyester	rough
Nomex®	windproff
Double-knit fabric	sharp military appearance
Gore-Tex®	wrinkles easily
Denim	hot
Nylon	itchy
Ideal fabric for combat	good in the desert
Polypropylene	soft
Spandex®	fire-resistant
Synthetic	shrinks easily
Polyester/cotton blend	lightweight
Wool	silky
Silk	rips/tears easily
Nylon/cotton blends	uncomfortable
	easy to care
	wicks moisture from body
	durable
	non-absorbent
	stiff
	cool
	good in the jungle
	water resistant
	strong
	stretchable
	dries quickly
	clingy
	scratchy

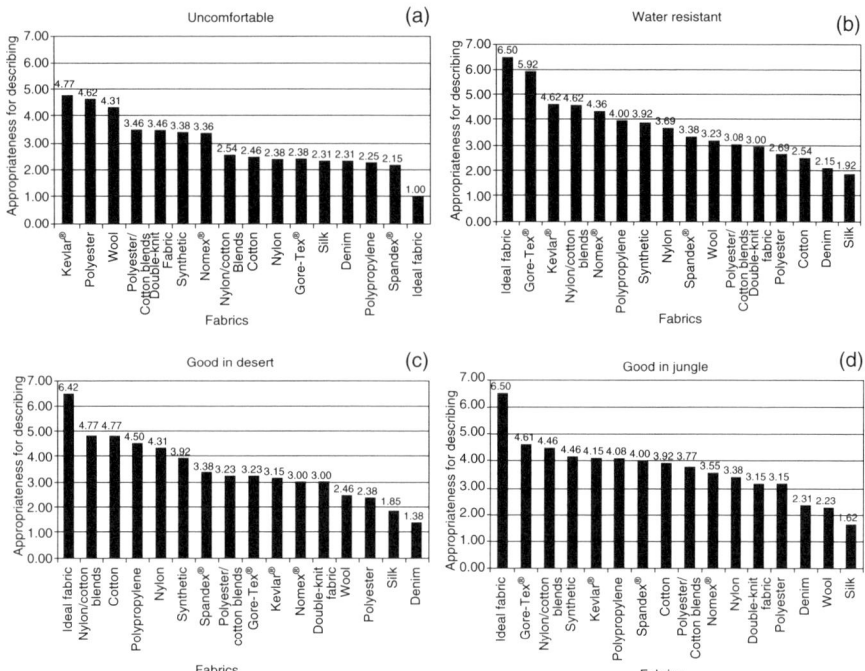

4.7 Mean appropriateness ratings for (a) 'uncomfortable', (b) 'water resistance', (c) 'good in the desert', and (d) 'good in the jungle' for each of 16 fiber/fabric names (from Schutz *et al.*, 2005).

Figure 4.7 shows the mean appropriateness ratings for all fibers/fabrics on selected characteristics/uses. Panel (a) shows data for the characteristic 'uncomfortable'. As might well be expected, the term 'ideal fabric' had the lowest mean rating for this characteristic. The other 15 fiber/fabrics fell into three groups. One group included 'Kevlar®', 'polyester' and 'wool', all of which had high ratings for 'uncomfortable'. 'Polyester/cotton blends', 'double-knit fabrics', 'Nomex®', and 'synthetic fabrics' had the next highest mean ratings, while the other fibers/fabrics had lower and roughly similar ratings. Fabrics used in military battle dress uniforms, which are primarily nylon/cotton blends, fell within this latter group. Panel (b) shows data for 'water resistant'. The 'ideal fabric' and 'Gore-Tex®' have the highest ratings on this characteristic, while the other fibers/fabrics show a more continuous decline, starting with 'Kevlar®' and 'nylon/cotton blends' and moving through 'denim' and 'silk', which were rated the least appropriate.

Panels (c) and (d) show the data for the characteristics/uses of 'good in the desert' and 'good in the jungle'. Except for the 'ideal fabric', 'cotton' (and 'nylon/cotton blends') rated highest for 'good in the desert'. Interestingly, however, cotton was rated much lower than nylon/cotton blends for 'good in the jungle'. The likely reason for this difference can be seen in panel (b), where cotton is at the bottom of the fabric list for its appropriateness for water resistance. This poor rating of cotton for water resistance makes it far less desirable for use 'in the jungle' than 'in the desert'. Clearly, item-by-use appropriateness data can provide important insights into the mind of the consumer and how he/she conceptualizes the important characteristics and situational/environmental uses for different fabrics.

4.6.2 Conjoint analysis of garment comfort and satisfaction

In a related study (Schutz *et al.*, 2005), a conjoint analytic approach was used to assess the importance of both fabric and garment characteristics to comfort and satisfaction with two military field uniforms. The comfort- and satisfaction-related factors of importance to soldiers were first identified in focus groups (group interviews) conducted on three separate days when the soldiers wore either their temperate weather, hot weather, or desert 'Battle Dress Uniform' (BDU). During these interviews, clothing attributes important to the comfort of the garment were generated. The attributes included 'abrasiveness', 'clinginess', 'absorbency', 'softness', 'breathability', 'thickness of the material', 'weight of the material', 'stiffness', 'coarseness' and 'thermal aspect (hot–cold)'. A similar list of factors generated for overall garment satisfaction resulted in the factors: 'fit', 'protection', 'thermal comfort', 'appearance', 'durability' and 'feel'.

The respondents for the hot weather uniform study were 97 soldiers stationed at Fort Lewis, WA, who had been wearing the hot weather BDU as their duty uniform for several months prior to the test. For the temperate weather BDU, 98 soldiers from Fort Hood, TX, who had worn that uniform served as respondents. For both studies, the questionnaires were identical, except for the reference garment. Each group completed two conjoint questionnaires – one for the relative importance of fabric properties to perceived *comfort* and one for the relative importance of garment factors to overall *satisfaction*. Soldiers rated the perceived comfort (or satisfaction) of a number of uniform concepts that were comprised of different combinations of two levels of comfort-related factors. The levels for each factor were light–heavy, clingy–not clingy, scratchy–not scratchy, soft–hard, absorbent–non-absorbent, breathes–doesn't breathe, thin–thick, stiff–not stiff, smooth–rough, and hot–cold. Two levels of the satisfaction-related factors were constructed using the adjectives 'good' or 'poor', e.g. 'fit' was either 'good fit' or 'poor fit', etc.). The dependent measure for the comfort questionnaire was a rating made on the CALM scale. The dependent measure for the satisfaction questionnaire was a rating on a seven-point satisfaction scale (1 = completely dissatisfied, 7 = completely satisfied). Figure 4.8 shows an

Please rate your likely satisfaction or dissatisfaction with a BDU
(battle dress uniform) that has the following characteristics:

```
Good fit
Poor durability
Good military appearance
Good protection
Poor feel
Poor thermal properties
```

Please make your satisfaction/dissatisfaction rating by circling
one of the numbers/phrases below:

7   Completely satisfied
6   Very satisfied
5   Somewhat satisfied
4   Neither satisfied nor dissatisfied
3   Somewhat dissatisfied
2   Very dissatisfied
1   Completely dissatisfied

4.8 A military clothing concept used in the conjoint analytic study of
clothing satisfaction.

example conjoint clothing concept from the satisfaction portion of the
study, along with the question posed to soldiers and the rating scale.

The data from the questionnaires were analyzed using a general linear
model that calculates the part-worths or 'utility values' for each level of
each factor. These utility values index the influence of each factor level on
the respondent's ratings, and an 'averaged importance' value is calculated
that indexes the relative range of utility values for the levels within each
factor. The results of these analyses showed that the data were well fit by
the statistical model for both questionnaires and for both subject groups
(Hot weather Comfort: $R^2 = 0.97$, Temperate weather Comfort: $R^2 = 0.99$,
Hot weather Satisfaction: $R^2 = 0.88$, Temperate weather Satisfaction: $R^2 = 0.55$).

Relative importance of fabric factors to comfort

The top panel (a) of Fig. 4.9 shows the average importance values for both
uniforms on each conjoint factor related to 'comfort'. ANOVAs showed
significant differences among the importance values for each factor for both
the hot weather ($F = 20.68$; df $= 9.846$; $P < 0.001$) and temperate weather
($F = 8.77$; df $= 9.846$; $P < 0.001$) garments. *Post-hoc* analyses revealed that
the 'thermal aspect' was significantly more important than all other factors
for the hot weather garment and more important to the comfort of the hot

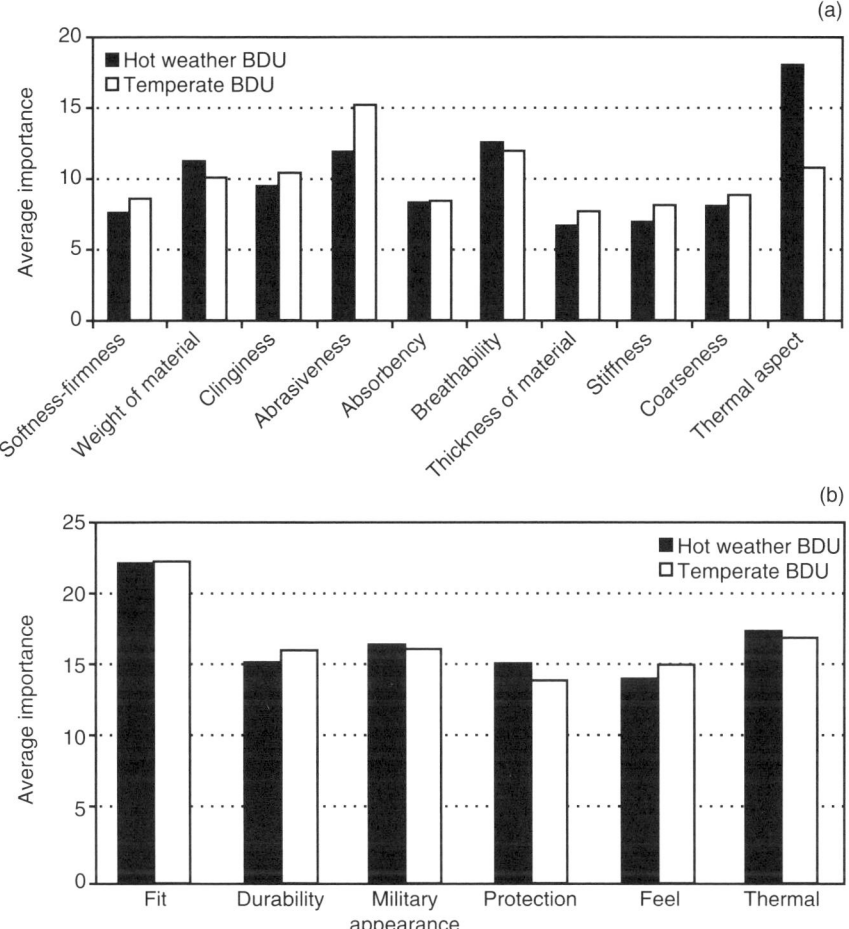

4.9 Average importance values for (a) comfort-related and (b) satisfaction-related factors for the hot weather and temperate weather BDUs (from Schutz *et al.*, 2005).

weather garment than the temperate. For the temperate weather garment, 'abrasiveness' was significantly more important than all other factors and was more important to this garment than to the hot weather uniform. All other fabric feel factors were similar in their contribution to comfort for both garments, suggesting that military clothing designers should focus more attention on the thermal properties of uniforms worn in hot weather and more attention on the feel characteristics or abrasiveness of those worn in temperate climates.

Relative importance of uniform factors to satisfaction

The bottom panel (b) of Fig. 4.9 shows the average importance values for both uniforms for 'overall satisfaction'. Although there are large differences in importance among the six satisfaction characteristics, there are remarkable similarities between the two uniforms. This consistency in the data is impressive and speaks to a high degree of construct validity in the ratings, especially since these data were obtained from two different groups of soldiers at two different geographic locations.

ANOVAs conducted on the 'satisfaction' data showed highly significant differences among the average importance values for both the hot weather ($F = 4.13$, df $= 5.440$, $P < 0.001$) and temperate weather ($F = 4.30$, df $= 5.440$, $P < 0.001$) uniforms. Post-hoc analyses showed almost the same pattern of significant differences for the two garments. 'Fit' was significantly more important than all other characteristics for both uniforms, which is consistent with previous data obtained on athletic uniforms (Casselman-Dickson and Damhorst, 1993; Feather *et al.*, 1996; Wheat and Dickson, 1997). The importance of other factors showed only minor differences within or between garments, even though protection, durability and thermal properties might well be expected to have outweighed military appearance and feel (tactile) factors in importance. However, studies on athletic uniforms also have shown that appearance attributes have high importance, even relative to the clothing's performance attributes (Casselman-Dickson and Damhorst, 1993; Wheat and Dickson, 1997), suggesting that clothing appearance is important, regardless of the functionality of the clothing. Since sharp military appearance is an important aspect of military personnel inspections and combat uniforms are now worn on a daily basis in garrison for a wide variety of non-combat tasks, it is not surprising that appearance has high importance relative to protection and durability for these garments.

4.7   The role of clothing comfort on military performance

4.7.1 Comfort and performance: a naturalistic study

A number of studies have shown that clothing factors can influence the cognitive performance of the wearer (Brooks and Parsons, 1999; Hancock and Vasmatzidis, 2003; Gordon *et al.*, 1989). However, these studies have focused primarily on the effects of heat stress and fit. Little data are available on the role of skinfeel comfort on cognitive performance. However, in a recent study conducted by NSRDEC, the influence of clothing comfort on cognitive performance was examined in a naturalistic setting (Bell *et al.*, 2005). Eighty-eight graduate students at a local university who were taking a one-hour statistics exam self-reported the type of clothing that they were wearing ('formal work clothes', 'casual work clothes', 'casual leisure clothes' or 'dress for comfort clothes') and their perceived level of comfort in this clothing using the CALM scale. To control for non-clothing related variables that might influence exam scores, the students also self-reported the total time that they spent studying for the exam, their confidence going into the exam, and their degree of perceived social support.

A multivariate analysis of the data that included comfort level, hours spent studying, level of confidence, perceived social support and gender resulted in a model that accounted for approximately 50% of the variability in exam scores. Controlling for all variables, both level of comfort and confidence level were positively associated with exam scores. Figure 4.10 shows the relationship between clothing comfort ratings on the CALM scale and exam score. For each 3% increase in self-reported clothing comfort, there was a 1% increase in exam score. Thus, if clothing comfort were increased by 30%, a 10% or full-grade increase in exam performance would be predicted.

Although these data pertain to civilian clothing, they point to the potential influence of military clothing comfort on performance, especially in tasks related to cognition and reasoning. To the extent that military clothing is uncomfortable, it may well interfere with cognitive attention by refocusing conscious awareness from the task at hand to the source of discomfort.

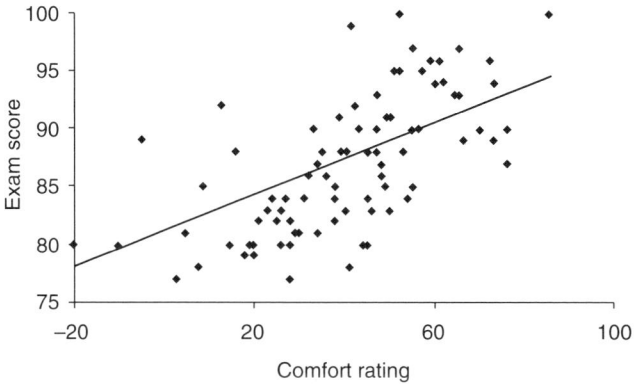

4.10 Plot of exam scores against clothing comfort ratings controlling for other variables (from Bell *et al.*, 2005).

4.7.2 Clothing comfort and performance: a laboratory study

In another study conducted by NSRDEC, comfort was directly manipulated through the use of different fabrics during a standardized test of cognitive performance (Bell *et al.*, 2003). Forty male and female subjects participated in a computerized visual vigilance task during test sessions in which they wore different clothing fabrics in contact with their neck and arms. The fabrics in contact with their skin were 80% cotton/20% stretch nylon, 85% wool/15% nylon, or they wore no material on their neck and arms. All sessions were randomized among subjects and comfort was rated four times during each session.

Table 4.8 shows the results obtained during the first and second half of the test sessions. As can be seen, the uncomfortable 85% wool fabric significantly increased reaction time during both halves of the test period versus either the 80% cotton fabric or no fabric. Although mean accuracy was also lowest for the wool material throughout the test, the difference with the control only reached significance during the second half of the test period. Correlations of comfort ratings with reaction time ($r = 0.34$) and with percent accuracy ($r = 0.46$) were both significant ($P < 0.001$).

The above data demonstrate the important role that fabric comfort can have on cognitive performance. To the extent that military clothing is *un*comfortable to the soldier, these data suggest that both reaction time and accuracy on important military tasks, like those that involve vigilance, can be adversely affected. Both studies by Bell *et al.* (2003, 2005) point to the

Table 4.8 Mean percent accuracy and reaction times in the first and last half of the cognitive trials (from Bell *et al.*, 2003)

Variable and condition	First half of task		Second half of task	
	M	SD	M	SD
Percent accuracy				
Control	97.2[a]	1.8	96.8[a]	1.7
Cotton	96.2[ab]	1.4	94.4[ab]	1.7
Wool	94.6[ab]	1.7	93.1[bc]	1.6
Reaction time (msec)				
Control	539[a]	28.1	511[a]	25.0
Cotton	544[a]	22.6	530[a]	30.5
Wool	597[b]	22.2	615[b]	27.6

Means with different superscript letters are different at $P < 0.05$.

need to focus greater attention on improving military clothing comfort as a means to improve combat effectiveness.

4.8   Conclusions

The comfort of military clothing is composed of a complex mix of sensory, cognitive, and affective variables. From the studies conducted to date it is clear that a judicious application of advanced sensory, psychophysical, and cognitive methods to the problem of military clothing comfort can lead to a better understanding of the factors that control the comfort of our fighting men and women. In addition, the application of many of these methods to laboratory assessments of military fabrics can lead to better predictive relationships with data obtained from wear trials of garments manufactured with these fabrics. Lastly, data are now becoming available that show the important relationship that exists between military clothing comfort and the performance of the soldier. Future research focused on the optimization of military clothing comfort may well enhance not only the overall comfort and morale of the Future Warrior, but also his/her ability to win on the battlefield.

4.9   Acknowledgment

The author wishes to acknowledge the important contributions of research collaborators who were involved in the joint conduct of the research studies described herein and who were co-authors of the original research reports upon which this chapter has been based. These collaborators include Howard Schutz, Carole Winterhalter, Rick Bell, Larry Lesher, William Santee and Larry Berglund.

4.10   References

BELL, R., CARDELLO, A.V. and SCHUTZ, H.G. (2003), Relations among comfort of fabrics, ratings of comfort and visual vigilance. *Perceptual and Motor Skills*, 97, 57–67.

BELL, R., CARDELLO, A.V. and SCHUTZ, H.G. (2005), Relationship between perceived clothing comfort and exam performance. *Family and Consumer Sciences Research Journal*, 31, 1–13.

BINNS, H. (1926), The discrimination of wool fabrics by the sense of touch. *British Journal of Psychology*, 16, 237–247.

BORG, G. (1982), 'A category scale with ratio properties for intermodal and interindividual comparisons', in Geissler, H-G. and Petxoid P. (Eds), *Psychophysical judgment and the process of perception*. Berlin, VEB Deutscher Veriag der Wissenschaften, 25–34.

BRAND, R.H. (1964), Measurement of fabric aesthetics. *Textile Research Journal*, 34, 791–804.

BRANSON, D.H. and SWEENEY, M. (1991), 'Conceptualization and measurement of clothing comfort: Toward a metatheory', in Kaiser, S. and Damhorst, M.L. (Eds.), *Critical linkages in textiles and clothing: Theory, method and practice*. Monument, CO, International Textile and Apparel Association, 94–105.

BROOKS, J.E. and PARSONS, K.C. (1999), An ergonomics investigation into human thermal comfort using an automobile seat heated with encapsulated carbonized fabric (ECF). *Ergonomics*, 42(5), 661–673.

BYRNE, M.S., GARDNER A.D.W. and FRITZ, A.M. (1993), Fiber types and end-uses: A perceptual study. *Journal Textile Institute*, 84 (2): 275–288.

CARDELLO, A.V., SCHUTZ, H.G. and WINTERHALTER, C. (2002), *Development and application of new psychophysical methods for characterization of the handfeel and comfort properties of military clothing fabrics*. US Army Soldier and Biological Chemical Command, Soldier System Center Technical Report NATICK/TR-02/022, Natick, MA August, 2002.

CARDELLO, A.V., WINTERHALTER, C. and SCHUTZ, H.G. (2003), Predicting the handle and comfort of military clothing fabrics from sensory and instrumental data: Development and application of new psychophysical methods. *Textile Research Journal*, 73 (3), 221–237.

CASSELMAN-DICKSON, M.A. and DAMHORST, M.L. (1993), Female bicyclists and interest in dress: Validation with multiple measures. *Clothing and Textiles Research Journal*, 11 (4), 7–17.

CIVILLE, G.V. and DUS, C.A. (1990), Development of terminology to describe the handfeel properties of paper and fabrics, *Journal of Sensory Studies*, 5, 19–32.

CROWN, E.M. and BROWN, S.A. (1984), Consumer trade-offs among flame retardance and other product attributes: A conjoint analysis of consumer preferences. *The Journal of Consumer Affairs*, 18(2), 305–316.

DELONG, M.R., MINSHALL, B.C. and LARNTZ, K. (1986), Use of schema for evaluating consumer response to an apparel product. *Clothing and Textiles Research Journal*, 5, 17–26.

ECKMAN, M. (1997), Attractiveness of men's suits: The effect of aesthetic attributes and consumer characteristics. *Clothing and Textiles Research Journal*, 15 (4), 193–202.

FEATHER, B.L., FORD, S. and HERR, D.G. (1996), Female collegiate basketball players' perceptions about their bodies, garment fit and uniform design preferences. *Clothing and Textiles Research Journal*, 14 (1), 22–29.

FORSYTHE, S.M. and THOMAS, J.B. (1989), Natural, synthetic and blended fabric contents: An investigation of consumer preferences and perceptions. *Clothing and Textiles Research Journal*, F (3): 60–64.

FOURT, L. and HOLLIES, N.R.S. (1970), *Clothing: Comfort and function*. New York: Marcel Dekker.

GAGGE, A.P., STOLWIJK, J.A.J. and HARDY, J.D. (1967), Comfort and thermal sensations and associated physiological responses at various ambient temperatures. *Environmental Research*, 1, 1–20.

GORDON, C.C., BRADTMILLER, B., CHURCHILL, T., CLAUSER, C.E., MCCONVILLE, J.T., TEBBETS, I.O. *et al.* (1989), *Anthropometric survey of U.S. Army personnel: Methods and summary statistics*. US Army Natick Soldier Center Technical Report NATICK/TR-89/044, AD A225 094, Natick, MA.

GREEN, B.G., SHAFFER, G.S. and GILMORE, M.M. (1993), Derivation and evaluation of a semantic scale of oral sensation magnitude with apparent ratio properties. *Chemical Senses,* 18, 683–702.

GREEN, P.E. and SRINIVASAN, V. (1978a), A general approach to product design optimization via conjoint analysis. *Journal of Marketing,* 45, 17–37.

GREEN, P.E. and SRINIVASAN, V. (1978b), Conjoint analysis in consumer research: Issues and outlook. *Journal of Customer Research,* 5, 103–124.

GRIDGEMAN, N.T. (1961), A comparison of some test methods. *Journal of Food Science,* 26, 171–177.

GUILFORD, J.P. and DINGMAN, H.F. (1955), A modification of the method of equal-appearing intervals. *American Journal of Psychology,* 68, 450–454.

HANCOCK, P.A. and VASMATZIDIS, I. (2003), Effects of heat stress on cognitive performance: The current state of knowledge. *International Journal of Hypothermia,* 19(3), 355–373.

HOLLIES, N.R., CUSTER, A.G., MORIN, C.J. and HOWARD, M.E. (1979), A human perception analysis approach to clothing comfort. *Textile Research Journal,* 49, 557–564.

JACOBSEN, M., FRITZ, A., DHINGRA, R. and POSTLE, R. (1992), A psychophysical evaluation of the tactile qualities of hand knitting yarns. *Textile Research Journal,* 62, 557–566.

KAWABATA, S. and NIWA, M. (1975), Analysis of hand evaluation of wool fabrics for men's suit using data of thousand samples and computation of hand from the physical properties, in *Proceedings of the 5th International Wool Textile Research Conference,* 5, 413–424.

KAWABATA, S., (1980), *Standardization and Analysis of Hand Evaluation* (2nd ed.), Osaka: The Textile Machinery Society of Japan.

MEISELMAN, H.L. and CARDELLO, A.V. (2002), Soldier-centric product development: quantifying the sensory and comfort properties of CB clothing, in: *Proceedings of the NATO human factors in medicine panel/symposium on operational medical issues in chemical and biological defense.* RTO-MP-075, AC/323 (HFM-060), Nevilly-sur-Seine Cedex, FRANCE: NATO/RTO, 23/1–23/15,.

PIERCE, F.T. (1930), The 'handle' of cloth as a measurable quantity. *Journal of the Textile Institute,* 21, T377–416.

PHILIPPE, F., SCHACHER, L., ADOLPHE, D.C. and DACREMONT, C. (2004), Tactile feeling: Sensory analysis applied to textile goods. *Textile Research Journal,* 74, 1066–1072.

PONTRELLI, G.J. (1977), 'Partial analysis of comfort's gestalt', in N.R.S. Hollies and R. F. Goldman (Eds.), *Clothing comfort: Interaction of thermal, ventilation, construction and assessment factors,* Ann Arbor, MI: Ann Arbor Science Publishers Inc., 71–80.

ROHLES, F.H. (1971), Psychological aspects of thermal comfort. *American Society of Heating, Refrigeration and Engineering Journal,* 13, 86–90.

ROBINSON, K.J., GATEWOOD, B.M. and CHAMBERS, E. (1994), Influence of domestic fabric softeners on the appearance, soil release, absorbency, and hand of cotton fabrics, in *Book of Papers, 1994 International Conference and Exhibition,* American Association of Textile Chemists and Colorists, Charlotte, NC, 58–66.

ROBINSON, K.J., CHAMBERS, E. and GATEWOOD, B.M. (1997), Influence of pattern design and fabric type on the hand characteristics of pigment prints. *Textile Research Journal,* 67 (11), 837–845.

SANTEE, W.R., BERGLUND, L.G., CARDELLO, A.V., WINTERHALTER, C.A. and ENDRUSICK, T.L. (2006), *Physiological and comfort assessments of volunteers wearing battle-dress uniforms (BDU) of different fabrics during intermittent exercise*. United States Army Research Institute for Environmental Medicine Technical Report T06–06, Natick, MA.

SCHUTZ, H.G. and CARDELLO, A.V. (2001), A labeled affective magnitude (LAM) scale for assessing food liking/disliking. *Journal of Sensory Studies*, 16, 117–159.

SCHUTZ, H.G., CARDELLO, A.V. and WINTERHALTER, C. (2005), Perceptions of fiber and fabric uses and the factors contributing to military clothing comfort and satisfaction. *Textile Research Journal*, 75(3), 223–232.

SCHUTZ, H.G. and PHILLIPS, B.A. (1976), Consumer perception of textiles. *Home Economics Research Journal*, 1: 2–14.

SLATER, K. (1977), Comfort properties of textiles. *Textile Progress*, 9, 1–71.

SLATER, K. (1985), *Human comfort*. Springfield, IL: Charles C. Thomas.

SLATER, K. (1996), 'Comfort or protection; the clothing dilemma'. In: J.S. Johnson and S.Z. Mansdorf (Eds), *Performance of Protective Clothing*. STP 1237, West Conshohocken, PA: American Society for Testing and Materials, 486–497.

STEVENS, S.S. (1957), On the psychophysical law. *Psychological Review*, 64, 153–181.

STEVENS, S.S. and GALANTER, E.H. (1957), Ratio scales and category scales for a dozen perceptual continua. *Journal of Experimental Psychology*, 54, 377–411.

SWEENEY, M.M. and BRANSON, D.H. (1990), Sensorial comfort. Part II: A magnitude estimation approach for assessing moisture sensation. *Textile Research Journal*, 60, 447–452.

WAGNER, J., ANDERSON, C. and ETTENSON, R. (1990), Evaluating attractiveness of apparel design: A comparison of Chinese and American consumers. In: P.E. Horridge (Ed.), *ACPTC Proceedings of National Meeting*, Monument, CO: Association of College Professors of Textiles and Clothing, 97.

WHEAT, K.L. and DICKSON, M.A. (1997), Uniforms for collegiate female golfers: Cause for dissatisfaction and role conflict? *Clothing and Textiles Research Journal*, 17 (1), 1–10.

WINAKOR, G., KIM, C.J. and WOLINS, L. (1980), Fabric hand: Tactile sensory assessment. *Textile Research Journal*, 50, 601–610.

WINSLOW, C-E.A., HERRINGTON, L.P. and GAGGE, A.P. (1937), Relations between atmospheric conditions, physiological reactions and sensations of pleasantness. *American Journal of Hygiene*, 26, 103–115.

WORKMAN, J.E. (1990), Effects of fiber content labeling on perception of apparel characteristics, *Clothing and Textiles Research Journal*, 8 (3): 19–24.

YICK, K.L. CHENG, K.P.S. and HOW, Y.L. (1995), Subjective and objective evaluation of men's shirting fabrics. *International Journal of Clothing Science and Technology*, 7 (4), 17–29.

5

Testing and analyzing comfort properties of textile materials for the military

F. S. KILINC-BALCI and Y. ELMOGAHZY,
Auburn University, USA

5.1   Introduction

The critical importance of testing and analyzing protective clothing stems from the need for a high-level performance under many major constraints including: predictable and unpredictable sources of life threats, inevitable sacrifice of some human comfort as a result of extreme protection, regulatory requirements, and liability costs. These constraints require reliable, precise, and better simulating techniques of testing and characterizing protective systems.

Today, the protective clothing industry provides protection products to a wide range of users including mineworkers, firemen, medical personnel, police officers, industrial workers, and military personnel. In addition, convenient protective systems that can be used by all people in case of emergency situations will potentially become an essential product in the years to come as a result of the various threats from Mother Nature, industrial contamination, or other hazardous sources. In the United States, over 25 million people wear uniforms in the workplace.[1] The cost of clothing and individual equipment each year by the United States Department of Defense (DoD) has reached over one billion dollars now. A large portion of these purchases are battle dress uniforms (BDU), the two-piece, camouflage uniforms worn by troops in combat, training, and garrison situations.[2]

An extensive review of the numerous literatures on protective fabric systems clearly suggests that, with few serious exceptions, there exists an obvious parallel in which research efforts of physical, chemical, and biological protection are on one side and those of human comfort are on the other. When protection is a crucial anticipated goal, total protection is ideally equal to total insulation (a space suit being the ultimate example) in which a protective fabric system must be totally impermeable to heat, radiation, gases, liquids, chemicals, dirt and dust, and bacteria. This objective in itself represents a fundamental violation of the whole concept of a traditional fabric system, which is to provide a portable breathable environment to the wearer and allow the performance of basic human functions from simple movement to specific physical tasks. For military applications, those tasks can be extended way beyond the normal human daily tasks. To make matters additionally complicated, the concept of comfort has not been fully resolved as a result of the complex interaction between thermo-physiological, neuro-physiological (tactile), and psychological factors, that make the best comfort analysis merely an issue of resolving the awareness/fuzziness trade-off of the comfort phenomenon.[3] In this trade-off, extreme awareness of fabric/body interaction also implies extreme discomfort; minimum or no awareness of fabric/body interaction often implies a pleasant feeling; and in-between there is a wide range of fuzziness imposed by the wide variability of human reactions. Indeed, with all the research effort made in this area, the best a human can do is to identify discomfort. Many of these challenges have been addressed in various independent research efforts. What has never been made is an integrated effort in the form of comprehensive modelling of the interactive nature of protection and comfort factors, supported by consideration of the psychological effects that can be independent when one crosses the issue of protection to that of comfort.

In this chapter, different approaches used for characterizing comfort-related attributes are discussed, and a new patented test method of fabric hand is described. Experimental results of comfort and protection related factors tested for military clothes are also reported. In addition, a design-oriented model is described in which both protection and comfort-related factors are integrated into a single index.

5.2   The multiplicity of characterization methodologies of comfort

In the context of protective clothing systems, comfort may be defined in many different ways:

- A state of satisfaction with the protective clothing system in terms of human body interaction with the system
- The presence of a friendly environment provided by the protective clothing system in terms of heat and moisture transfer from and to the body
- A state of unawareness of the protective clothing system by the user

The first definition implies physical effects, the second implies thermal effects, and the third implies psychological effects. The above definitions were carefully chosen by Elmogahzy and his co-workers in a long study of clothing comfort[3] so that an expectation of a neutral state, rather than the more active state of pleasure found in some fashionable clothing, is achieved. Indeed, it is often preferable to determine the discomfort state with most protective clothing systems rather than the comfort state as a result of the special functions of these systems that often impose some inconvenience resulting from heavier weight or pore closeness. Psychologically, comfort being a natural state makes it easier for the wearer to describe discomfort using common terms such as 'too prickly', 'too stiff', 'too hot', or 'too cold'. These characterizations are easier to quantify using some psychological scales.[2,4]

Among all aspects associated with human feelings and desires, comfort represents a central concern. Indeed, just about every activity a human performs in life involves a process of seeking comfort or relief from environmental and/or mental constraints. The comfort level of a human is driven by a host of factors, which may be divided into three main categories: environmental (air temperature, radiant temperature, humidity, etc.), physical (human minimum inherent status, health and physical condition, etc.), and psychological (human psychological condition) (see Fig. 5.1). To make matters additionally complex, these factors typically interact in a nonlinear manner. Furthermore, a human hardly ever experiences a still

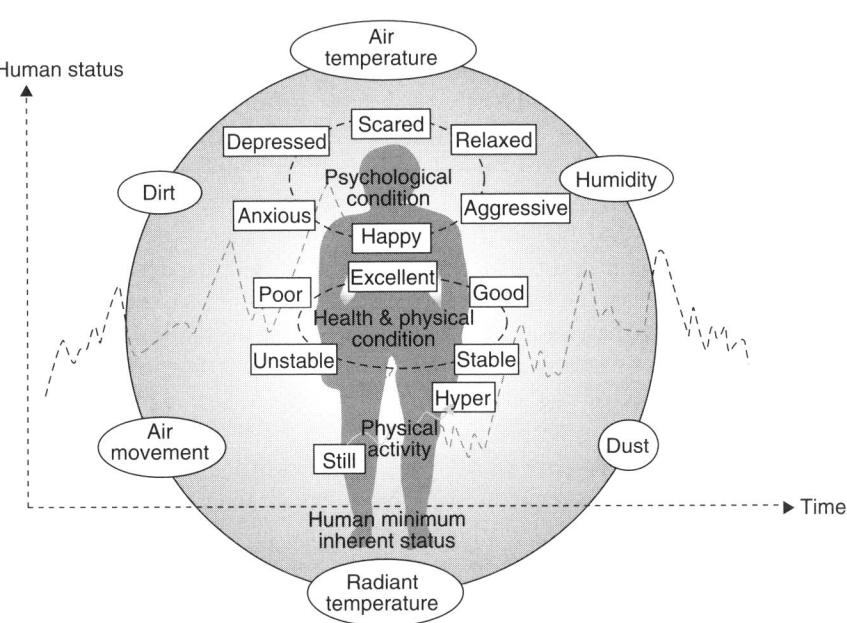

5.1 Primary factors influencing human comfort.[5]

environment or body conditions. In other words, there is a continuous change over time that leads to transitional effects.[5]

The multiplicity and complexity of factors influencing the comfort phenomenon has resulted in numerous research investigations dealing with different comfort-related aspects. Indeed, since Peirce's well-known study in 1930,[6] there have been hundreds of studies conducted with numerous experimental and analytical approaches, each providing an insight into the nature of the comfort phenomenon. However, a complete evaluation of this critical phenomenon faces many challenges, particularly cost and time related, and requires a substantial multidisciplinary involvement.

Comfort analysis can be divided into three main categories:

(i)   Objective analysis in which quantitative measures characterizing the comfort status can be determined (tactile and thermal parameters).
(ii) Subjective analysis in which psychological evaluation is made using surveys, ratings and scales.
(iii) Correspondence analysis in which the above two categories are combined to develop quantitative measures.

Comfort or discomfort is a well-realized integrated mental status by human individuals. However, there are no objective output parameters that can fully describe this realization. Instead, there are hundreds of parameters, each emphasizing one comfort-related aspect, yet none truly reflects the whole comfort or discomfort realization. The problem with relying totally on subjective evaluation is that humans are different in their perceptions of comfort and some may have different views than those of the expert evaluator. In addition, comfort-related factors typically interact in a very complex, non-linear fashion that makes traditional linear, discrete, or bi-polar physiological scaling automatically deficient. More importantly, most subjective analyses rely on descriptors developed by a few experts, each of which deals with a single aspect of comfort. This makes comfort analysis a form of multiple choice questionnaire rather than a collective analysis leading to an integrated index of comfort.[3] In addition, and in contrast with hand or the initial feel of comfort, a reliable evaluation of the discomfort status of a clothing system may require a significant amount of time and experience with the system being tested. In the absence of full consideration of these difficult factors, any comfort study will be limited by many constraints including: people attitude, familiarity with the clothing system being tested or similar systems, external influences, expectations, prejudices, quality assumptions, and stereotypes. In some studies, these constraints can mask the particular factors determining the comfort status of a particular clothing system and lead to misleading results.

5.3   The trade-off between protection and comfort

Avoiding situations that involve discomfort and difficulty is human nature. Due to this nature, some people do not wear a seat belt because of the mild constraint that it imposes on the human body, even though they are well aware that it is for their safety and it might avoid a fatal car accident. One can then expand this simple example to more complicated situations in which protective clothing can truly impose real constraints on human body movement. Imagine a mine worker, a fireman, a policeman, or a military officer in action and under the typical environmental conditions in which they work. Among all these people, the primary concern with protective clothing systems is undoubtedly the parallel discomfort. This concern is often expressed in common words such as 'unbearable', 'suffocating', 'too heavy', 'too bulky', and 'too hot'. These are subjective descriptors that are useful in determining the psychological status of the individuals, but do not translate to precise values of design parameters that engineers can use to improve the performance of protective clothing systems. Even if these systems are tolerated by virtue of their necessity, efficiency and long-lasting effectiveness will certainly be in question.

The decision of suitability of certain protective clothing is largely based on the protective features of the product (e.g. fire resistance, tear resistance, durability, chemical detection, sharp objects and projectile resistance, environmental protection, and dust or dirt protection). After some use, two critical aspects of protective clothing surface very quickly: care and comfort (see Fig. 5.2). If these two criteria are not met, a negative wearer's view of the product will develop and it may grow to the point that even under the most risky situations, protective clothing may be disregarded. It is critical therefore that the design of protective clothing accounts for these two aspects as well as providing protection.[7] Given the fact that, for protective systems, protection is the primary design focus, the key design question is to what extent comfort and care should be emphasized. This question will depend on many factors including: the nature of the protection mission, the duration of use of the protective systems, and the extent of training by the wearers of protective clothing.

5.4   The comfort trilobite: Tactile, thermal, and psychological

As indicated earlier, comfort is a result of three basic aspects: tactile, thermal, and psychological. These aspects are discussed in the following sections. Before proceeding with this discussion, it should be pointed out that realization of these aspects should be based on three coexisting factors: body,

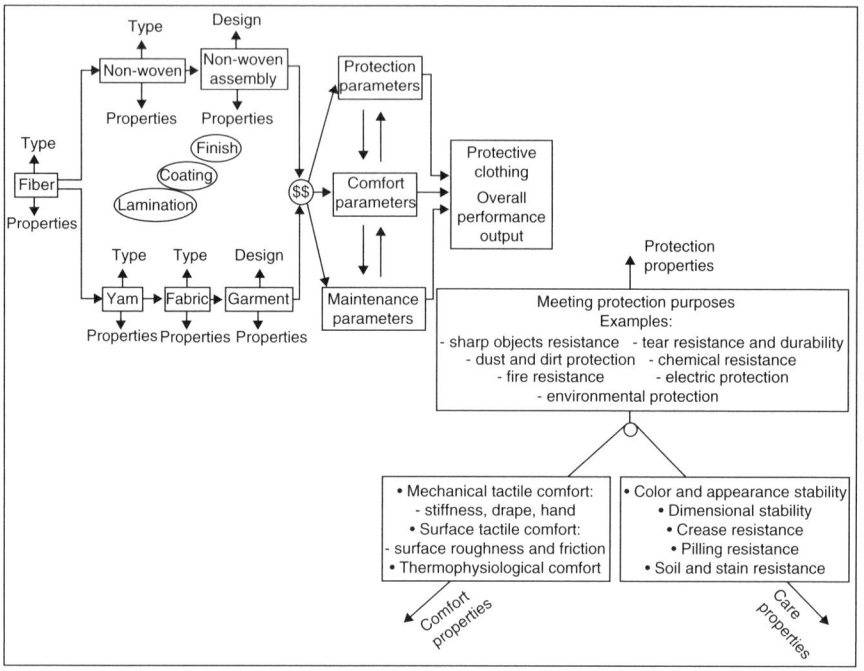

5.2 Basic design components and criteria and related sub-criteria of protective clothing.[7]

clothing, and environment. With regard to the human body, the key factor is the level of physical activity being performed (from a relaxed status to extremely harsh action). The clothing system consists of constituents each of which can contribute to the comfort status. These constituents are listed in Fig. 5.3. Environment implies the portable environment between the skin and the clothing system and the surrounding environment.

5.4.1 Tactile (neurophysiological) aspects of fabric comfort

Tactile, or neurophysiological, comfort reflects the feel of fabric against the skin. This feel is triggered by sensory receptors in the skin, which are connected to the brain by a network of nerve fibers (see Fig. 5.4). Three basic sensory nerve sensations are realized: the pain group, the touch group of pressure and vibration, and the thermal group of warmth and coolness.

The skin/fabric interaction is normally stimulated by many factors, some of which are mechanically related (bending rigidity, surface roughness, etc.), and others are thermally related (warm and cool sensation). Since the classic work by Peirce in 1930,[6] numerous investigations have been devoted

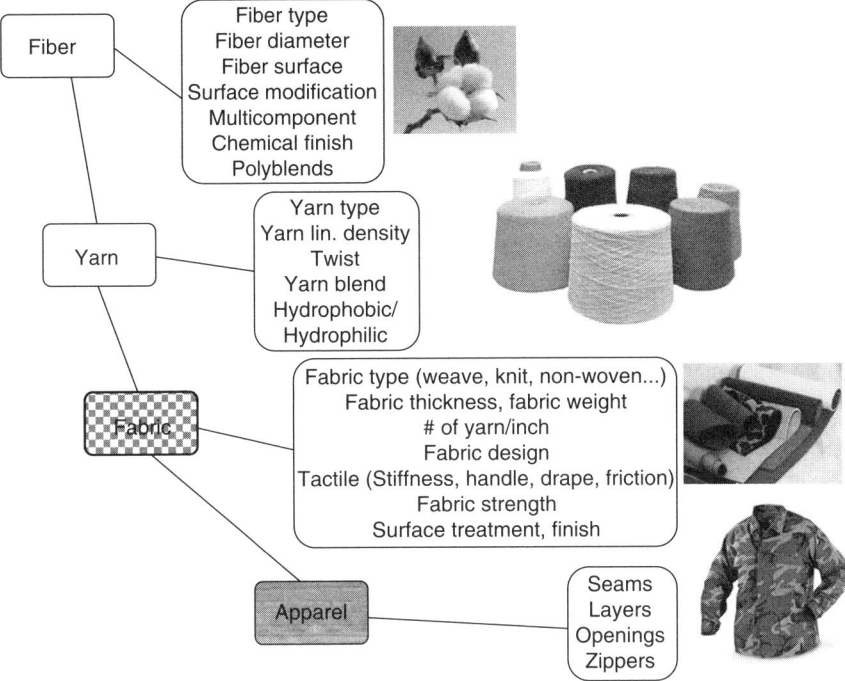

5.3 Constituents of fabric comfort from clothing point of view.

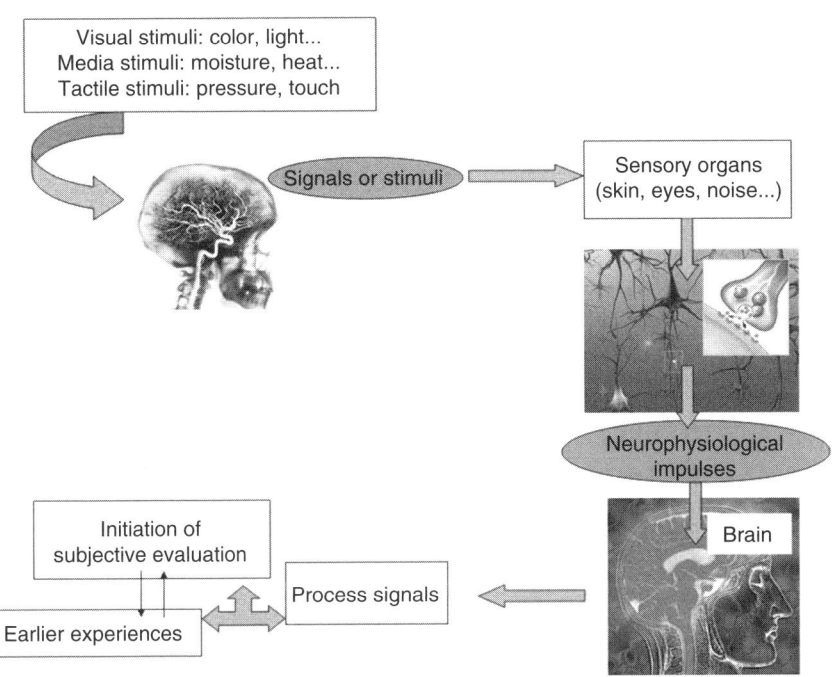

5.4 The mechanism of subjective evaluation in the human body.

to analyze various neurophysiological aspects of comfort, particularly hand-related factors.

The most notable outcomes of the neurophysiological comfort studies are systems to test fabric tactile properties (e.g. bending, handle, friction). A review of the different developments reveals two main categories of fabric hand (handle) evaluation systems:

• Indirect systems of fabric hand (handle) evaluation
• Direct methods of fabric hand (handle) evaluation

The difference between these two categories lies in the types of parameters produced by each category and their associated interpretations. Indirect systems do not characterize handle in a direct fashion. Instead, they produce instrumental parameters that are believed to represent basic determinants of fabric handle such as fabric stiffness, fabric roughness, and compressibility. Only through parallel subjective assessment and cross-correlations, some parameters that are believed to simulate fabric handle are estimated. The two common methods of this category are the Kawabata system (KES®) and the FAST® (Fabric Assurance by Simple Testing) system.

Direct methods of fabric hand (handle) evaluation represent creative techniques that are intended to simulate two or more aspects of hand evaluation and produce quantitative measures that are labelled as hand force or hand modulus. These methods include: the ring method and the slot method. It should be pointed out that the term 'direct' does not necessarily mean more representative or more accurate in comparison with the indirect systems. Brief descriptions of these systems are given below.

Indirect handle evaluation systems: The Kawabata (KES®) system and the Fabric Assurance by Simple Testing (FAST®) system

The Kawabata and FAST systems are commercial systems that are available in many fabric testing laboratories around the world. Reviews of these systems have been discussed in many literatures.[8–10,27] These systems belong to the correspondence analysis defined earlier, in which objective and subjective evaluations are combined to develop quantitative measures of fabric hand.

Kawabata et al.[8] developed a system that was later called KES-F (The Kawabata Evaluation System for Fabrics) for measuring the fabric mechanical properties and an objective method for the fabric handle measurement in 1972. The system, manufactured by Kato Tech. Co. of Kyoto, measures physical, mechanical and surface properties of fabrics using four separate instruments: Shear/Tensile Tester, Bending Tester, Compression Tester and Surface Tester.

In order to effectively use the Kawabata's system, it is typically important to get experts to agree on what aspects of handle are important and the relative contribution of each aspect with respect to the fabric under consideration. In this regards, the Kawabata system establishes the so-called 'primary hand' as a measure characterized by properties such as stiffness, smoothness, fullness/softness, crispness, anti-drape stiffness, scrooping, flexibility with soft feeling, and soft touch. Given the fact that these descriptors may have interpretive differences, particularly when translated to other languages, Kawabata decided to use Japanese descriptors corresponding to these properties. However, the terms used exhibit a great deal of overlap and have their share of confusion. In addition, they are certain to be different from one fabric category to another. Indeed, it has been found that there are differences between countries in their perception of what truly constitutes fabric handle with respect to a particular application. The end result is assessed through a correlation of tested parameters with the subjective assessment of handle using linear regression equations.

In a recent study, Cardello and Winterhalter[2] used sound psychophysical principles for assessing both qualitative and quantitative aspects of sensory handle and comfort of military clothing fabrics. They checked the sensitivity and reliability of a standardized hand evaluation methodology combined with Kawabata data. They found a high degree of predictability of comfort responses from a combination of sensory and Kawabata parameters.

The FAST (Fabric Assurance by Simple Testing) system was designed with a more global view of fabric handle in the late 1980s. It was developed by CSIRO for use by goods manufacturers to detect and diagnose problems associated with the process of conversion from fabric to garment. As a result, the system aims at distinguishing loosely constructed fabrics which are easily deformable, from tightly constructed fabrics. The system consists of three instruments: compression meter, bending meter, and extension meter.

Direct hand evaluation systems

Direct hand methods include the ring test, the slot test, and the Elmogahzy-Kilinc handle measurement system. These methods were developed to measure the handle properties of fabric through simulations of the various mechanisms reflecting fabric hand.[11–14] The first two methods were based on pulling a fabric sample through a ring (the ring method) or pushing a fabric sample through a slot (the slot method) and measure the resistances to the pull-through or push-through mechanisms. This, in part, simulates how a person tends to handle a piece of fabric when he/she is attempting to evaluate it.

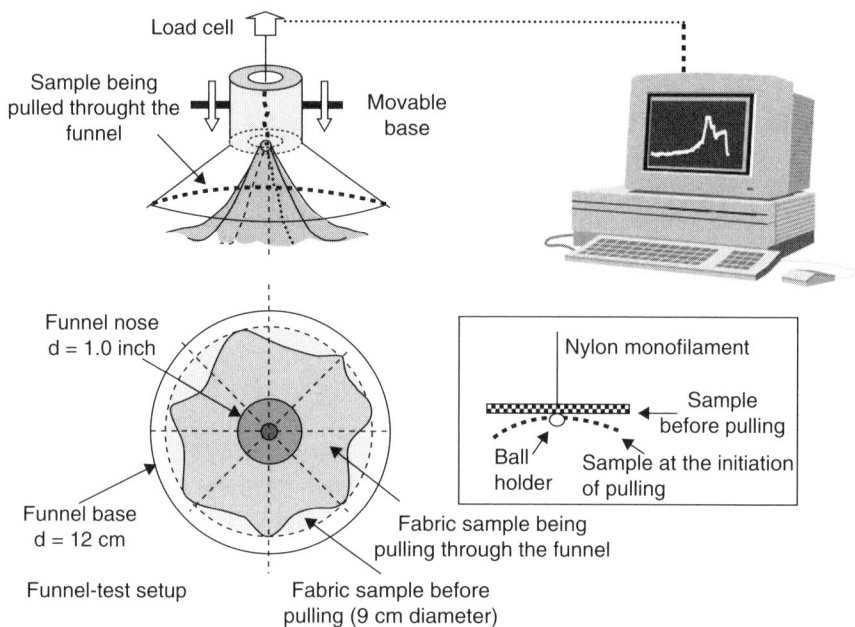

Load cell

Sample being pulled throught the funnel

Movable base

Funnel nose
d = 1.0 inch

Funnel base
d = 12 cm

Funnel-test setup

Nylon monofilament

Sample before pulling

Ball holder

Sample at the initiation of pulling

Fabric sample being pulling through the funnel

Fabric sample before pulling (9 cm diameter)

5.5 The Elmogahzy–Kilinc fabric hand method.[14]

A more recently patented method is Elmogahzy–Kilinc hand method.[14,28,29] Basic components used in this method are shown in Fig. 5.5. The underlying concept of this method was inspired by the theoretical and experimental efforts made by many previous investigators. In addition, the method aimed at overcoming some of the problems associated with the statistical reproducibility and characterization parameters found in previous methods. A flexible light funnel is used to represent the media through which the fabric sample is pulled. The idea of using a funnel media instead of a ring or a slot arrangement is to provide multiple configurations of fabric hand that closely simulate the various aspects of the hand phenomenon. The contoured flexible surface of the light funnel simulates anticipated hand modes such as mild stretch, drape, compression, lateral pressure, and surface friction. These modes are achieved both simultaneously and sequentially. In addition, the funnel media allows both constrained and unconstrained fabric folding or unfolding; a key aspect of fabric handle.

In general, as the movable head of the AU® mechanical tester moves downward, the funnel moves downward and the rounded fabric sample, which is connected to the load cell at its center, is pulled through. During the duration of the fabric pull through the funnel, a force–time profile is generated, which is termed the 'handle profile'. For most apparel fabrics, this profile takes the common shape shown in Figure 5.6. From this profile,

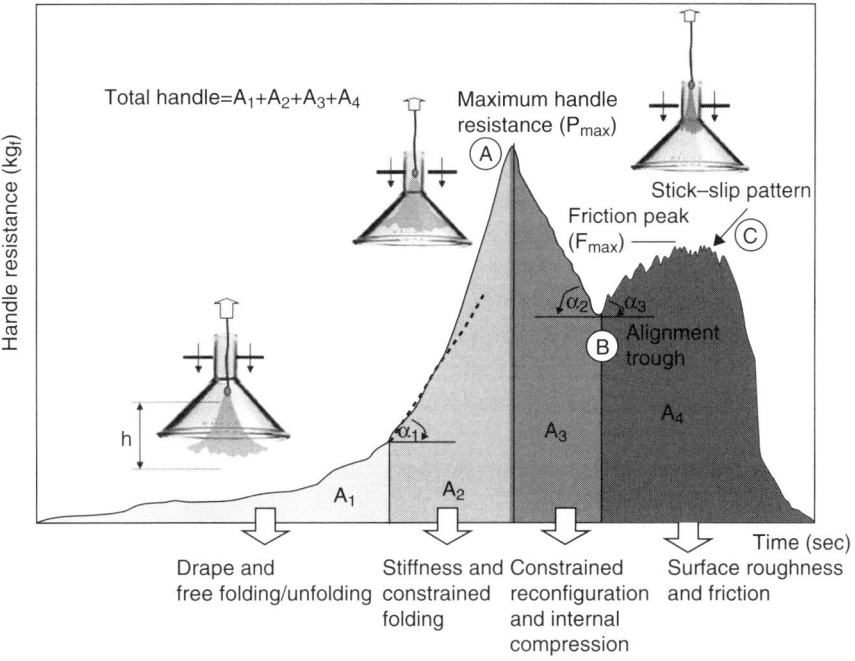

5.6 Elmogahzy–Kilinc fabric hand profile.[14]

different handle parameters can be obtained.[15] Three categories of handle parameters are obtained:

(i)   Handle modulus parameters represented by the slopes at the portions of the profile (i.e. α_1, α_2 and α_3).

(ii) Handle work parameters represented by the areas under the profile curves (i.e. A_1, A_2, A_3, and A_4).

(iii) Handle resistance parameters represented by the forces (i.e. first handle peak, first handle drop, and second handle peak).

The handle profile reflects most possible deformational modes involved in a hand trial. In addition, each zone of the profile reflects a specific mechanism of fabric hand. This point is important particularly when an enhancement of a particular hand-related parameter is required in the process of fabric design. Also, the total area under the hand profile provides an integrated parameter of fabric hand, which can be termed 'objective total hand', or OTH. This parameter is the sum of the four handle work areas, A_1, A_2, A_3 and A_4, discussed above. A detailed study in which this parameter was evaluated[5] proved that it is highly correlated to subjective hand assessments of many woven and knit fabrics.

5.4.2 Thermophysiological comfort

Thermal comfort has been defined by the American Society of Heating Refrigeration and Air Conditioning Engineers (ASHRAE) Standard 55-66 as 'that condition of mind, which expresses satisfaction with the thermal environment'. It is also defined as the 'attainment of a comfortable thermal and wetness state; involves transport of heat and moisture through a fabric'. People reach this type of comfort when they do not need to add or remove clothing in order to be satisfied with the temperature. This aspect of comfort becomes particularly important for military clothing and is enormously difficult to achieve since military personnel are exposed to several different thermal environments during their duties. Binns,[17] Peirce,[6] Houghton and Yaglou,[18] Winslow,[19–20] Fourt and Hollies,[21] Slater[22] and Rohles[23] are among the many researchers who have analyzed the thermal comfort.

Fanger[16] identified six variables, which influence the condition of thermal comfort. These variables are:

- air temperature,
- mean radiant temperature,
- relative air velocity,
- water vapour pressure in the ambient air,
- activity level (heat production), and
- thermal resistance of clothing (clo).

The effect of clothing on thermal comfort depends mainly on such factors as:

- physical properties of fabric, and
- air spaces between the body and the fabric (or between the fabrics themselves), and
- characteristics of the ambient environment.

In addition to heat transfer, water-absorbing properties play an important role in comfort and warmth. In general, if the clothing becomes wet, the insulating ability of the fabric will be lowered.

In 1970, Fanger[16] developed a mathematical model to define the neutral thermal comfort zone of men in different combinations of clothing and activity levels. Based on Fanger's study, ASHRAE developed comfort charts and indices of thermal sensations. By using these indices, it is possible to predict the comfort acceptance under different combinations of clothing insulation, metabolic level, air temperature, and wet-bulb temperature (or radiant temperature). An international thermal comfort standard (ISO7730) was developed based on Fanger's comfort model. This standard is based upon the predicted mean vote (pmv) and predicted percentage of dissatisfied (ppd) thermal comfort indices.[16] It also provides methods for the assessment of local discomfort caused by draughts, asymmetric radiation and temperature gradients.

Even though comfort in military clothing has been studied for a long time, the major focus was on thermal rather than tactile comfort, since thermal stress is a major factor contributing to the human performance. Focus has turned to tactile properties in recent years since battle dress uniforms (BDUs) are also worn by military personnel on a daily basis in garrison situations; thus the tactile comfort can be as important as thermal comfort for combat uniforms.[2]

5.4.3 Psychological comfort

The trade-off between comfort and protection should be realized on the basis that comfort itself is a form of protection from environmental changes. Clothing, being an intermediate environment between the human body and surrounding media, plays a critical role, particularly for military personnel, in providing protection at minimum physical hindrance to body functions and maximum mobility freedom. By virtue of the fact that clothing does not represent a natural element in human life (unlike fur or wool for animal inherent protection), the key issue of comfort becomes an issue of a trade-off between accommodation with the surrounding media and adaption with human skin and body movement, especially when the clothing is used mainly for protection purposes, as in the military. The degree of protection required in traditional clothing is much less than that required in military clothing, as shown in Fig. 5.7. As the degree of protection increases, discomfort and cost also increase. When protection implies total isolation and

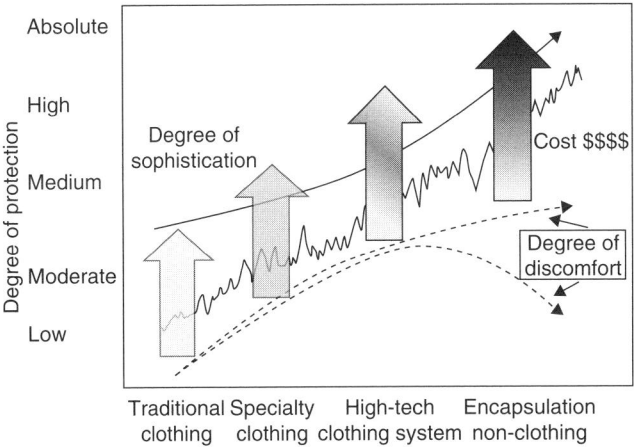

5.7 Basic criteria of protective clothing and related sub-criteria.

insulation, discomfort becomes inevitable. This trade-off adds to the complexity of the comfort phenomenon since it requires in-depth analysis of the nature of interaction between clothing and surrounding media (the human body and its external environment).

Despite the challenges associated with developing a unified comfort output, one can assume that there exists a neutral state at which optimum satisfaction with fabric/garment is felt and realized by a human. The term 'optimum' here implies 'a status at which a moderately pleasant feeling is achieved with respect to the situation/application in question'. Table 5.1 gives examples of products, with associated general applications and general anticipated optimum comfort criteria.[5] As demonstrated in this table, the anticipated comfort criteria may vary substantially with different product types and applications.

A key point, also revealed in the Table 5.1, is that, although the comfort reference varies significantly from one application to another, the two aspects that are commonly revealed in most responses are *awareness* and *protection*. This is a result of the fact that humans invented clothing primarily for protection, not for added body comfort. Humans would be more comfortable in the nude state if the environment and the surrounding would cooperate.

One conceptual way to illustrate the meaning of optimum comfort status is to examine possible relationships between the level of human awareness and human realization of the comfort/discomfort status. In a previous study conducted by El Mogahzy *et al.*,[3] many fabric types were examined and a generalized view of these relationships was developed. This view is illustrated in Fig. 5.8, which indicates that both awareness level and comfort status were determined on a 0 to 100% ranking scale. An optimum comfort status was defined as the status at which more than 80% of subjects responses give a score value of 40 to 60 on a 0 to 100% comfort scale (with 0 being extremely comfortable and 100% being totally unbearable or extremely uncomfortable). It should be pointed out that this scale is developed under normal standard environmental conditions (70 °F and 65% RH) and normal (low to moderate) physical activity. This means that only clothing-related factors are considered.

As we go further away from the neutral comfort state, uncomfortable or more comfortable feelings are progressively felt. Along with this, both the level of awareness and the extent of fuzziness (or the extent of clearly identifying the comfort/discomfort status) will also change, as shown in Fig. 5.8. The underlying assumption associated with the concept illustrated in Fig. 5.8 is that at extreme discomfort (e.g. bulky garment, an under shirt made from very stiff fibers, or a totally closed hydrophobic fabric structure against the skin), the level of awareness is very high and the level of fuzziness is very low. On the other hand, an extremely comfortable

Table 5.1 Examples of optimum comfort status[5]

Clothing type	Application	Anticipated optimum comfort criteria
Fire-resistant	Specialty product	• Minimum awareness at low physical activity level • Moderate awareness at moderate to high physical activity level • Moderate breathability • Durability
Military clothing	Transitional physical activity and environmental conditions:	• Minimum to moderate awareness at both resting mode and at high physical mode • Maximum physical protection • Maximum environmental protection • Great breathability • Coolness/lightness at high level of physical activities and in warm/hot environment • Warmth/heaviness in cold environment • Dust-free
Children clothing	Simple apparels	• Minimum awareness at all activity modes • Self-supportive/protective/light • Environmental adaptive/dirt-resistant • Safety oriented
Girdle	Specialty product	• Maximum intimacy with the body at reasonable awareness • Maximum physical protection/ High breathability • Coolness/lightness at high level of physical activities and in warm/hot environment
Night gown	Relaxing mode: Sitting Sleeping Easy walking	• Pleasant awareness associated with intimacy with the fabric (cuddling) • Nice touch and feel inside/out/smoothness • Coolness/lightness in hot environment • Warmth/moderate heaviness in cold environment

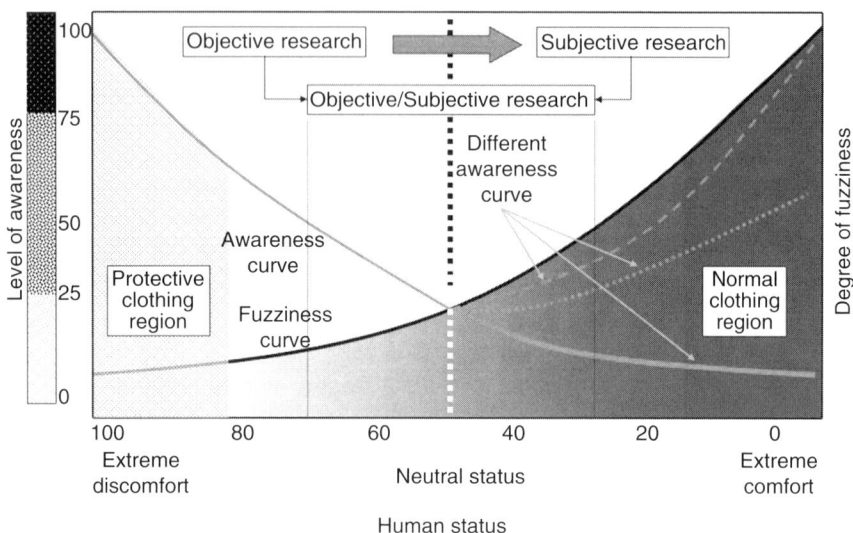

5.8 The interactive effects of awareness and fuzziness.[3]

status cannot be defined or identified with a high degree of objectivity; it is highly variable and immensely relative. Thus, at this state, which is almost imaginative, the degree of fuzziness is at its highest level by virtue of subjectivity, variability and relativity, and the level of awareness may vary from very low to very high depending on a host of factors that are primarily psychologically oriented. Obviously, different clothing products will result in different awareness curves, as indicated in Table 5.1.

Fuzziness curves, on the other hand, will depend on the subjects' ability to express how they feel and the level of anticipated comfort. In this regard, it should be pointed out that almost all people can identify a discomfort status (simply by coming into contact with the objects) and only a few can agree on an extreme comfort status. In general, an optimum comfort status can be identified at the transition stage from extreme discomfort to neutral comfort. Beyond the neutral comfort status and toward the maximum comfort side, a great deal of subjectivity will be encountered.

In the light of the above discussion, comfort/discomfort can be defined as a status of the level of awareness of clothing. In this regard, the main factors influencing the level of awareness are:

- physical activity level,
- fabric tactile behavior, and
- fabric thermal behavior.

The relationship between the level of awareness and each one of these factors will obviously vary, depending on the product type and application considered. However, in most cases, the physical activity exhibits an exponential relationship with the level of awareness. Thermal and tactile factors typically exhibit power functions with the level of awareness, with indices varying according to the application and garment type.[3,5]

5.5   Modeling the comfort phenomena: The ultimate challenge

In the context of protective/military clothing, different products can be compared in the marketplace on the basis of their impacts on the level of awareness. Figure 5.9 shows hypothetical examples of products compared on the basis of the level of awareness.[3] A key design and acceptance criterion in this regard is the initial slope of the power relationship between the level of awareness and the required degree of protection ($\theta_{A/P}$) (A: Awareness and P: Protection). Products with high $\theta_{A/P}$ values typically perform poorly in real applications by virtue of the fact that they have lower acceptability and convenience, and lower use duration (t_u).

The definition of comfort discussed above, coupled with the protection aspect, yield a state of relative comfort defined as 'the state at which the fabric/garment has a minimum mechanical interaction with the skin, and an optimum positive interaction with the environment; the environment here being the surrounding media and the localized against-skin media'. This definition was established by the present authors as the basis for a design-oriented comfort model.[5,15] In this model, the common factor that can truly tie all aspects associated with fabric comfort is fabric/skin interaction. In

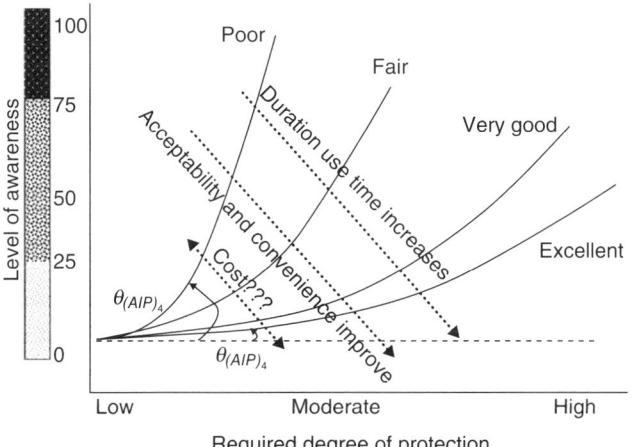

5.9 The conceptual relationship between awareness and protection.[3]

the context of pure comfort, this interaction should be minimized to provide maximum unawareness of clothing contact with the skin. In the context of protection, this interaction should be optimized so that it is not too high, to avoid irritation and loss of mobility, and it is not too low, to insure acceptable levels of protection both physiologically and psychologically. The question now is how to translate this factor into a characterization index of fabric comfort and how to establish design parameters of textile fabrics and garments that reflect this interaction. This question has led to the concept of Area Ratio as a unique index to characterize fabric/skin interaction. The Area Ratio is defined as the ratio between the true area of fabric/skin contact and the corresponding apparent area:

$$Area\ Ratio\ (AR) = \frac{True\ area\ of\ contact}{Apparent\ area\ of\ contact} = \frac{A_t}{A_a} \qquad [5.1]$$

5.5.1 The concept of Area Ratio

One of the fundamental structural differences between textile fabrics and other non-fibrous structures (e.g. paper) stems from the fact that what is perceived as a flexible flat sheet (the fabric) actually never exhibits a complete flatness when it comes into contact with other solid surfaces (as shown in Fig. 5.10a). In other words, the apparent area of contact between a textile

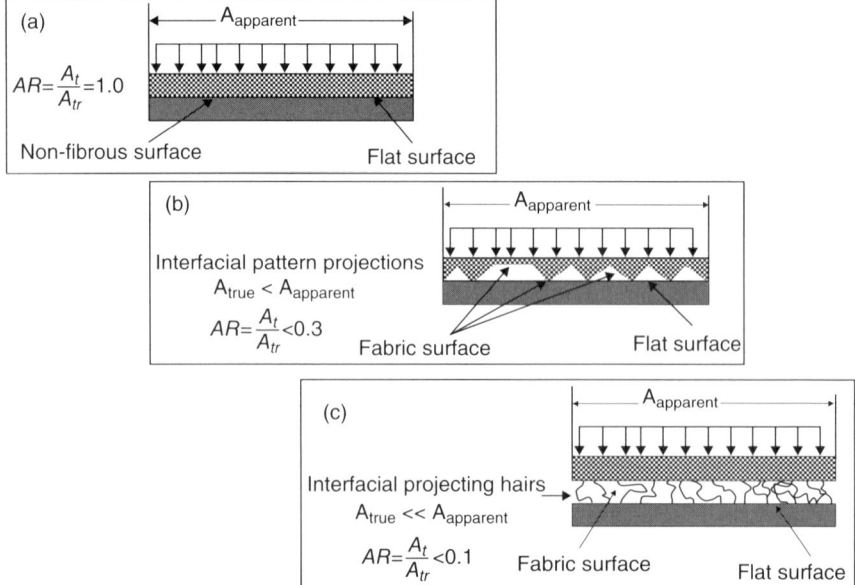

5.10 The difference between the apparent and the true area of fabric/other surface contact.[5]

fabric and another surface is typically much greater than the actual area of contact. This is a direct result of structural and deformation effects inherent to fibrous structures.[24,25] As a result, garment conformity to body contours is achieved at the expense of a substantial gap between the skin and the fabric in contact. In the context of comfort, this can be considered as a unique added-value phenomenon by virtue of the advantage that it provides with respect to the level of wearer awareness of the fabric/skin contact. In the context of protection, the level of awareness is not a uni-directional effect; instead, an optimum level of awareness (not too high and not too low) is required to achieve an optimum comfort/protection trade-off.

On a macroscopic level, fabric natural irregularities created by the fabric pattern (valleys and troughs) prevent the fabric from a pure contoured contact (see Fig. 5.10b). On a microscopic scale, surface disturbances such as projecting hairs or pills (Fig. 5.10c) further decrease the true contact. When one considers lateral effects, such as applied pressure, multiple fabric layers (weight and thickness), and human dynamics, one will see that the Area Ratio represents a key variable that can determine fabric perfor-mance in different applications.

The fact that the true area of fabric/skin contact is much smaller than the apparent area of contact represents a unique structural feature of textile fabrics that has been largely overlooked in previous comfort studies. In extreme cases, such as fabric puckering and dimensional instability, this feature is undesirable as it results in a great difficulty in handling fabrics. However, under normal wearing conditions, the large difference between the apparent contact area and the true contact area can provide many merits in several applications. For example, in relation to fabric comfort one can think of two extreme situations of fabric/skin contact: (a) an Area Ratio of approximately one, and (b) an Area Ratio of zero. At an Area Ratio of one, the fabric is in total (complete) contact with the body skin. This virtual situation can only occur if the contact area is under extremely high pressure or if the fabric is virtually cemented to the body skin, as it can be imagined in non-breathing, total isolation protective clothing. This situation is totally undesirable as it completely hinders body mobility and closes the fabric/skin gaps required for critical effects of clothing such as thermal insulation, air permeability, and breathability. On the other hand, if the Area Ratio is close to zero, a complete unawareness of the fabric will be felt. This situation can be positive if the body does not require any protection as a result of the surrounding environment being cooperative. In the case of pro-tective clothing, where the human body requires protection from the sur-rounding environment, an intermediate (optimum) fabric/skin contact will be desirable.

In the light of the above discussion, it follows that an optimum Area Ratio between the fabric and the skin should be maintained to achieve an

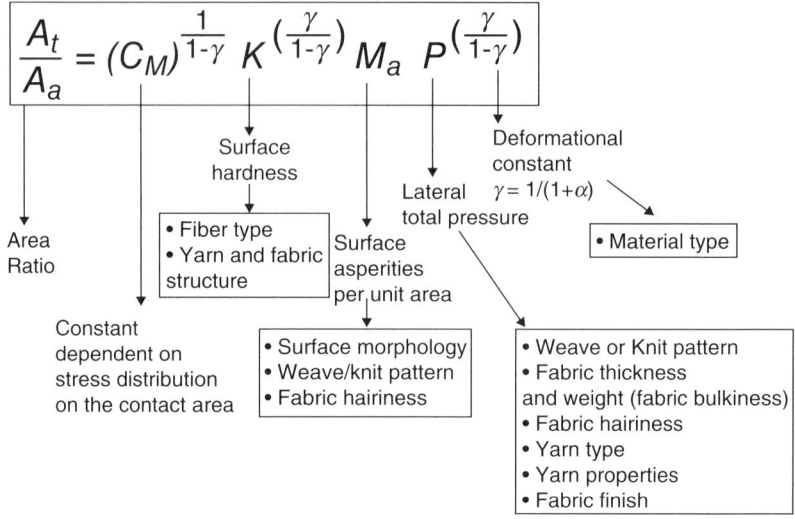

$$\frac{A_t}{A_a} = (C_M)^{\frac{1}{1-\gamma}} K^{(\frac{\gamma}{1-\gamma})} M_a \ P^{(\frac{\gamma}{1-\gamma})}$$

Area Ratio

Surface hardness
• Fiber type
• Yarn and fabric structure

Constant dependent on stress distribution on the contact area

Lateral total pressure
Surface asperities per unit area
• Surface morphology
• Weave/knit pattern
• Fabric hairiness

Deformational constant $\gamma = 1/(1+\alpha)$
• Material type
• Weave or Knit pattern
• Fabric thickness and weight (fabric bulkiness)
• Fabric hairiness
• Yarn type
• Yarn properties
• Fabric finish

5.11 Structural model of Area Ratio.[5]

optimum trade-off between comfort and protection. Although the Area Ratio is defined by purely geometrical parameters, it is actually a function of a complex interaction between a host of parameters, some of which are geometrical and others mechanical (see Fig. 5.11). This point is demonstrated by the general expression of the Area Ratio:[5]

$$\frac{A_t}{A_a} = (C_M)^{\frac{\alpha+1}{\alpha}} K^{-\frac{1}{\alpha}} M_a P^{\frac{1}{\alpha}} \qquad [5.2]$$

where
A_t/A_a = the ratio between the true area of fabric/skin contact and the corresponding actual area, K = Surface resilience, or hardness, constant (SI units = N/m$^{2(\alpha+1)}$), α = A constant which lies between 0 for purely plastic behavior and 1.0 for purely elastic behavior, M_a = the number of contacting asperities at the fabric/skin interface (an index of surface roughness), C_M = a parameter determined from the type of load distribution applied on the interface, and P = the lateral pressure applied on the area of fabric/skin contact

Based on the relationship between the lateral pressure and fabric thickness, a more detailed expression of Area Ratio has also been derived:[5]

$$\frac{A_t}{A_a} = (C_M)^{\frac{\alpha+1}{\alpha}} K^{-\frac{1}{\alpha}} M_a [K_t \cdot E \cdot V^3 \cdot (\frac{1}{t^3} - \frac{1}{t_o^3})]^{\frac{1}{\alpha}} \qquad [5.3]$$

where K_t is a constant, E is the Young's modulus of fibers, V is the volume of fibers, and t and t_o are thicknesses of the fiber mass at two different levels of pressure.

The above expressions clearly indicates that the Area Ratio is related to many deformational and surface parameters that collectively reflect the tactile mechanical and surface behavior of fabrics. Empirical analysis performed in Kilinc's study[5] also revealed that the Area Ratio is related to non-tactile parameters such as thermal insulation, thermal absorpitivity, air permeability, wicking parameters, and pore size distribution.

In Kilinc's study, the Area Ratio was measured using a reference method in which fabric samples were coated with high-viscosity ink and the true imprint area against a flat metallic surface was determined using Image Analysis.[5] Measures were taken at different levels of lateral pressure. In addition, two empirical indices of area ratio, namely 'a' and 'm', were estimated for each fabric from the general power function:

$$\frac{A_t}{A_a} = aP^m \qquad\qquad [5.4]$$

More efficient thermo-photographic and 3-D optical methods were also examined to determine the Area Ratio between the fabric and human skin (see Fig. 5.12).

Based on the experiments conducted in Kilinc's study, it was found that different fabric structures exhibited different values of Area Ratio (see Fig. 5.13). In general, most traditional garments had an Area Ratio ranging from 0.01 to 0.05, depending on the lateral pressure applied (typical pressure applied 0.1 to 2.6 g/cm^2). Some structures exhibited exceptionally higher Area Ratios (e.g. AR for some woven satin fabrics was as high as 0.25, and AR for some knit interlock fabrics was as high as 0.15). With protective clothing, it was found that a wide range of AR does exist, from 0.001 (e.g. some hospital uniforms) to 0.46 (e.g. some military and police uniforms).

The concept of area ratio introduced above primarily stems from the need for a unique parameter that can ultimately represent a single-index of garment comfort based on the justifications discussed earlier and the numerous preliminary results obtained in Kilinc's study. To some, this may seem a radical approach considering the multiplicity of factors associated with fabric comfort. However, we all can agree that a phenomenon that has no output parameter, to uniquely define it, is a phenomenon that no one can design for. Indeed, it is often the case that an improvement in the tactile aspect of comfort comes at the expense of the thermal aspect or vice versa. When protection is the primary goal, this situation becomes more complicated as protection is often associated with a great deal of garment awareness reflected in loss of mobility, heat stress and other adverse human-related effects.

Extensive empirical analysis of the Area Ratio parameter proved that it did relate to all tactile (mechanical and surface) comfort parameters and many thermal comfort parameters.[5] An exploratory partitioning model of

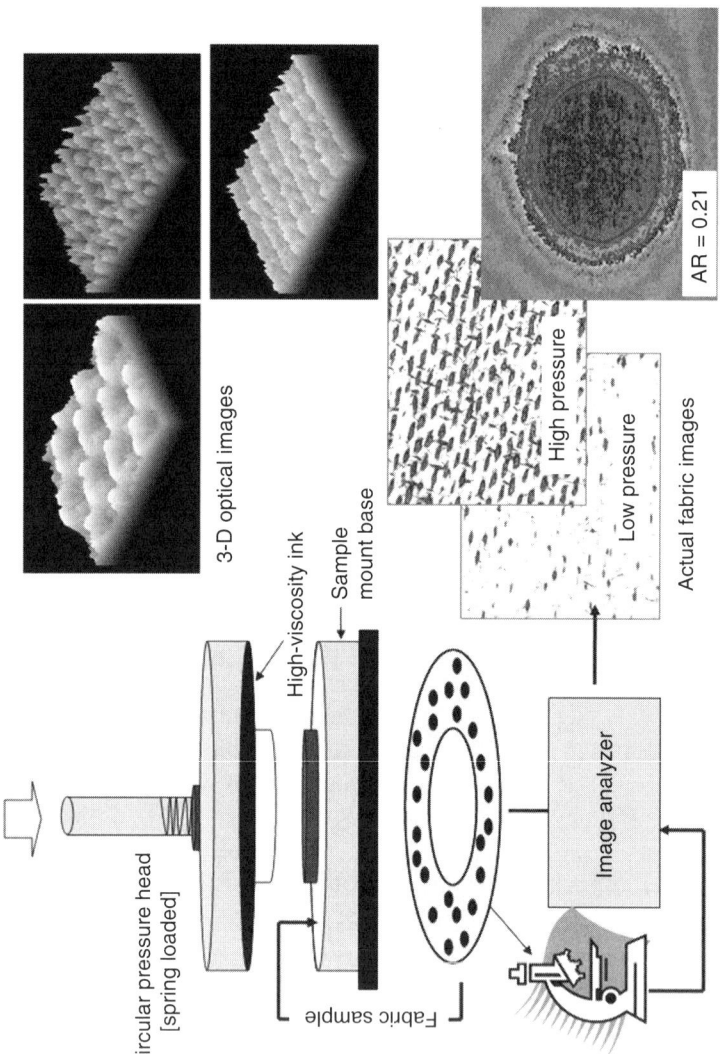

3-D optical images

Thermo-photographic images

AR = 0.21

High pressure

Low pressure

Actual fabric images

High-viscosity ink

Sample mount base

Image analyzer

Fabric sample

Circular pressure head [spring loaded]

5.12 Methods of measuring the Area Ratio.[5]

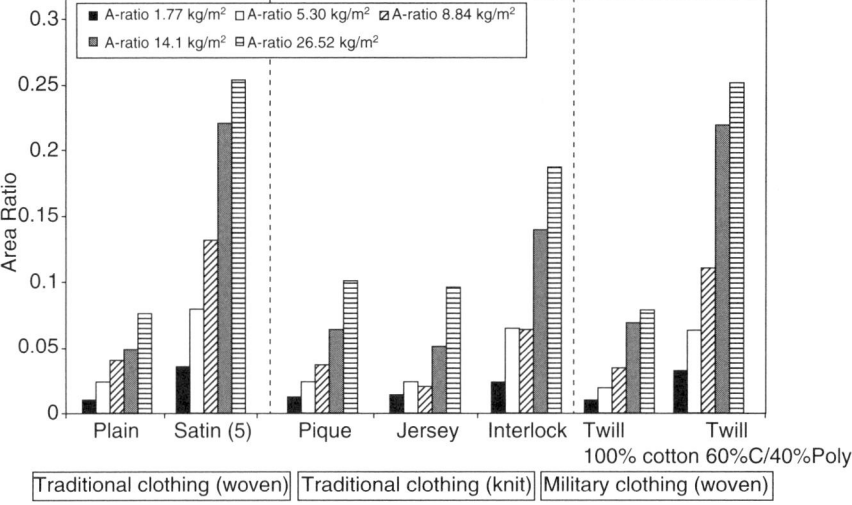

5.13 Area Ratio at different pressures of different fabrics.

Area Ratio developed over different fabric patterns and a wide range of tactile and thermal properties clearly proved that it is significantly related to most of these factors.

5.5.2   Design capability of fabrics

Designers of protective clothing should realize the design capacity of fibrous assemblies; more specifically, the range of parameter values that can be obtained by changing basic structural features of the fabric such as weight, thickness, fiber fraction (FF) or fiber/air ratio, yarn type, yarn structure, fabric count, and fabric balance. It is this realization that will allow designers to achieve the protection task at the full potential of comfort capability of the fabric used. Figures 5.14 and 5.15 provide actual values of design capability for knit and woven fabrics. These radar diagrams clearly illustrate the incredible design capabilities of fibrous structures that are often overlooked in the design process of protective clothing. As an example, air permeability can change by over 150% using different knit structures made from the same fibers and the same yarns. For woven fabrics, air permeability can be changed by over 500% using different woven structures made from the same fibers and the same yarns. Changes in thermal insulation for knit and woven fabrics can be as high as 50% and 100%, respectively. Changes in stiffness for knit and woven fabrics can be as high as 50% and 1500%, respectively, and so on.

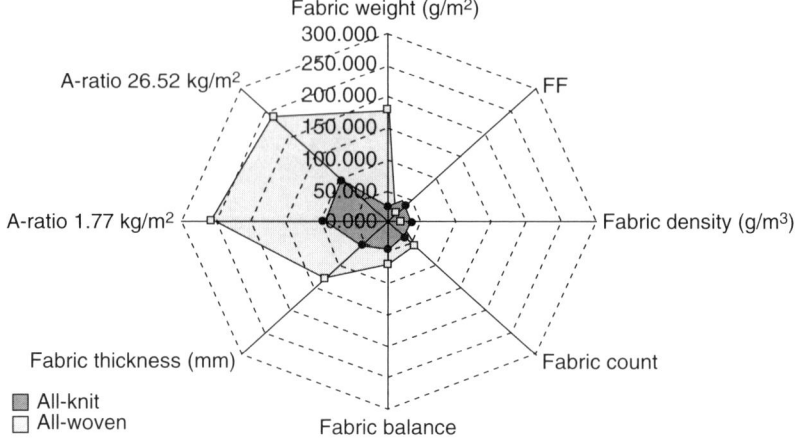

5.14 Design capacity profile: Structural factors and Area Ratio.[5]

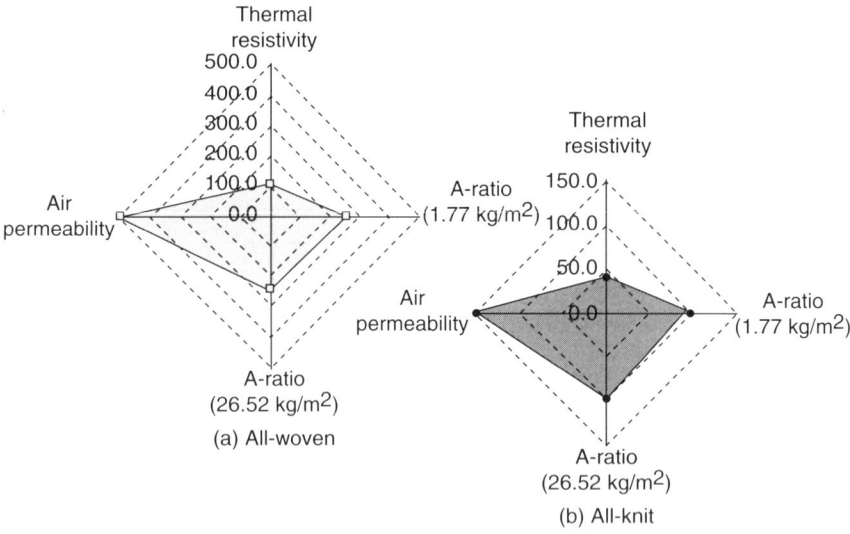

5.15 Design capacity profile: Air and heat flow factors and Area Ratio.[5]

5.6   Comfort and protection in military clothing

Figures 5.16 through 5.19 show the comparison of tactile and thermal prop-erties of two different fabrics used for military clothing. The physical prop-erties of these two different fabrics are very similar as shown in Fig. 5.16. The only difference is in the fiber type used (100% cotton and 60/40% cotton/polyester). As can be seen from the figures, thermal, tactile and mechanical properties of these two military fabrics are significantly different.

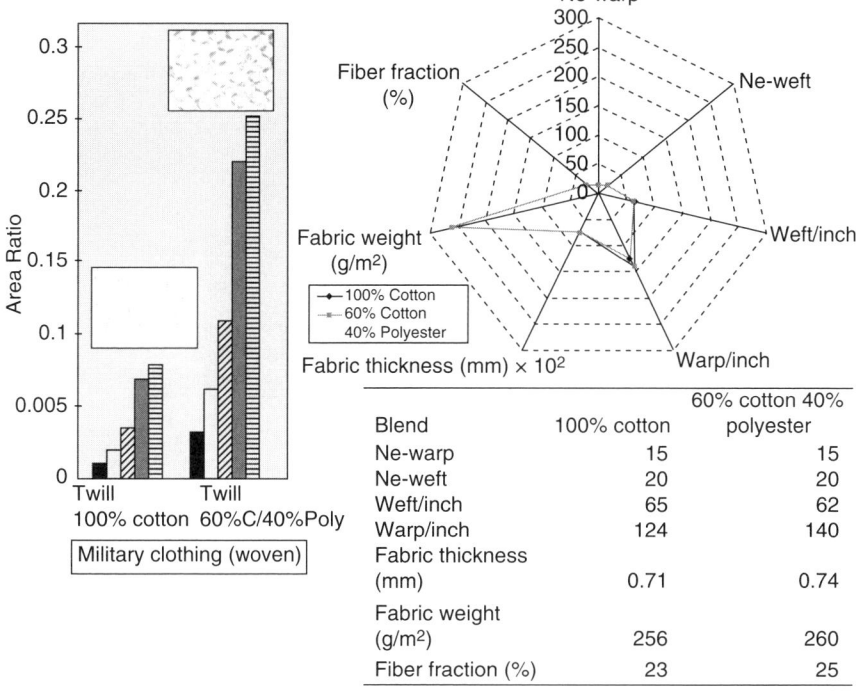

Blend	100% cotton	60% cotton 40% polyester
Ne-warp	15	15
Ne-weft	20	20
Weft/inch	65	62
Warp/inch	124	140
Fabric thickness (mm)	0.71	0.74
Fabric weight (g/m²)	256	260
Fiber fraction (%)	23	25

5.16 Area Ratio and structural parameters of two different military uniform fabrics.[5]

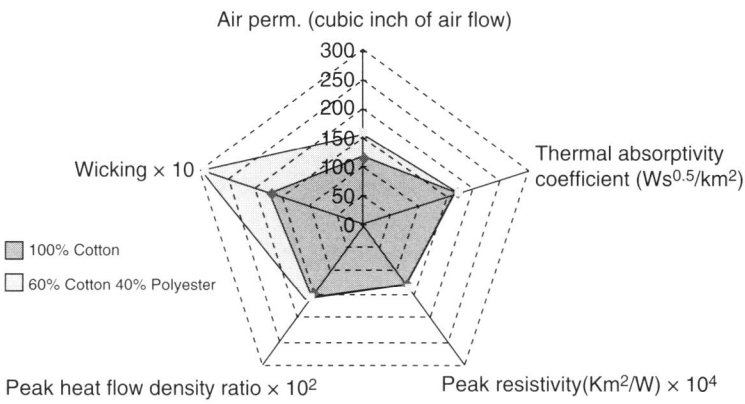

Blend	100% cotton	60% cotton 40% polyester
Air permeability (cubic feet of air flow)	9.5544	13.104
Thermal absorptivity coefficient (Ws^0.5/km²)	172.475	171.725
Thermal resistivity (km²/W)	0.0128575	0.0134475
Peak heat flow density ratio	1.48475	1.54325
Wicking max weight (mg) (max absorbed water on 1 mm fabric)	16.59	29.05

5.17 Thermal properties, air permeability, and wicking of two different military uniform fabrics.[5]

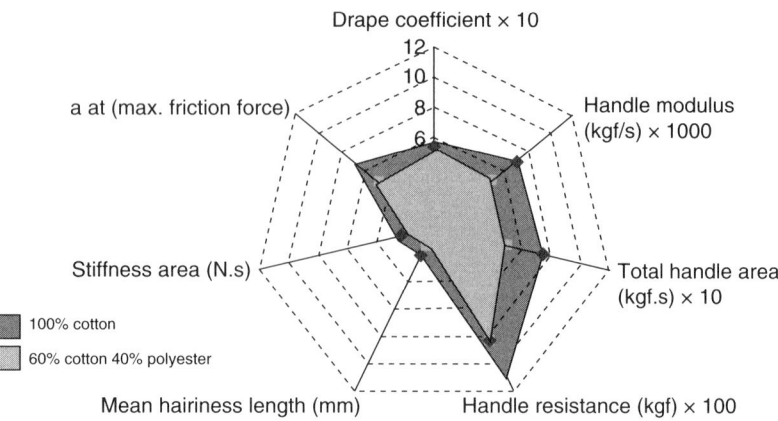

Blend	100% cotton	60% cotton 40% polyester
Drape coefficient	0.55	0.506498131
Total handle area (kgf.s)	0.73073695	0.5179327
Handle resistance (kgf)	0.10161	0.088657
Handle modulus (kgf/s)	0.00703695	0.005255732
Stiffness area (N.s)	2.303666667	1.533333333
a at (max. friction force)	6.742174676	4.704354057
Mean hairiness length (mm)	2.083505866	1.503636364

5.18 Mechanical and surface tactile parameters of two different military uniform fabrics.[5]

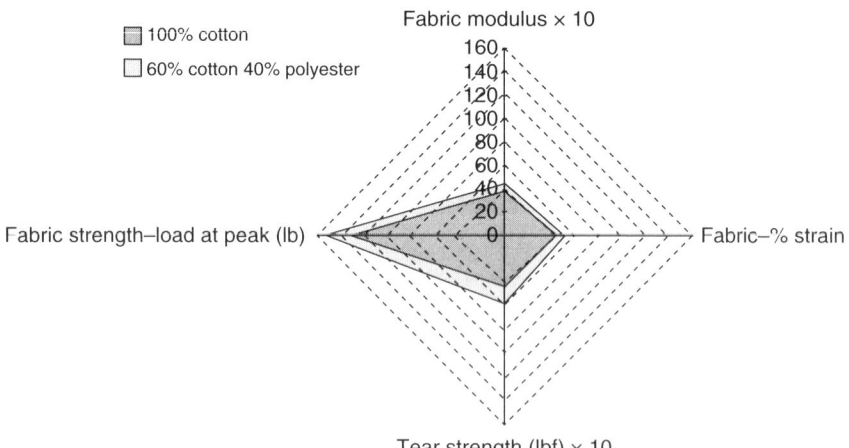

Blend	100% cotton	60% cotton 40% polyester
Fabric strength–load at peak (lb)	138.3	148.9
Fabric strength–% strain	40.8	42.3
Fabric modulus	338.9705882	352.0094563
Tear strength (lbf)	4.2	4.5

5.19 Durability parameters of two different military uniform fabrics.[5]

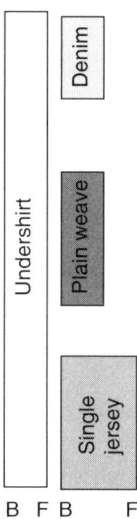

5.20 Multi-layer fabric arrangement.[5]

5.7   Multiple-layer systems

Multiple-layer garments are commonly used in military applications. One of the interesting experiments conducted in Kilinc's study[5] was the analysis of multi-layer fabrics. The purpose of this experiment was twofold: (i) to determine the effects of multi-layering on comfort characteristics and (ii) to simulate actual wearing situations. Accordingly, pairs of fabric layers were used, with the arrangements shown in Fig. 5.20. Tables 5.2 and 5.3 summarize the results of these experiments.

The results in Tables 5.2 and 5.3 reveal the following points:

- Tactile parameters typically multiply as a result of multi-layering. This is primarily due to the increase in thickness and fiber fraction (FF) (fiber/air ratio) imposed by wearing multilayer garments.
- The multiplication effect in tactile parameters is not linear and different types of outer fabric will yield different multiplication effects.
- Multi-layering typically increases thermal resistivity, reduces thermal absorptivity, and reduces air permeability.
- Different outer fabric types will have different levels of thermal or airflow values.

5.8   Future trends

After almost a 100 years of extensive research on comfort, and millions of dollars spent on comfort R&D, we are left with more questions than answers regarding this critical phenomenon. When protection is the driving force,

Table 5.2 Comfort characteristics with woven fabrics[5]

Multi-layer condition	Undershirt	Denim–undershirt	Plain weave–undershirt	Jersey cotton shirt–undershirt
Thermal absorptivity	177	167	147	145.3400
Thermal resistivity	0.0099	0.0217	0.0186	0.0210
Air permeability (ft^3/ft^2.min)	243	5	85	74.5500
Thickness (mm)	0.5247	1.2175	0.8175	1.0725
Weight (g/m^2)	143	460	248	325.4438
Stifness area (N.s)	0.1985	7.9400	0.7363	0.8270
Stifness max load (N)	0.4550	10.2100	1.2867	1.3800
Handle work (kg_f.sec)	0.0688	1.4676	0.2472	0.2840
Handle resistance (kg_f)	0.0113	0.2072	0.0397	0.0355
Handle modulus (kg_f/sec)	0.0007	0.0130	0.0024	0.0021
Area ratio	0.1400	0.4000	0.1200	0.1500
FF (%)	17.9868	24.8508	19.9482	19.9634

Table 5.3 Comfort characteristics of woven fabrics alone[5]

Multi-layer condition	Denim	Plain shirt	Jersey cotton shirt
Thermal absorptivity	195.7111	208.8000	186.2500
Thermal resistivity	0.0130	0.0064	0.0096
Air permeability (ft^3/ft^2.min)	4.3022	126.7000	141.0000
Thickness (mm)	0.7000	0.3048	0.5400
Weight (g/m^2)	316.4444	104.4327	182.0000
Stifness area (N.s)	4.2980	0.2497	0.2603
Stifness max load (N)	7.1800	0.4800	0.4733
Handle work (kg_f.sec)	1.2288	0.1072	0.1327
Handle resistance (kg_f)	0.1868	0.0151	0.0196
Handle modulus (kg_f/sec)	0.0116	0.0008	0.0012
Area ratio	0.4000	0.1200	0.1500
FF (%)	29.7410	22.5412	22.1735

other factors must be taken into consideration. In addition to fabric durability, air permeability, pore size, and moisture transfer, factors such as dust infiltration and dirt or sand resistance are expected to play a vital role in both comfort and protection.[26,27] Furthermore, treatments such as anti-odor, germ-killing, and antibacterial finish will become increasingly important.

In the United States, as in other countries, individuals who serve in the military wear a utility uniform as part of their issued clothing. In the US Army, these utility uniforms are called battle dress uniforms (BDU). They consist of three different types, designed for three climates – hot weather, temperate weather, cold weather and desert. Although designed for use in field training and in combat situations, BDUs are also worn by soldiers for routine tasks in offices, maintenance facilities, and other Homeland Defense situations. It is our opinion that the design of these clothing systems should be revisited in view of the concepts and the models discussed in this chapter.

5.9   References

1 SCHUTZ, H., CARDELLO, A., and WINTERHALTER, C., 'Perceptions of Fiber and Fabric Uses and the Factors Contributing to Military Clothing Comfort and Satisfaction', *Textile Res. J.*, 75(3), pp. 223–232, 2005.

2 CARDELLO, A., and WINTERHALTER C., 'Predicting the Handle and Comfort of Military Clothing Fabrics from Sensory and Instrumental Data: Development and Application of New Psychophysical Methods', *Textile Res. J.*, 73(3), pp. 221–237, March 2003.

3 ELMOGAHZY Y., KILINC F.S., HASSAN M., FARAG R., and KAMEL A., 'Protective Clothing. The Unresolved Ultimate Trade-Off Between Protection and Comfort: The Concept of Area Ratio', *4th International Conference on Safety & Protective Fabrics*, IFAI, October 2004.

4 HATCH, K. L., *Textile Science*, West Publishing Company, MN, USA, 1993.

5 KILINC-BALCI, F. S., *Developing a Design-oriented Comfort Model*, Ph.D. Thesis, Auburn University, Auburn, AL, 2004.

6 PEIRCE, F. T., 'The Handle of Cloth as a Measurable Quantity', *J. Text Inst.*, 21, p. T377, 1930.

7 ELMOGAHZY, Y., KILINC, F.S., and KAMEL, A., 'Tug of War: The Unresolved Ultimate Trade-off between Protection and Comfort in Protective Clothing', *Industrial Fabric Products Review*, pp. 30–32, July 2004.

8 KAWABATA, S., and NIWA, M., 'Fabric Performance in Clothing and Clothing Manufacture', *J. Text. Inst.*, 80, pp. 19–43, 1989.

9 MINAZIO, P.G., 'FAST–Fabric Assurance by Simple Testing', *International Journal of Clothing Science and Technology*, 7(2/3), pp. 43–48, 1995.

10 PAN, N., ZERONIAN, S.H., and RYU, H.S., 'An Alternative Approach to the Objective Measurement of Fabrics', *Text. Res. J.*, 63, pp. 33–43, 1993.

11 KIM J.O., and SLATEN, B.L., 'Objective Assessment of Fabric Handle in Fabrics Treated with Flame Retardants', *J. of Testing and Evaluation*, 24(4), pp. 223–228, 1996.

12 KIM, J.O., and SLATEN, B.L., 'Objective Evaluation of Fabric Hand: Part I: Relationships of Fabric Hand by Extraction Method and Related Physical and Surface Properties', *Text. Res. J.*, 69(1), pp. 59–67, 1999.

13 ALLEY, V.L., JR., and MCHATTON, A.D., 'Quantitative Measurement of the Feel of Fabric', *NASA Tech. Brief. LAR-12147*, 1977.

14 ELMOGAHZY, Y., KILINC, F., and HASSAN, M., 'Ch 3: Developments in Measurement and Evaluation of Fabric Hand', in *Effect of Mechanical and Physical Properties on Fabric Hand*, Ed. Hassan Behery, Woodhead Publishing, pp. 51–57, 2005.

15 KILINC, F.S., ELMOGAHZY, Y.E., HASSAN, M., FARAG, R., EBIELY, R., EL DIEB, A.S., and TOLBA, A., 'The Tactile Behavior of Textile Materials: New Perspectives – Part I: A Study on the Nature of Fabric Handle', *Proceedings of Cotton Beltwide Conference*, U.S. Cotton Council, January 2004.

16 FANGER, P.O., *Thermal Comfort*, Danish Technical Press, Copanhagen, Denmark, 1970.

17 BINNS, H., 'The Discrimination of Wool Fabrics by the Sense Touch', *Br. J. Psychol.* 16, pp. 237–247, 1926.

18 HOUGHTON, F.C., and YAGLOU, C.P., 'Determining Lines of Equal Comfort', *ASHRAE Trans.* 28, pp. 163–176, 361–384, 1923.

19 WINSLOW, C-E.A., HERRINGTON, L.P., AND GAGGE, A.P., 'Physiological Reactions of the Human Body to Varying Environmental Temperature', *Am. J. Physiol.* 120, pp. 1–20, 1937.

20 WINSLOW, C-E.A., HERRINGTON, L.P., and GAGGE. A.P., 'Relations between Atmospheric Conditions, Physiological Reactions and Sensations of Pleasantness', *Am. J. Hyg.* 26, pp. 103–115, 1937.

21 FOURT, L., and HOLLIES, N.R.S., *Clothing: Comfort and Function*, Marcel Dekker, NY, 1970.

22 SLATER, K., 'Comfort Properties of Textiles', *Textile Progress*, Vol. 9 No. 4, pp. 1–91, 1977.

23 ROHLES, F.H., 'Psychological Aspects of Thermal Comfort', *ASHRAE J.* 13, pp. 86–90, 1971.

24 ELMOGAHZY, Y., and GUPTA, B.S., 'Friction in Fibrous Materials, Part II: Experimental Study of the Effects of Structural and Morphological Factors', *Text. Res. J.* 63(4), pp. 219–230, 1993.

25 GUPTA, B.S., and ELMOGAHZY, Y., 'Friction in Fibrous Materials, Part I: Structural Model', *Text. Res. J.* 61(9), pp 547–555, 1991.

26 STEPHENS, R.A., 'From North Dakota to Mars', *Industrial Fabric Products Review*, pp. 22–24, August 2006.

27 BISHOP, D.P., 'Fabrics: Sensory and Mechanical Properties', *Textile Progress*, Textile Institute, 26, pp. 1–62, 1996.

28 ELMOGAHZY, Y., KILINC-BALCI, F., and FARAG, R., 'Developments in Measurement and Evaluation of Fabric Hand', *Provisional US Patent Application 60/962835*, filed August 1, 2007.

29 ELMOGAHZY, Y., KILINC-BALCI, F., and FARAG, R., 'Elmogahzy–Kilinc Fabric Hand Method and Apparatus: Method and Apparatus for Testing and Evaluating Total Fabric Hand and Tactile Aesthetics', *Provisional US Patent Application 60/918885*, filed March 19, 2007.

5.10   Bibliography

BEHERY, H.M., 'Comparison of Fabric Hand Assessment in the United States and Japan', *Text. Res. J.*, 56, pp. 227–240, 1986.

FANGER, P.O., 'Thermal Environment–Human Requirements', *Sulzer Technical Rev.*, 67, pp. 3–6, 1985.

GROVER, G., SULTAN, M.A., and SPIVAK, S.M., 'A Screen Technique for Fabric Handle', *J. Text. Inst.*, 84(3), pp. 1–9, 1993.

LI, Y., 'The Science of Clothing Comfort', *Textile Progress*, Textile Institute, 31(1/2), 2000.

PAN, N., YEN, K.C., ZHAO, S.J., and YANG, S.R., 'A New Approach to the Objective Evaluation of Fabric Handle from Mechanical Properties, Part II: Objective Measure for Total Handle', *Text. Res. J.*, 58, pp. 444–483, 1988.

6

Sweat management for military applications

N. PAN, University of California, USA

6.1 Introduction: Body/clothing/environment – the microclimate

Supplementing coverage in Chapters 4 and 5, this chapter will focus on the influences of sweat and heat, and their coupling effect. Because of the complex physical and physiological interactions constantly taking place within the microclimate – the climate in the space between clothing and human skin – a careful examination of these interactions is a critical step towards a better understanding of the microclimate's dynamic nature and its effect on clothing comfort.

6.1.1 The major components of the microclimate

Figure 6.1 illustrates this dynamic climate system in the space formed by the triad of body/clothing/environment. This is called a dynamic system because:

- There is a constant exchange of energy (mainly heat) and mass (fluids) between the three components;
- This constant exchange and the nature of the interactions dictate that any equilibrium of the system is temporary and unstable, both temporally and spatially.

Of the three components forming the microclimate system, as illustrated in Fig. 6.1, the human body can adjust body comfort to a certain degree through the skin, by constantly exchanging heat and moisture with the environment. According to Arens and Zhang (2006), the body's heat exchange mechanisms include sensible heat transfer at the skin's surface (via conduction, convection, and radiation – longwave and shortwave); latent heat transfer (via moisture evaporating and diffusing through the skin and sweat evaporation on the skin surface); and sensible plus latent heat exchanges via respiration from the lungs. Liquid sweat dripping from

137

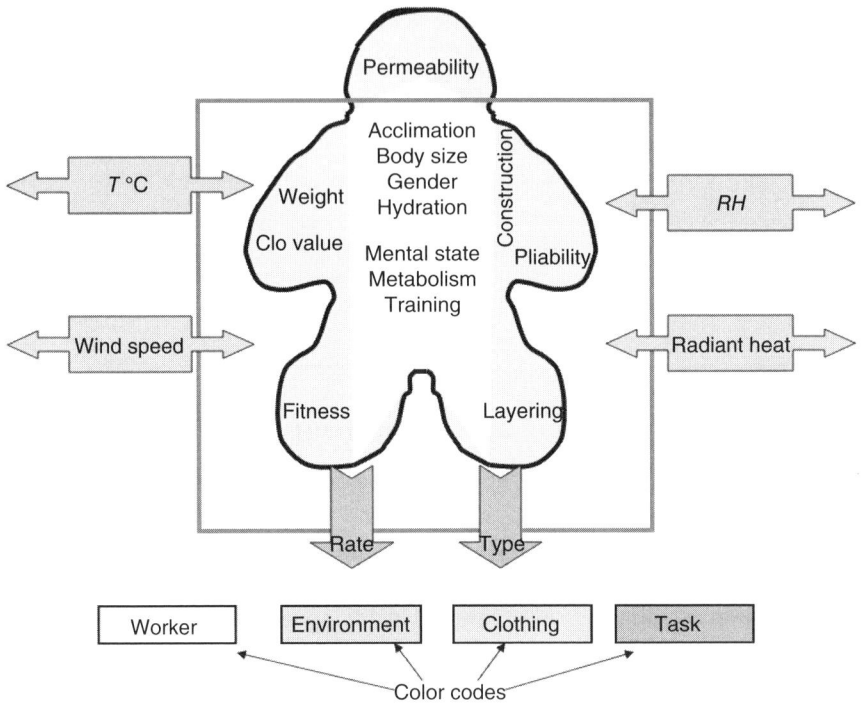

6.1 Dynamics in the microclimate.

the body or discharge of bodily fluids cause relatively small amounts of heat exchange, but exposure to rain and other liquids in the environment can cause high rates of heat loss or gain.

Depending on the situation, our surrounding environment can vary drastically. For this book, we are mainly concerned with an extreme climate or extreme conditions, and will therefore focus only on some extreme numbers and facts (Elert, 2006):

- At present, although the global average annual surface temperature is about 15°C, the highest temperature ever recorded is 57.7 °C (135.9 °F) at Al 'Aziziyah, Libya, on 13 September 1922;
- The lowest temperature ever recorded on Earth was –89.4 °C (–129 °F), recorded on Thursday, 21 July 1983 at the Russian (formerly Soviet Union) Vostok Station in Antarctica;
- Temperature fluctuation on Earth, therefore, can be as great as 147.1 °C (264.9 °F).

Such huge temperature fluctuations are obviously far greater than our body's thermal regulatory system can manage. To compensate for this

Table 6.1 Thermal conductivity of polymer materials used in textile fibers. Adapted from Morton and Hearle (1997)

Material	Thermal conductivity ($Wm^{-1}K^{-1}$)
Poly(vinyl chloride)	0.16
Cellulose acetate	0.23
Cotton	0.10
Nylon	0.25
Polyester	0.14
Polyethylene	0.34
Polypropylene	0.12
PET	0.14
Still air	0.026
Water	0.6
Silver	418

deficiency in the body's self-adjustment, we choose clothing as a means of buffering these disturbances from the wider environment. Clothing reduces sensible heat transfer, while in most cases allowing evaporated moisture (latent heat) to escape. Some clothing resists rain penetration to prevent the rain from cooling the skin directly and also to prevent a reduction in the clothing's effectiveness as insulation. Wet clothing has a higher heat transfer than dry, and depending on design, this could range from almost no difference to a nearly 30-fold increase (See Table 6.1 for thermal conductivity data). Clothing is nearly always designed to allow the wearer's breath to enter and exit freely, in order to keep the temperature and humidity of inhaled air low, and avoid moisture condensation within the clothing (Arens & Zhang, 2006).

However, as a material system with its own physical properties, clothing is not completely passive, but also interacts inwards with the skin and outwards with the ambient environment. Furthermore, considering the three components of the microclimate, very little can be done about the body's thermal regulatory physiology, and not much can be done – or not efficiently – about the ambient conditions. Clothing, therefore, is the only variable where active and pre-design measures can be taken to improve, compensate, alleviate or at least delay any adverse changes that occur within the microclimate during the course of interactions in the system.

6.1.2 The complexity of the microclimate's dynamics

The microclimate is a heat–moisture-coupled field created by inserting a layer of clothing between the living mechanism of a human body, with its own temperature and moisture regulation system, and the ambient environment, which is often highly variable. The clothing, therefore, has to be designed in order to establish a much more favorable and manageable climate or dynamic equilibrium between the clothing and skin, so that the body can function normally. More importantly, this equilibrium has to be established within the narrow confines between the body's complex metabolic physiology and movement, and a porous multiple-layered barrier system in contact with the complex ambient environment. It is a daunting challenge to attempt to create such an intricate assembly. For military personnel, the problems also have additional dimensions, including extreme ambient conditions, intense physical and physiological burdens (weight carried, demanding body movement, etc.) and chemical, biological and other war-related hazards.

6.2   Heat, moisture and interactions within the microclimate

6.2.1   Analysis of the dynamics within the microclimate

The major variables involved in establishing and influencing the microclimate are listed in Fig. 6.1. The body's heat transfer mechanisms include radiation, conduction, convection and evaporation of perspiration. One critical fact is that, except for evaporation, the heat transfer direction of all the other mechanisms is determined by the temperature gradient between the environment and the naked human body. If the temperature gradient is outward (i.e. in cold weather), body heat will be leaked out collectively via conduction, radiation, convection (if in movement) and evaporation (if liquid is involved). Consequently, over-cooling becomes a concern. However, if the ambient temperature exceeds body temperature, the body will be overloaded by all the heat transferred to it from the environment. As a result, virtually all this excessive heat has to be dissipated by the *evaporation of sweat alone*, or by lowering the temperature gradient – turning on the air conditioning! This reversible heat transfer highlights the paramount importance of perspiration and its subsequent evaporation because of the body's self-adjusting temperature function.

6.2.2   Temperature regulation of the human body

In their widely used textbook, Guyton and Hall (1996) presented a thorough coverage of metabolism and temperature regulation, and provided a wealth of related information. Our skin temperature is not constant and changes with, among other factors, the ambient temperature, as indicated in Fig. 6.2. In our discussions that follow, unless otherwise specified, we assume an unclothed person at rest at a room temperature of 23 °C. Hot weather is set at 45 °C. The typical skin temperature of 34 °C is derived from physiology textbooks, whereas the normal core body temperature is 37 °C. Table 6.2 summarizes all the numerical results mentioned below in order to facilitate the discussion.

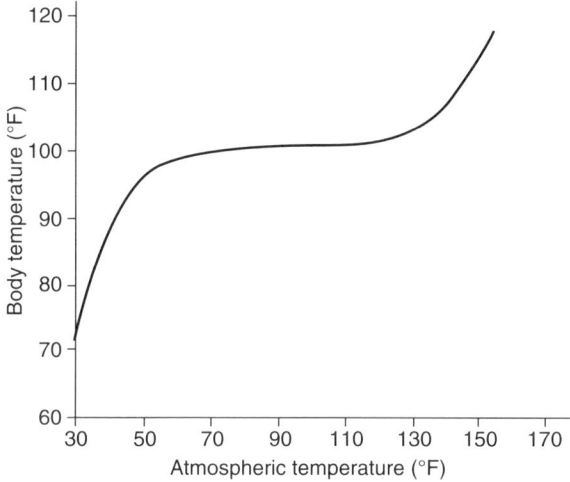

6.2 Correspondence between ambient and body temperatures (Guyton and Hall, 1996).

Table 6.2 Comparison of heat exchange in hot and cool conditions for a bare body

Case:	Hot	Cool
Skin temperature (°C)	37	34
Ambient temperature (°C)	45	23
Temperature difference (°C)	**−8**	**11**
Conduction (watt)	−8	11
Convection (watt)	0*	0*
Radiation (watt)	−109	133
Perspiration (watt)	207	17
Basal production (watt)	|90|	|90|
Net heat to environment (watt)	0	71
Sweat generated (kg/day)	**7.08**	**0**

Note: (i) +heat released from the body *to* the environment.
(ii) −heat absorbed by the body *from* the environment.
(iii) the basal heat is the minimum to be dissipated by the body to the environment.
*no air flow and still air convection (boundary layer effect) included in conduction.

Heat generated by the body

(i)   Basal metabolic rate of all the body's cells;
(ii)   Extra metabolic rate produced by muscle activity, including muscle contractions caused by shivering;
(iii) Extra metabolism caused by the action of thyroxine and other hormones on the cells;
(iv)   Extra metabolism caused by other biochemical cell activity, especially when the cell temperature increases.

Heat transferred between the ambient environment and the body via

(i)   Radiation – accounting for about 60% of total heat loss. To tackle such a huge loss of heat in extremely cold weather, a coated layer of clothing, reflecting radiant heat back to the body, is a highly effective measure for thermal protection.
(ii)   Conduction – conduction heat loss as a result of contact with other solid objects is very limited because of the rapid reduction of temperature difference, but the loss to the surrounding air is significant (about 15%). However, conduction of heat between the body and fluid is self-limiting, because once a temperature equivalence is established, heat conduction ceases unless the heated air moves away from the body, allowing unheated air to move in – as, for instance, when using a fan to cool down in the summer. In the opposite situation, however, if for instance a naked person is in cold water, it is better for him not to move around too much in order to avoid disturbing the newly established heat equilibrium boundary layer between the body and water that has been heated by his own body temperature.
(iii) Convection – in this case, heat loss by convection is associated with conduction. Air heated by the body via conduction tends to rise and gives way to unheated air, allowing air flow or convection to take place and thus carry away the heat.
(iv)   Evaporation – as mentioned above, evaporation is the most important mechanism in body temperature regulation because:
   • It cools the body down in hot weather, unlike the other mechanisms which always work against the body's needs;
   • Much higher energy is required for evaporation, so therefore it is the most effective way to reduce heat.

6.2.3 Perspiration and evaporation

Because of the importance and complexity of the evaporation process, it is discussed in detail in this separate section.

Moisture from perspiration

Sweating (also called perspiration or sometimes transpiration) is the production and evaporation of *a watery fluid*, consisting mainly of sodium chloride in solution, that is excreted by the sweat glands in the skin of mammals. In humans, sweating is primarily a means of temperature regulation. Evaporation of sweat from the skin's surface has a cooling effect due to the latent heat of the evaporation of water. Hence, in hot weather, or when an individual's muscles heat up due to exertion, more sweat is produced. Sweating is decreased when the body is cold.

As part of the physiological regulation of body temperature, the skin begins to sweat almost precisely at 37 °C and perspiration will increase rapidly with increasing skin temperature, as indicated in Fig. 6.3. It is reported (Guyton and Hall, 1996) that a normal rate of maximum perspiration is about 1.5 liters/hr, but that, after 4 to 6 weeks of acclimatization in a tropical climate, this rate can reach 3.5 liters/hr, corresponding to a maximum cooling power of almost 2.4 kilowatts! Even though one may be unaware of perspiration, physiology textbooks quote an amount of about 600 grams per day of 'insensate loss' of liquid moisture from the skin.

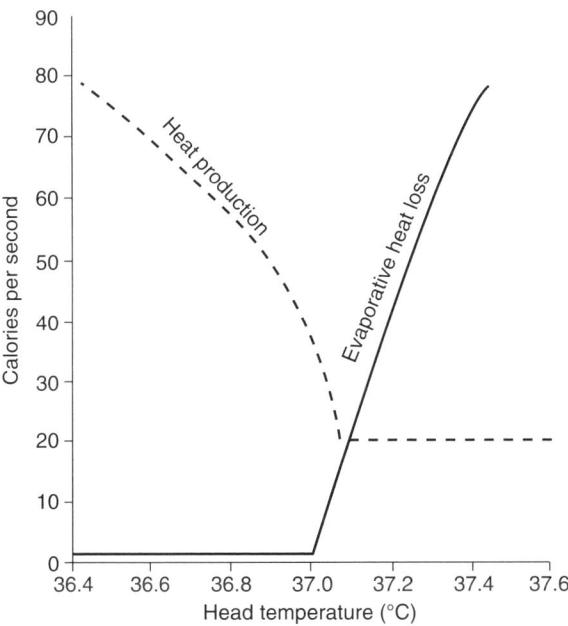

6.3 Body heat generation and heat evaporation as a function of body temperature (Guyton and Hall, 1996).

Heat of perspiration

The energy required to change a gram of liquid into a gaseous state at boiling point is called the 'heat of vaporization' of that liquid. Once evaporated, for an ideal gas, there is no longer any potential energy associated with the intermolecular forces. So the internal energy is entirely in the form of molecular kinetic energy. This energy in water breaks down the intermolecular attractive forces and must also provide the necessary energy to expand the gas (the $P\Delta V$ work): a significant feature of the vaporization phase-change of water is the accompanying large change in volume. Once a mole of water is evaporated into steam at 100 °C, its volume will expand by a factor of 1700! This is a physical fact known to firefighters, because the 1700-fold increase in volume when water is sprayed on a fire or hot surface can be explosive and dangerous.

If part of a liquid evaporates at a temperature below the boiling point, it must extract the necessary heat of vaporization from the remaining liquid in order to make the phase change to the gaseous state. The cooling effect of perspiration is due to this very great heat loss from the vaporization of water: *539 calories/g* at boiling point, but even larger, *580 cal/g*, at normal skin temperature. The heat of vaporization is greater at body temperature than at boiling point because the binding energy of the water molecules is higher at lower temperature, and it therefore takes more energy to break them apart into the gaseous state (Nave, 2005).

6.2.4 Temperature regulation of the human body

The temperature of the body is regulated by neural feedback mechanisms which operate primarily through the hypothalamus (Guyton & Hall, 1996). The hypothalamus contains not only the control mechanisms, but also the key temperature sensors. Under the control of these mechanisms, sweating begins almost precisely at a skin temperature of 37 °C and increases rapidly as the skin temperature rises above this value. The body's heat production remains almost constant under these conditions as the skin temperature rises. Hypothalamic body temperature regulation is illustrated in Fig. 6.3. Even when inactive, an adult male loses heat at a rate of about 90 watts as a result of his basal metabolism. The additional heat to be dispersed depends on the temperature difference between the body and the ambient environment.

Warming the body at low ambient temperature

The human body is equipped with some thermal insulation functions of its own for protection in cold conditions. The skin, the subcutaneous tissues, and the fat of the subcutaneous tissues are heat insulators themselves, especially the fat, whose thermal conductivity is only one third that of the muscles. If skin temperature drops below 37 °C, a variety of responses are initiated to conserve body heat and increase heat production. These include:

- Vasoconstriction to decrease the flow of heat to the skin;
- Cessation of sweating;
- Shivering to increase heat production in the muscles;
- Secretion of norepinephrine, epinephrine, and thyroxine to increase heat production;
- In lower animals, erection of hairs and fur to increase insulation.

Since this process to warm up a cold body to normal skin temperature involves very little perspiration or moisture and associated phase change and heat-moisture coupling, it is a relatively straightforward matter. Furthermore, as indicated in Table 6.2, on a cold day and when inactive, an adult male loses heat at a rate of about 90 watts as a result of basal metabolic needs. In this case, the largest heat loss to the environment is through radiation at a rate of 133 watts or 83% of the total heat loss, and the losses by conduction and perspiration are at 11 and 17 watts, respectively. A net heat loss rate of 161 − 90 = 71 watts will certainly make an unclothed person at rest at room temperature of 23 °C feel uncomfortably cool.

Cooling the body at high ambient temperature

A problem arises when the ambient temperature is above body temperature, because all three standard heat transfer mechanisms work against the need for body heat loss and instead transfer heat into the body. Our ability to survive in conditions of high ambient temperature derives from efficient cooling through the evaporation of perspiration, as well as evaporative cooling from exhaled moisture. Because of the body's temperature regulation mechanisms, skin temperature would be expected to rise to 37 °C, at which point, as shown in Fig. 6.3, perspiration is initiated and increases until evaporation cooling is sufficient to hold skin temperature at 37 °C, if possible. With an ambient temperature of 45 °C, according to Stefan–Boltzmann's law, for an estimated body surface area of 2 m² and emissivity 0.97, there is a net input of power to the body of 109 watts, as seen in Table 6.2. Cooling perspiration must overcome that heat gain, in addition to an outflow of 90 watts for body equilibrium. Dissipating body heat at this minimum rate of 207 watts through perspiration requires 7081 g or roughly 7 kg/day of sweat!

6.3   Heat and moisture interactions in the microclimate

6.3.1   Introduction

So far, we have yet to add clothing to that naked body. Once we do that, several things happen (Guyton and Hall, 1996; Pan and Gibson, 2006; Pan and Zhong, 2006):

(i)   Due to the clothing, a barrier is formed between the body and the environment, as well as boundary layers on both sides of the clothing;

(ii)   Since clothing is a good thermal insulator, the heat transfer discussed above will be shielded significantly in both inward and outward body directions;

(iii)   The barrier will also interact with perspiration, mainly through wetting and wicking processes. The extent of the influence caused by these processes is dependent on the nature of the clothing – whether it is hydrophilic or hydrophobic, for instance. Sorption/desorption in turn involves sorption heat – the energy exchange during the phase-change between liquid and gaseous states. Moisture condenses back into liquid water by releasing the sorption heat, while liquid water evaporates if this same amount of energy is made available. In theory, this sorption heat can serve as a thermal buffer, for evaporation of sweat from a hot body absorbs the heat to more or less chill the body; and once cold, moisture condensation releases the sorption heat to warm up the body. In practice, sweat often blocks the airflow channels in the clothing, also causing fiber swelling, which in turn reduces the free pores in the clothing, both of which inhibit the 'breathability' of the clothing. Furthermore, the sorption heat can be a safety hazard for materials storage. The collective sorption heat can raise the temperature to burning point!

(iv)   The properties of the clothing will often be significantly affected because firstly the clothing turns into a mixture of fiber, air and water whose behavior is dictated by three media with drastically different thermal conductivities, as indicated in Table 6.1. In addition, once hydrophilic fibers are wetted, they change their geometric dimensions in both thickness and length, with the result that the entire garment seems more or less to turn into one made of different fibers!

Ghali *et al.* (2006) have summarized the influences of all these factors, acting individually or collectively. The structure of the fabric system consists of a solid fiber and entrapped air. The amount of entrapped air contributes largely to the fabric's ability to transport heat, while the volume and arrangement of the solid fiber mainly determine the fabric's ability to transport water vapor through the fabric system. The solid fiber represents an obstacle for the moving water vapor molecule and therefore always increases the fabric's evaporative resistance. However, the fiber is not an inert solid whose sole function is to impede the mobility of moisture. As discussed above, it can also adsorb or de-adsorb moisture, depending on the relative humidity of the entrapped air in the microclimate, and also on the type of solid fiber. For instance, wool fiber can take up to 38% of moisture relative to its own dry weight. The fabric's moisture sorption/desorption capability influences heat and moisture transport across the fabric and thus affects both its dry and evaporative resistance.

Furthermore, the moisture phase change in fabric is not limited to the sorption/de-sorption of the fiber property; it can also result from moisture condensing in the void space or the evaporation of moisture present in the void volume. The fabric's water content does not only include the adsorbed water in the solid fiber and the water vapor in the entrapped microclimate, but also liquid water that can be present in the void space. This liquid water could originate from a moist source from which the liquid water is wicked, or it could result from condensation when water vapor continues to diffuse through a fully saturated solid fiber. Like the sorption/desorption of moisture, liquid condensation and evaporation influence the flow of heat and moisture across the fabric by releasing or absorbing the heat of vaporization and thus acting as a heat source or sink in the heat transfer process. In addition, condensation has a high relevant effect on thermal comfort because of the uncomfortable sensation of wetness (Ghali *et al.*, 2006).

6.3.2 Modeling of coupled heat and moisture

To analyze the heat-moisture coupling process, Jones *et al.* (2006) have derived the following equations in finite difference form:

$$\Delta R_i \rho = \varphi \frac{P(\phi_{i-1}, T_{i-1}) + P(\phi_{i+1}, T_{i+1}) - 2P(\phi_i, T_i)}{\Delta x^2} \Delta t \qquad [6.1]$$

$$\Delta T_i c_i \rho = Q_s(\phi_i) \Delta R_i \rho + k \frac{T_{i-1} + T_{i+1} - 2T_i}{\Delta x^2} \Delta t \qquad [6.2]$$

where
  ΔR is the change in moisture regain R (%),
  ΔT is the change in temperature T (°C),
  Δt is the integration time step (s),
  Δx is the distance step in the x direction (m),
  c is the heat capacitance of the clothing (kJ/kg °C),
  ϕ is the local relative humidity (fraction),

ρ is the bulk density of the dry clothing (kg/m^3),

i refers to a specific discrete location in the x direction,

$P(\phi, T)$ is the equilibrium vapor pressure for the fibrous media at the local relative humidity and temperature (kPa), and

$Q_s(\phi)$ is the sorption heat for clothing at local relative humidity (kJ/kg).

According to Jones *et al.* (2006), Equations 6.1 and 6.2 show a clear coupling between moisture and heat in clothing. In particular, Equation 6.2 shows that any increase in regain ΔR results in an increase ΔT in temperature and vice versa. As the sorption heat $Q_s(\phi)$ is large, consequently small changes in regain can result in large temperature changes. Since heat flows are driven by the temperature gradients, the adsorption and desorption of moisture by the clothing has a large impact on the heat fluxes through the clothing as well.

It has been known for many years that moisture sorption and desorption can impact on body heat loss and affect perceptions of the thermal environment (Rodwell *et al.*, 1965). This effect has been modeled for clothing systems using the above equations and has been measured experimentally as well (de Dear *et al.*, 1989; Jones and Ogawa, 1992). The effect is so great that a person wearing clothing made of highly adsorptive fibers, such as wool or cotton, can experience a short-term change in heat loss from the body of the order of 50 W/m^2 when going from a dry environment (e.g. 25% RH) to a humid environment (e.g. 75% RH), i.e. experiencing a chilling effect even when the temperatures of both environments are identical. This effect is relatively short-lived and may last only for 5–10 minutes, but it is sufficient to elicit a strong change in thermal sensation and plays an important role in the perceived effect of humidity on comfort in many situations. A lesser, but still important, effect can persist for 30–60 minutes for moderately heavy indoor clothing made of highly adsorptive fibers.

This interaction is particularly important for the drying of porous media. The transport of adsorbed moisture from a porous media is driven by the vapor pressure gradient. A negative vapor pressure gradient from the media to the surroundings will result in transport of water vapor from the media to the surroundings. The source of this water vapor is moisture adsorbed on the fibers. As the moisture is released and the regain decreases, there is a cooling effect on the media, as quantified by Equations 6.1 and 6.2. A very small decrease in regain results in a large cooling effect. This small decrease in regain has a minimal impact on the local equilibrium relative humidity. However, the large change in temperature has a major impact on the saturation pressure. The net result is a large decrease in local vapor pressure. The ultimate consequence is that the cooling effect nearly eliminates the partial pressure gradient that is driving the moisture removal and, in the absence of a heat source, drying proceeds at a very low rate. The drying of a porous media is almost always limited by heat transfer, explaining why thick media can take hours or even days to dry.

For hydrophobic fibers such as polypropylene or polyethylene that adsorb very little moisture, the interaction of heat and moisture is minimal unless conditions are such that condensation occurs. When condensed moisture is present but still relatively immobile, the equations given previously still apply and strong interaction between heat and moisture will be present.

A more detailed discussion of all these factors is clearly beyond the scope of this chapter. Although more information can be found in Pan and Gibson, 2006, a great deal still remains to be understood.

6.4   Sweat management for military apparel applications

It is obvious that, since clothing is an extra barrier and a load carried by the body, it will, in general, physically impede the body's movement and functions. Although this physical obstruction is part of the adverse influences of clothing on body comfort, it is nonetheless an ergonomic and design issue (Pissiotis *et al.*, 1997), and is thus beyond the present discussion. Instead, we will focus on how clothing influences the moisture (and thus heat) transfer in the microclimate system.

6.4.1 Moisture condensation and deterioration of clothing performance

As discussed, a wet fabric loses not only its bodily thermal protection in cold conditions (because water's almost 30-fold higher thermal conductivity (Table 6.2) drastically accelerates the body's heat loss), but it also causes an unpleasant sensation in contact with the skin. It is, therefore, critical in extreme conditions to maintain clothing's dryness by preventing or alleviating any external water penetration and/or condensation of body sweat onto clothing.

Condensation is a process where the moisture turns into liquid water once it encounters a cooler temperature. In an ideal situation, the watery sweat from perspiration would evaporate and leave the human body dry. However, clothing covering the body severely blocks this channel of escape for the sweat. Instead, as soon as the sweat at skin temperature comes into contact with the clothing at cooler ambient temperature, condensation takes place. This leads, firstly, to a vast reduction of the clothing's thermal insulation, and secondly, to dissipation of condensation heat, as discussed above, increasing the discomfort in cold weather.

Also as discussed above, in a hot environment where the ambient temperature is higher than body temperature, all the other heat transfer mechanisms work against clothing comfort: they all transfer external heat to the body. Evaporation of perspiration becomes the only heat-dissipating channel. This capacity for evaporative heat loss is, however, reduced by the use of protective clothing (PC). Clothing creates an additional resistance to vapor transport, which results in increased thermal strain from elevated body temperatures, thus reducing worker tolerance and, in extreme cases, posing the risk of overheat sickness (Joy and Goldman, 1968; Kenny et al., 1988; Bishop et al., 1988; Faff and Tutak, 1989; Nunneley, 1989; Sun et al., 2000; White et al., 1989). It should also be added that, according to Iampietro and Goldman (1965), rectal temperature is not a good indicator of body tolerance to heat, but skin temperature and heart rate are.

6.4.2 The body's physical state versus psychological perceptions

Human beings have no known sensors that detect humidity directly, but they are sensitive to skin moisture caused by perspiration, and skin moisture is known to correlate with warm discomfort and unpleasantness. The mechanisms for discomfort are thought to be related to the swelling of the epidermis as it absorbs moisture (Berglund and Cunningham, 1986). Luo et al. (1998) also suggested that the swelling of the skin may stimulate its tactile mechanoreceptors in some way, and this is perceived as uncomfortable.

Not surprisingly the relationship between bodily comfort and the physical conditions of clothing is not simple and monotonic. Li and Yi have studied the effect of clothing material on the thermal responses of the human body (Li & Yi, 2005). One unique aspect of their research is their investigation into the relationship between psychological perceptions of temperature and moisture, and objectively measured skin and fabric temperatures and relative humidity in the clothing microclimate (Li, 2005; Wong and Li, 2004). According to their research, perception of dampness appears to follow Fechner's law more closely than Stevens' power law with a negative correlation with skin temperature, and it is non-linearly and positively correlated with relative humidity in the clothing microclimate. The perception of comfort is positively related to the perception of warmth and negatively to the perception of dampness. This perception of comfort is positively related to skin temperature, which appears to follow both Fechner's law and Stevens' law, and is also non-linearly and negatively related to relative humidity in the clothing microclimate.

In addition, mechanoreceptors are clearly stimulated by the friction of clothing moving across the skin surface. With moisture absorption, the stratum corneum outer layer softens, allowing clothing fibers to dig in and hence increasing the friction. The additional friction is perceived as fabric coarseness and cling (Grosdow et al., 1986). A moisture-induced cling effect also occurs with architectural and furniture surfaces, particularly smooth, hydrophobic materials (ASHRAE, 2005).

Fabrics with different moisture absorptance properties are potentially perceived differently, but there is little experimental evidence for this at present. Toftum *et al.* (1998) studied knitted and woven cotton and polyester clothes at controlled levels of skin relative humidity ranging from 10 to 70%, and found that fabric type had no effect on comfort or on perceived humidity of skin or fabric. They found that acceptance of skin humidity decreases as the skin's relative humidity increases, and produced a predictive model for this. It should be noted that for normal environments, where air temperature is lower than skin temperature, the effect of the *air's* relative humidity is much less than that of the skin's relative humidity. For cool environments, very high air relative humidity produces almost no perceived skin comfort effect. Even at the warm limit of the comfort zone, 70% air relative humidity causes less than 15% of subjects to perceive discomfort due to skin humidity.

Havenith *et al.* have also conducted a series of studies in these areas. They investigated moisture accumulation in sleeping bags at subzero temperatures (2004); the resultant clothing insulation as a function of body movement, posture, wind, clothing fit and ensemble thickness (1990); and clothing's evaporative heat resistance (1999); the effects of moisture absorption in clothing on the human heat balance (Lotens and Havenith, 1995); and regional microclimate humidity of clothing during light work as a result of the interaction between local sweat production and ventilation (Ueda *et al.*, 2006).

6.4.3 Measures of thermal resistance and permeability of clothing

The *tog*, a unit of thermal resistance, was defined by Peirce and Rees (1946) as the thermal resistance that is able to maintain a temperature gradient of $0.1\,°C$ with a heat flux of $1\,W/m^2$; accordingly, a light summer suit offers 1 tog insulation. In 1941, Gagge *et al.* proposed (1941) the unit *clo* to measure the thermal insulation of clothing where 1 clo defines the insulation of a clothing system that maintains an average sitting–resting man comfortably in a normally ventilated room (0.1 m/s air velocity) at the air temperature of $21\,°C$ and relative humidity less than 50%. It is roughly the metabolism of a person resting comfortably in a sitting position, or the value of the insulation of one's everyday clothing. Woodcock and Breckenridge (1957) presented a so-called moisture index as a new variable in predicting cloth comfort, and continued this methodology in their subsequent studies (Breckenridge *et al.*, 1960; Woodcock, 1962a, 1962b; Woodcock and Breckenridge, 1965).

Very recently, Huang undertook some investigations (2006) firstly by summarizing and analyzing the existing parameters in describing the heat–moisture comfort of clothing, i.e. *clo* and *tog* units, permeability index, evaporative transmissibility, permeation efficiency factor, index of water permeability. He then outlined their applications for the calculation of heat exchange between the human body and its environment, and for the prediction of the physiological variables under heat stress conditions.

Currently, ISO9920 (1995) and ISO15831 (2004) refer to two methods to determine the evaporative resistance of clothing ensembles:

- The use of F_{pcl}, a reduction factor for evaporative heat loss with clothing compared to the nude person.
- The use of i_m, the permeability index of clothing, which provides a relation between evaporative and dry heat resistance of clothing items or systems.

Havenith *et al.* (1999) have reviewed these two commonly used parameters as well as other alternative approaches to the problem. The different approaches were evaluated for their accuracy and usability. The old approach using F_{pcl} was shown to lead to serious under-estimations of the reduction in evaporative heat resistance due to wind and movement. Another empirical description of the relation between the clothing permeability index (i_m) and the changes in clothing heat resistance due to wind and movement was selected as the most promising method of deriving clothing vapor resistance. For this method, the user needs to know the static heat resistance, the static i_m value of the clothing and the wind- and movement-speed of the wearer. This method results in a predicted maximal decrease in clothing vapor resistance by 78% when clothing heat resistance is reduced by 50%, which is consistent with theoretical expectations and available data.

There is a trend to develop an overall thermal parameter to incorporate the above variables. Wang and Li introduced the relative thermal diffusion ratio (RTDR) and heat and moisture ratio (HMR) to assess the heat and moisture transfer properties of clothing materials during transient sweating conditions. The greater the RTDR and the more rapidly the moisture vapor evaporates, the more comfortable the body perceives itself to be. The larger the HMR, the less accumulation of moisture in the microclimate between the body and clothing, the less uncomfortable the body feels (Wang and Li, 2005). It is apparent that much more extensive research is needed before any conclusive comments can be made on the usefulness or effectiveness of these indices.

6.4.4 Clothing moisture management

Fabrics' liquid moisture transport properties in multiple dimensions, called moisture management properties, significantly influence human perceptions of moisture sensations. A new method and instrument called the moisture management tester (MMT) has recently been developed by Hu *et al.* (2005) and by Yao *et al.* (2006) in order to evaluate textile moisture management properties. In one step, this new method can be used to quantitatively measure multi-directional liquid moisture transfer in a fabric, where liquid moisture spreads on both sides of the fabric surface and transfers from one side to the other. Ten indexes are generated by the instrument to characterize the liquid moisture management properties of fabrics.

Bishop *et al.* (1995) have reported that protective clothing always obstructs the wearer's performance, and work tolerance was found to be decreased in all cases, even in cool (18 °C) environments. This review of the ergonomics of work in the US military chemical protective clothing included attempts to relate heat strain in the chemical defense ensemble to the ambient environment, and comparison of a military chemical suit with an industrial usage vapor barrier suit across two thermal environments. The paper also pointed out that, although PC seriously reduces worker productivity, minimal guidance is available to PC users. Part of the difficulty in providing general guidance is that the research is complicated by the variety of types of PC, the variety of subject profiles, and the variety of work loads and environments.

Clearly, fabric moisture management is crucial in extreme weather conditions, both hot and cold. In cases of high temperature, such as firefighting or, to a lesser degree, during a scorching summer, two conflicting requirements compete with each other. On the one hand, protective clothing with high insulation is indispensable to prevent the body burning from the external heat source. On the other hand, this highly insulative protective clothing virtually shuts down the channels for sweat evaporation, and thereby the body's release of heat stress.

In extremely cold conditions, however, the key issue is to keep the body warm by insulating it from the environment. However, such strong protection once again impedes basal sweat evaporation and the cumulated sweat will sooner or later condense onto the clothing, thus causing its thermal protection to deteriorate.

In both cases, maximizing heat protection (against external heat penetration or body heat leakage) and minimizing the blockage of sweat evaporation due to this protection pose a constant and major challenge to clothing designers, and perhaps more importantly call for *more effective* materials.

6.4.5 The transient process

Another issue in this area is the transient behavior of the whole system. The human body is constantly subject to changes in environmental variables, clothing and activities, and therefore the transient thermal response of the human–clothing system becomes more relevant during these changes.

Woodcock has discussed the 'after-exercise chill' (Woodcock, 1962a, 1962b) and Spencer and Smith have studied the buffering effect of hygroscopic clothing at the onset of sweating (Spencers-Smith, 1966). Both effects involve the simultaneous transport of heat and moisture vapor through fabrics or fiber assemblies. Dent (2001) investigated other transient phenomena, such as the analogous buffering effect due to changing ambient conditions (both temperature and humidity) and the initial 'cold feel' of fabrics, raising questions about the role played by fiber moisture regain in all transient cases.

Jones and Ogawa (1993) pointed out that moisture adsorption and desorption by the fabric are the major factors affecting the transient response of clothing. During transients, the mix of latent and sensible heat flow from the skin may differ considerably from the corresponding heat flows from the clothing surface to the environment. Furthermore, the heat exchange between the body and the environment during the transient period involves a change in moisture vapor pressure within the clothing, surface temperature and heat loss from the body. The comfort performance of a clothing system in dynamic conditions needs to be re-evaluated and further investigated on the basis of all the variables mentioned above, rather than just a few of them.

6.5   Conclusions

(i)   Evaporation of perspiration cools the body down in hot weather, unlike other thermal transfer mechanisms that always work against the body's needs: in a hot summer they place an additional heat-load on the body, but leak heat away from the body in a cold winter.

(ii)   On a cold day, our body loses heat mainly due to radiation (more than 80%). But in a hot summer, the body cools down solely through the evaporation of perspiration.

(iii)   In an extreme heat situation (e.g. fire fighting), it is not wise to rely solely on a textile to shield heat from the body because that would require clothing so thick that it would severely block perspiration and therefore add great heat stress to the body. Using a radiative or reflective surface to fend off the external heat is recommended.

(iv)   In Antarctica or other super-cold locations, wearing clothing with an interior coated irradiative layer to retain body heat is most effective, since, as indicated above, radiation accounts for more than 80% of total heat loss.

(v)   Transient processes taking place in the microclimate play a very important role in clothing comfort, but research progress in this area is still slow, due mainly to the complexity of the related problems.

6.6   References

ARENS, E., and ZHANG, H. (2006). The skin's role in human thermoregulation and comfort. In N. Pan and P. Gibson (Eds.), *Thermal and Moisture Transport in Fibrous Material* (pp. 560–597). Cambridge, UK: Woodhead Publishing Ltd.

ASHRAE. (2005). Chapter 8, *Handbook of Fundamentals*: American Society of Heating, Refrigerating and Air-conditioning Engineers.

BERGLUND, L. G., and CUNNINGHAM, D. J. (1986). Parameters of human discomfort in warm environments. *ASHRAE Transactions 92*, 732–746.

BISHOP, P., RAY, P., and RENEAU, P. (1995). A Review of the Ergonomics of Work in the US Military Chemical Protective Clothing. *International Journal of Industrial Ergonomics, 15*(4), 271–283.

BISHOP, P. A., NUNNELEY, S. A., GARZA, J. R., and CONSTABLE, S. H. (1988). Comparisons of air *vs* liquid microenvironmental cooling for persons performing work while wearing protective clothing. In *Trends in Ergonomics/Human Factors V* F. Aghazedeh (ed) (pp. 433–440). Elsevier, Amsterdam.

BRECKENRIDGE, J. R., PRATT, R. L., and WOODCOCK, A. H. (1960). Effect of Clothing Color on Heat Load from Solar Radiation. *Federation Proceedings, 19*(1), 178.

DE DEAR, R. J., KNUDSEN H.N., and FANGER, P. O. (1989). Impact of Air Humidity on Thermal Comfort during Step Changes. *ASHRAE Transactions, 95*, 129.

DENT, R. W. (2001). Transient comfort phenomena due to sweating. *Textile Research Journal, 71*(9), 796–806.

ELERT, G. (2006). *http://hypertextbook.com/facts/*. Retrieved February 10 2007.

FAFF, J., and TUTAK, T. (1989). Physiological responses to working with fire fighting equipment in the heat in relation to subjective fatigue. *Ergonomics, 32*, 629–638.

GAGGE, A. P., BURTON, A. C., and BAZETT, H. C. (1941). A practical system of units for the description of the heat exchange of man with his environment. *Science, 94*, 428–430.

GHALI, K., GHADDAR, N., and JONES, B. (2006). Phase Change in Fabrics. In N. Pan and P. Gibson (Eds.), *Thermal and Moisture Transport in Fibrous Materials* (pp. 402–423). Cambridge, UK: Woodhead Publishing.

GROSDOW, A. R., STEVENS, J. C., and STOLWIJK, J. A. J. (1986). Skin friction and fabric sensations in neutral and warm environments. *Textile Research Journal, 56*, 574–580.

GUYTON, A. C., and HALL, J. E. (1996). *Textbook of Medical Physiology* (9th ed.). New York: W.B. Saunders Company.

HAVENITH, G., DEN HARTOG, E., and HEUS, R. (2004). Moisture accumulation in sleeping bags at −7 degrees C and −20 degrees C in relation to cover material and method of use. *Ergonomics, 47*(13), 1424–1431.

HAVENITH, G., HEUS, R., and LOTENS, W. A. (1990). Resultant Clothing Insulation – A Function of Body Movement, Posture, Wind, Clothing Fit and Ensemble Thickness. *Ergonomics, 33*(1), 67–84.

HAVENITH, G., HOLMER, I., DENHARTOG, E. A., and PARSONS, K. C. (1999). Clothing evaporative heat resistance–proposal for improved representation in standards and models. *Ann. Occup. Hyg., 43*, 339–346.

HU, J. Y., YI, L., YEUNG, K. W., WONG, A. S. W., and XU, W. L. (2005). Moisture management tester: A method to characterize fabric liquid moisture management properties. *Textile Research Journal, 75*(1), 57–62.

HUANG, J. (2006). Thermal parameters for assessing thermal properties of clothing. *Journal of Thermal Biology, 31*(6), 461–466.

IAMPIETRO, P. F., and GOLDMAN, R. F. (1965). Tolerance of men working in hot, humid environments. *J. Appl. Physiol., 20*, 73–76.

ISO9920. (1995). *Ergonomics of the Thermal Environment – Estimation of the Thermal Insulation and Evaporative Resistance of a Clothing Ensemble:* International Organization for Standardization.

ISO15831. (2004). *Clothing – Physiological Effects – Measurement of Thermal Insulation by Means of a Thermal Manikin:* ISO.

JONES, B., GHALI, K., and GHADDAR, N. (2006). Heat–moisture interactions and phase change in fibrous material. In N. Pan and P. Gibson (Eds.), *Thermal and Moisture Transport in Fibrous Materials* (pp. 424–436). Cambridge, UK: Woodhead Publishing.

JONES, B. W., and OGAWA, Y. (1992). Transient Interaction Between the Human Body and the Thermal Environment. *ASHRAE Transactions, 98*, Part 1.

JONES, B. W., and OGAWA, Y. (1993). Transient Response of the Human–Clothing System. *Journal of Thermal Biology, 18*(5–6), 413–416.

JOY, R. J. T., and GOLDMAN, R. F. (1968). A Method of Relating Physiology and Military Performance – A Study of Some Effects of Vapor Barrier Clothing in a Hot Climate. *Military Medicine, 133*(6), 458–470.

KENNY, W. L., LEWIS, D. A., ARMSTRONG, C. G., HYDE, D. E., DYKSTERHOUSE, T. S., FOWLER, S. R., and WILLIAMS, D. A. (1988). Psychometric limits to prolonged work in protective clothing ensembles *American Industrial Hygiene Association Journal, 49*, 390–395.

LI, F. Z., and YI, L. (2005). Effect of clothing material on thermal responses of the human body. *Modelling and Simulation in Materials Science and Engineering, 13*(6), 809–827.

LI, Y. (2005). Perceptions of temperature, moisture and comfort in clothing during environmental transients. *Ergonomics, 48*(3), 234–248.

LOTENS, W. A., and HAVENITH, G. (1995). Effects of Moisture Absorption in Clothing on the Human Heat-balance. *Ergonomics, 38*(6), 1092–1113.

LUO, H. A., FISHBEIN, M. C., BAR-COHEN, Y., NISHIOKA, T., BERGLUND, H., KIM, C. J., *et al.* (1998). Cooling system permits effective transcutaneous ultrasound clot lysis in vivo without skin damage. *Journal of Thrombosis and Thrombolysis, 6*(2), 125–131.

MORTON, W. E., and HEARLE, J. W. S. (1997). *Physical Properties of Textile Fibers.* Manchester, UK: The Textile Institute.

NAVE, C. R. R. (2005). *http://hyperphysics.phy-astr.gsu.edu/.* Retrieved Jan., 26, 2007, 2006.

NUNNELEY, S. A. (1989). Heat stress in protective clothing. *Scandinavian Journal Work Environmental Health, 15*, 52–57.

PAN, N., and GIBSON, P. (2006). *Thermal and Moisture Transport in Fibrous Material.* Cambridge, UK: Woodhead Publishing.

PAN, N., and ZHONG, W. (2006). *Fluid Transport Phenomena in Fibrous Material.* Cambridge, UK: Woodhead Publishing. .

PEIRCE, F. T., and REES, W. H. (1946). The transmission of heat through textile fabrics, Part 2, *J. Text. Inst.*, T181–T204.

PISSIOTIS, C. A., KOMBOROZOS, V., PAPOUTSI, C., and SKREKAS, G. (1997). Factors that influence the effectiveness of surgical gowns in the operating theatre. *European Journal of Surgery, 163*(8), 597–604.

RODWELL, E. C., REBOURN, E. T., GREENland, J., and KENCHINGTON, K. W. L. (1965). An Investigation of the Physiological Value of Sorption Heat in Clothing Assemblies. *Journal of the Textile Institute, 56*, 624–645.

SPENCERS-SMITH, J. I. (1966). Buffering Effect of Hygroscopic Clothing. *Textile Research Journal, 36*(9), 855–856.

SUN, G., YOO, H. S., ZHANG, X. S., and PAN, N. (2000). Radiant protective and transport properties of fabrics used by wildland firefighters. *Textile Research Journal, 70*(7), 567–573.

TOFTUM, J., JORGENSEN, A. S., and FANGER, P. O. (1998). Upper limits for indoor air humidity to avoid uncomfortably humid skin. *Energy and Buildings, 28*(1), 1–13.

UEDA, H., INOUE, Y., MATSUDAIRA, M., ARAKI, T., and HAVENITH, G. (2006). Regional microclimate humidity of clothing during light work as a result of the interaction between local sweat production and ventilation. *International Journal of Clothing Science and Technology, 18*(3–4), 225–234.

WANG, L. P., and LI, C. (2005). A new method for measuring dynamic fabric heat and moisture comfort. *Experimental Thermal and Fluid Science, 29*(6), 705–714.

WHITE, M. K., VERCRUYSSEN, M., and HOUDOUS, T. K. (1989). Work tolerance and subjective responses to wearing protective clothing and respirators during physical work. *Ergonomics, 32*, 1111–1123.

WONG, A. S. W., and LI, Y. (2004). Relationship between thermophysiological responses and psychological thermal perception during exercise wearing aerobic wear. *Journal of Thermal Biology, 29*(7–8), 791–796.

WOODCOCK, A. H. (1962a). Moisture Transfer in Textile Systems, Part 1 *Textile Res. J., 32*, 628–633

WOODCOCK, A. H. (1962b). Moisture Transfer in Textile Systems, Part 2 *Textile Res. J., 32*, 719–723.

WOODCOCK, A. H., and BRECKENRIDGE, J. R. (1957). Moisture Index, a New Clothing Variable. *Federation Proceedings, 16*(1), 139.

WOODCOCK, A. H., and BRECKENRIDGE, J. R. (1965). A Model Description of Thermal Exchange for the Nude Man in Hot Environments. *Ergonomics, 8*(1–4), 223–235.

YAO, B. G., LI, Y., HU, J. Y., KWOK, Y. L., and YEUNG, K. W. (2006). An improved test method for characterizing the dynamic liquid moisture transfer in porous polymeric materials. *Polymer Testing, 25*(5), 677–689.

7

Cold-weather clothing

C. THWAITES, W. L. Gore and Associates UK Ltd, UK

7.1 Introduction

The history of warfare warns us that many armies have been defeated by extreme weather rather than by the enemy. Cold weather is often referred to as the second foe. In addition, cold or cold wet winter conditions have sometimes caused more casualties than the battles themselves, because, for amongst other reasons, the soldiers' clothing was not suitable. Logistical problems, lack of training and the prevention of good hygiene by ongoing combat are also contributory reasons. Cold stress can lead to local and extremity cooling, which in turn can lead to cold injuries as well as hypothermia, which in turn can eventually result in death. Even in such a recent conflict as the Falklands, cold injuries, specifically trench foot, have been major issues, which have led to increased risks being taken by commanders to bring the war to a close (Evans, 2007).

In the last few decades, the textile industry has been able to engineer the properties of fabrics and laminates to substantially improve the properties of base-layers, insulative mid-layers and protective outer shells. This has increased the performance and wear comfort of clothing, particularly in cold weather, thus reducing the cold strain of the wearer. One of the biggest steps in the improvement of protective clothing items has been the invention of waterproof yet highly moisture vapour permeable laminates in the 1970s, reducing the problems that cold wet weather can cause.

The purpose of cold weather clothing is to protect the soldier from the adverse effects of rain, snow, wind, and cold. To do this, sufficient but not too much insulation is needed to balance the heat loss of the body with its heat production. Heat production can change by a factor of ten between rest and intense exercise. Insulation in clothing is created by air entrapped between the fibres and clothing layers. Since only still air provides effective insulation, the air in the clothing needs to be protected from wind penetration by a windproof outer layer. Moisture should also be kept out of the clothing since the insulation value of moist insulation is only about 30% of its dry value. There are two ways that clothing can get wet: by the ingress of rain or water from the outside or from the condensation of sweat internally. This latter problem is prevented or reduced by the use of waterproof but vapour permeable textile laminates.

For the design of a military clothing system we first of all need to know the range of climatic conditions in which it will be worn. Depending on the weather, different properties like waterproofness, windproofness, water vapour permeability, ventilation, and moisture management are of changing importance. If the different clothing layers are not designed as a system, only a part of the potential improvements can be perceived, e.g. quite often a clothing system is let down by a poor choice of base-layer, which can cause a wet, clammy feeling on the skin. Durable waterproofness is essential in cold wet conditions and windproofness is a must in extreme dry cold. Methodical testing of the garments as well as the textiles is important, to make sure that they are fit for purpose. These areas are covered in this chapter.

7.2   Cold weather

When designing clothing or deploying soldiers, the climate of the intended use is needed as well as the temperature extremes that may be encountered. Climates can be classified using Military Standards, e.g. DEF STAN 00-35 Part 4 (2006). World climatic data has been compiled by Wernstedt (1972). Historical weather data is also available on the World Wide Web, e.g. Weather Underground. It should be remembered that extremes of temperature and relative humidity do not often occur simultaneously.

Cold weather can be divided into three main categories;

Wet cold $+10\,°C$ down to $-10\,°C$ where we can expect thawing and refreezing,
Very cold $-10\,°C$ down to $-30\,°C$ and
Extremely cold $-30\,°C$ and below, the extreme lower limit being about $-60\,°C$.

7.3   Physiological responses to cold

For thermal comfort, a normal core body temperature is about $37\,°C$, varying about $0.5\,°C$ either side of this over a day for a non-active person. A comfortable mean skin temperature for a resting individual is in the range of 33 to $34.5\,°C$, with local skin temperatures variable over the body but generally in the range $32–35.5\,°C$. Blood flow to the feet and hands under these conditions will be high (Hardy, 1970), which explains why they are quite near core temperature. When we start to feel cold, the blood flow to the

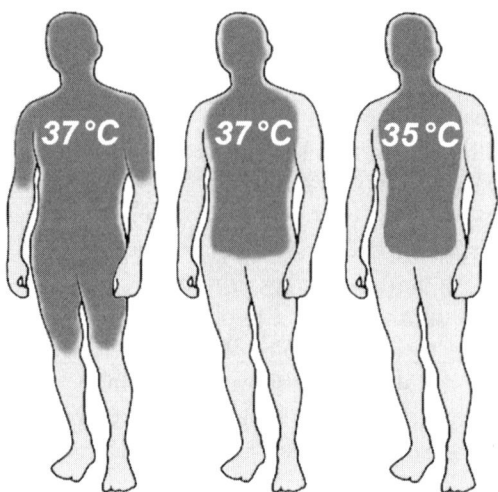

7.1 Approximate relation of core and shell compartments at different thermal states. After Lloyd (1994). Skin temperatures are decreasing from left to right.

extremities is reduced in order that the core temperature (that of the vital organs) is protected; see Fig 7.1. The left-hand depiction shows the shell/ core compartments for a comfortable person with relatively high skin blood flow, whilst that in the middle shows the shell for a person with a normal core temperature but cool skin and reduced skin blood flow; the depiction on the right shows a person about to become hypothermic with the core temperature at 35 °C.

Vasoconstriction starts when the skin temperature falls below about 34 °C, the reduction in skin blood flow being approximately 50% with a 2 °C drop in mean skin temperature. Due to this reduction in warm blood flow, the hands and feet start to lose more heat than they gain and cool down so that the temperature of the extremity is reduced.

The onset of cold pain begins when the skin temperature is about 15 °C, and at 7 °C cold pain ends and numbness sets in; 'At local temperatures of 7 °C both motor and sensory nerves are blocked, resulting in paralysis. When sensation is lost, there is little indication of any subsequent cold injury. Consideration of this fact and others pertaining to limits of tactile sensitivity suggest that the skin temperature should not fall below 15 °C for useful work and not below 8 °C for the maintenance of sensation and little likelihood of injury' (Hayes, 1989). For the US Military, Pimental (1991) has suggested the following physiological wear criteria for cold weather cloth- ing – the mean weighted skin temperature should be above 28 °C and local skin temperatures at any site equal or higher than 18 °C.

This reduction in skin blood flow is of great importance for inactive soldiers, since it is the main source of heat for the fingers and toes; deprived of this, they can reach dangerously low temperatures in a relatively short time. This cold induced vasoconstriction can sometimes lead to cold injury. When a person becomes cold and/or wet, the increased heat loss from the body has to be balanced by an increase in activity to keep the core temperature from falling into the hypothermic region, <35 °C. Another problem with muscles is that they become less efficient as they get colder and, as a consequence, 'exercise of the same intensity becomes increasingly more difficult as it demands more energy, and this might explain the earlier onset of fatigue when exercising in the cold' (Noakes, 2000). Noakes also cites Adolph and Molnar (1946) reporting that 'subjects with a core temperature of 36.9 °C but a mean skin temperature of about 16 °C became exhausted presumably when muscle temperature fell.'

A protective mechanism can kick-in when hands and feet get too cold – Cold Induced Vaso Dilation (CIVD). This releases blood through the arterio-venous anastomoses in the fingers and toes and re-warming occurs. This occurs periodically. This response is, however, blunted or absent in hypothermic people (Vanggaard, 1975).

Shivering is another source of heating to combat the cold. It usually begins in the torso and spreads to the limbs. The intensity of shivering varies according to the severity of the cold stress. Xu *et al.* (2003) have published an equation relating metabolic heat generation due to shivering based on core and skin temperatures as part of a model for predicting thermoregulatory responses during long-term cold exposure.

The effect of activity on toe temperature has been clearly demonstrated in a study by Mekjavic *et al.* (2005), who measured toe temperatures in two groups of people, one of which hiked for 3–4 hours whilst the other was on guard duty for 3 hours; the ambient temperature for both was 2 °C. In the hikers, the toe temperature increased from 28 °C to 31 °C whilst in those on guard duty, it dropped to 15 °C.

Cold injuries can be classed as either non-freezing cold injuries (NFCI), such as trench foot, or freezing cold injuries, e.g. frostbite. Trench foot or non-tropical immersion foot has been a major source of injuries in the Second World War (Whayne and DeBakey, 1958), in Korea (Orr and Fainer, 1952) and in the Falklands (Oakley, 1984). It still remains a source of injury in peacetime training (Tek and Mackey, 1993; DeGroot *et al.*, 2003; Schissel *et al.*, 1998; Taylor, 1992; Hawryluk, 1977; Candler and Ivey, 1997).

As previously mentioned, to re-warm cold feet, exercise is needed. The best type is one where the actual foot moves. It is not just the exercise itself that is beneficial but the actual physical movement of the foot. In a study of the ability of lower and upper body exercise to re-warm cold feet, it was found that step exercise was much better than cycling (Rintamaki *et al.*, 1992) This confirms what winter cyclists with cold feet know; only walking brings back the warmth. During the Second World War, General Patton wrote in one of his Third Army operation orders 'The most serious menace confronting us today is not the German Army, which we have practically destroyed, but the weather, which may well destroy us through the incidence of trench foot. Soldiers must look after themselves, particularly in wet or cold weather. This applies particularly to trench foot which can largely be prevented if the soldier will put on dry socks' (Whayne and DeBakey, 1958).

The head is one area that is not subject to vasoconstriction in the cold, so substantial amounts of heat can be lost if it is not insulated. At $-4\,°C$, about half the metabolic heat generation of an otherwise well-clothed person can be lost this way (Froese and Burton, 1957). To keep the feet warm, the advice is often given to put a hat on – this is a very sound proposal since any action that preserves heat loss, such as wearing of a hat and gloves, will improve the blood flow to the feet.

7.4   Clothing design principles

Layering – although it is good advice and practice to have more than one clothing layer, so that the clothing insulation can be varied to suit the environment and activity, two points should be borne in mind: (i) each clothing layer will for a given task increase the energy required. (Teitlebaum and Goldman, 1972); (ii) If you want a highly insulating clothing system, this cannot be achieved by continually adding relatively thin layers – thick insulation is necessary. The increase in energy expenditure caused by adding layers is in the order of 4% a layer (Lotens, 1988), but might be reduced by decreasing the inter-layer friction. Recent research has looked at this topic in more detail and the interaction of clothing bulk and weight with the number of layers made it difficult to separate out the individual effects of these parameters (Dorman, 2007). In this study, the effect of wearing four layers compared to a single layer when walking was a 5 to 8% increase in metabolic rate. When clothing friction was investigated, no significant difference was found between high and low friction fabrics.

An interesting approach to selecting clothing was suggested by Siple (1949), in which he divided the world into clothing zones and described them by the number of clothing layers worn in each zone. The largest number of clothing layers practical is four which he describes as needed for sub-Arctic and Arctic winter-type environments. This should still be considered the maximum number of layers to be worn.

Looseness of fit – one of the biggest benefits of traditional Inuit clothing design is the ability to ventilate when active, which is due to its looseness. They can also tighten the clothing with a belt when needed. Because they use animal skins and furs which deteriorate when wet, they have to keep their clothing as dry as possible (Stefansson, 1944). As argued by Fiennes (1993) in relation to the attempt to reach the South Pole, Amundson was correct to use Inuit fur clothing since he was dog-sledging, whereas Scott was right *not* to use it since he was man-hauling sledges and presumably the harness would restrict the ventilation needed for the fur clothing to work. With military clothing, when carrying a ruc-sac or wearing webbing, the ability to ventilate the torso is severely reduced but efforts should still be made to get as much ventilation as is possible without compromising waterproofness, both in the jacket and the trousers. To be effective, any vents allowing air in have to have a corresponding vent to allow the ventilating air out. Arms that can be pulled up to above the elbow allow increased heat loss from the lower arms. Full or partial length trouser zips can be useful to aid ventilation.

The acronym COLD sums up how cold weather clothing should be worn (Siple, 1949):

C is to keep the clothing clean so that it will not cling or matt and destroy the insulation
O is to open the clothing
L if for wearing the clothes loosely and in layers
D is to keep the clothing dry.

Here is a description about the onset of hypothermia in the Antarctic by a doctor who, at the time, worked for the UK Ministry of Defence advising on exercise performance and survival (Stroud, 1998). During an expedition across the continent he woke up one day with the sun shining, dressed accordingly and set off as the weather began to deteriorate. Although he realized that he was perhaps not wearing enough clothing, he thought that the hard work of pulling a heavy sledge would keep him warm. The weather got colder and his hands become numb with cold. He stopped but could not put on his spare pair of thicker mitts and his companion following behind him caught up, and in a 'wonderful gesture' swapped gloves. However having had to stop for ten minutes, he was deeply chilled and 'my muscles had cooled to the point where they no longer functioned properly. Unable to move fast, I could not generate enough heat to match my further losses. I was getting colder and colder'. He then stopped again to put on another fleece jacket. 'Dressing would also need some dexterity, which meant taking off my gloves again. Once I removed my gloves it was only moments before my fingers would not obey commands. This time it was the zips on the jacket that defeated them and failed to move. I was dismayed to see the tips of each digit turning chalky white . . .' Ranulph Fiennes, his companion caught up with him again, got his jacket and gloves on, but after half an hour Stroud could not warm up and had begun to wander off course. He was now hypothermic but Fiennes was there to put up their tent and re-warm him.

This example is interesting because some of the problems encountered with adjusting clothing are a consequence of poor design. Take, for instance, the need to remove an outer layer so that extra insulation can be put on underneath. In these severely cold conditions and in wet cold environments, it makes sense to avoid having to take the outer layer off by ensuring that insulative layers have some weather protection. In this case it would be to have a windproof insulation that could have been worn without removing the existing outer layer. In wet conditions it would mean having a water-proof or water-resistant covered insulation. In fact the Canadian Military have done such a thing with their Improved Environmental Clothing System IECS. As Frim (1995) states 'The most important attribute of the IECS is that it finally brings the layering principle into practicality, and it does this with no sacrifice, and possibly even some significant gains, in thermal protection against the cold'.

Gloves and mitts should be designed to be put on with cold hands. Zips for cold weather should have a zip pull extension fitted so that it can be used with gloves or cold hands. For use in freezing conditions, zips should have a cover to prevent icing up. The means of adjusting clothing should be simple and be possible even in the dark.

With multiple layers, the ease of going to the toilet has to be thought about and designed into all the clothing layers. With an increasing number of layers being worn, the problem of bulkiness at the waist needs to be addressed. Wearing salopettes rather than trousers may be the better option since, not only do they help reduce waist bulkiness, they also allow the possibility of increased ventilation if some of the lower trouser leg can be opened up. Salopettes also protect the midriff better in cold conditions.

During the Second World War, a counter-intuitive phenomenon was noticed when evaluating heavy aviators' clothing. The core temperature decrease in the cold was much larger than that observed in previous trials with more lightly clad subjects and it was postulated that the vasoconstriction response had been delayed. This resulted in the skin temperature remaining high with a concomitant increase of heat loss to the environment, hence the lower core temperature. This delay was not noted when the face-masks were removed and the face was exposed to cold air. It was thought that at least some of the body should be exposed to the cold to initiate the vasoconstriction needed to maintain the body core temperature (Winslow and Herrington, 1949).

Performance of the clothing

When designing clothing it should be designed for ease of movement, especially with regard to legs and arms. The amount that the garment needs to stretch can be measured on a soldier carrying out the important tasks that he needs to do. After making the prototype clothing, how does it affect the tasks? This can be tested by measuring the performance times when carrying out typical activities (Lotens, 1988).

The effect of cold weather clothing on body movement has recently been investigated (O'Hearn *et al.*, 2005). Comparing it with a temperate uniform, the cold weather clothing interfered with bending at the waist and shoulder movements. Wearers leaned forward more and moved their arms less. The cold weather boots also interfered with leg swing compared to regular combat boots.

Burn injury – there has been a trend in many countries to use synthetic base-layers in cold weather Military uniforms. To minimize burn injury, it is often stated that clothing should be non-melt. The evidence for this has been questioned by McLean (2001), who argued that there appeared to be little objective evidence that thermoplastic materials worsen burns. Even if it was true, for cold weather clothing which has more than one layer, any adverse effects would be expected to be much less than for hot weather clothing. It is, of course, important to have flame-resistant clothing that does not melt, but for combat soldiers in cold-weather clothing outside of vehicles, it is less of a danger than being inside a vehicle at warmer temperatures. However, for those soldiers using tents, flame resistance could be of vital importance.

7.5   Estimation of the clothing insulation required

A very useful standard has been developed which estimates the thermal insulation required for thermal neutrality in cold environments $IREQ_{neutral}$ (ISO/DIS 11079, 2005). The calculation requires as inputs: air temperature (Ta), velocity (v_{air}), relative humidity (rh), the air permeability of the outer fabric, and the mean radiant temperature (Tmrt). An estimate of the metabolic heat generation (M) has to be made – this can be obtained from ISO 8996 (1989) or calculated from equations developed by Pandolf *et al.* (1976, 1977). These can predict the metabolic heat generation for soldiers walking and running, carrying ruc-sacs, and different types of terrain can be selected. Most data on clothing has been obtained on standing manikins in relatively still air – this is the basic thermal insulation Icl. Movement and the effect of wind have been investigated and their effects have been incorporated into the prediction of the resultant thermal insulation under actual wear conditions. Clothing insulation as measured on a standing manikin in

Table 7.1 Required insulation and basic insulation at various temperatures, air permeability and activities. (1 clo = 0.155 m²K W⁻¹). Based on ISO DIS11079 (2005). Wind speed 5 m s⁻¹

Example	Ta, ambient temperature (°C)	Activity metabolic heat production (W m⁻²)	Air permeability L m⁻² s⁻¹ at 100 Pa	IREQ$_{neutral}$ (clo) Insulation needed in use	Icl Basic clothing insulation – (clo) Corresponding insulation on manikin in still air
1	10	175	1	0.8	1
2	0	175	1	1.3	1.6
3	−15	175	1	2	2.6
4	−30	175	1	2.6	3.5
5	−15	175	50	2	3.4
6	−15	250	1	1.3	1.7
7	10	100	1	1.7	2.2
8	−10	100	1	3.2	4.4
9	−30	100	1	4.8	6.6

relatively still air can be markedly reduced by walking and wind, especially if the outer fabric is air permeable – see Table 7.1. In the first four examples, we can see the effect of reducing ambient temperature on the insulation required; all other factors are constant. The activity is equivalent to walking at about 4.5 km h⁻¹. Example 5 can be compared to Example 3, the difference here being that the outer fabric has an air permeability of 50 L m⁻² s⁻¹, which is approximately the value for a woven fabric. The basic clothing insulation needed has been increased by 30%. This air permeability is typical for relatively tightly woven uniforms – standard fleece fabrics have air permeabilities of about 1000 L m⁻² s⁻¹. Example 6 can be compared to Example 3 and shows the reduction in insulation required when the activity is increased from 175 to 250 Wm⁻². This latter rate is equivalent to walking at 4.5 kph with a 30 kg ruc-sac. The insulation required has dropped by 30 percent. Examples 7 through 9 show the insulation required for a standing soldier increasing dramatically as the temperature is reduced. The standard also estimates the exposure time for a person wearing insufficient insulation for an 8 hour exposure. A program is available to do these calculations (IREQalfa, 2002), and all the necessary information and equations are given to solve them. One clo is equivalent to a business suit ensemble. The most insulating garments for Arctic use are about four to five clo. Some care needs to be taken when comparing insulation values since the methods used to calculate insulation can differ. The insulation value is for the whole body, so in addition we need to make sure that, when evaluating a clothing system, the extremities have the insulation they need to prevent excessive cooling.

7.6   Evaluation system for textiles and garments

NATO has published a five-stage guide for the testing of combat suits (ACCP 1, 1992). This is shown in Table 7.2. It is an excellent system to follow.

At the first level, the thermal and evaporative resistances of all the textiles used in the clothing system are measured using the skin model (ISO 11092, 1993). Using resistance values rather than permeabilities has one clear advantage – easy calculation: the total resistance of multiple layers is simply the sum of the single layers.

When comparing textiles or insulations with different thermal insulations, it is better to look at the water vapour permeability index i_{mt}, which is a weighted ratio of the thermal insulation and evaporative resistance.

$$i_{mt} = 60\frac{Rc_t}{Re_t}$$

Base layer testing can also be carried out at this level, by far the best and most comprehensive method having been developed by the Hohenstein Institute (Bartels and Umbach, 2001). Although this system has been designed to evaluate sportswear, we have found it to work well for base-layers worn as part of a clothing system. The base-layer parameters tested are: wet cling, the number of contact points, stiffness, water vapour permeability, a sorption index (wetting angle and time) and a surface index which estimates hairiness.

At the next stage the thermal resistance of the clothing system is measured on a heated thermal manikin. For this test the manikin surface

Table 7.2 Five-level system for clothing evaluation

Level 5	Large-scale fitness for use tests.
Level 4	Controlled field trials measuring ambient and physiological parameters.
Level 3	Human subject physiological tests conducted in a climatic chamber.
Level 2	Measurement of the thermal and evaporative resistances of the complete clothing system on a thermal manikin. Predictive modeling.
	Measurement of ventilation rate on humans.
Level 1	Measurement of thermal and evaporative resistances (ISO 11092, 1993). Base-layer testing (Umbach, 1992).
	Insulation testing.

temperature is normally kept at 35 °C, whist the chamber is set at a lower temperature, commonly 15 or 20 °C. The power needed to keep the manikin skin at 35 °C is measured and from this the thermal insulation of the clothing can be calculated. The evaporative resistance can be measured directly by covering the manikin with a wet skin, which is wetted at the start of the test, and increasing the chamber temperature to match the manikin skin temperature so that only evaporative heat loss occurs (Goldman, 2006). By use of tubes and pumps, the 'skin' can be rewet at will. Alternatively, the evaporative resistance can be estimated from skin model measurements of the textiles and the clothing thermal insulation. By accurate measurement of the areas of each clothing layer including overlaps, the contribution of the textile layers to the overall insulation can be calculated and the difference is the thermal resistance of the enclosed and boundary air layers. Once these air layers have been estimated, the evaporative resistance of the clothing ensemble can be calculated by adding the evaporative resistance of the air layers and the area-weighted textile evaporative resistances together (Umbach, 1984).

Clothing ventilation rates can also be measured, either on articulated thermal manikins or on humans. In the latter method, a trace gas is introduced into a clothing system through a distribution harness of small tubes. Another identical harness samples the microclimate. From the difference in concentration of the trace gas in the two flows the ventilation rate can be calculated (Lotens, 1993). Care has to be taken when comparing the ventilation in air permeable and non-air permeable clothing. In non-air permeable clothing the trace gas leaves the clothing solely by ventilation – however if the clothing is air permeable then the trace gas can also leave by an extra parallel route by diffusion through the clothing. Whilst the equations governing the two routes have been described, and a method for their solution given (Lotens, 1993), ventilation rates for air permeable clothing are rarely interpreted this way.

If fabric waterproofness is necessary then it is logical that the clothing needs to be tested for waterproofness too since poorly designed garments can leak through closures and unfinished seams. This can be done by shower testing and the tests can be tailored to suit the specific end use.

After the determination of thermal and evaporative resistances, and as a precursor to human subject testing, predictive modeling should be carried out before carrying on to the next stage. Typically, when evaluating clothing we are looking at a minimum of two types – a current and a prototype system. Predictive modeling can help in deciding whether to test at the next level. If we think the new clothing system will be warmer, will a difference in physiological responses, for example, core temperature, be predicted when wearing the new clothing? If the answer is no, then there is no point in doing the tests. Predictive models can be divided into physical or rational types (Fiala, 1998; Stolwijk and Hardy, 1966) and empirical or experimental (Umbach, 1984; Givoni and Goldman, 1972).

At the third stage, the clothing is worn by humans, typically four to eight subjects, in a climate chamber, for physiological evaluation under realistic and controlled conditions of temperature and relative humidity. Every source of unwanted variability should be minimized for these tests to be successful. In particular, the following recommendations should be followed (Levine *et al.*, 1995)

(i)   Each subject must serve as their own control.
(ii)   Prototypes must be compared to control garments in the same tests.
(iii)   Tests are counter-balanced not randomized.

When no difference is found between clothing ensembles, it is good practice to note what size of difference could have been detected in the tests, i.e. is the fact that no difference was found due to high variability of the data.

The next stage, Level 4, is the physiological evaluation of the clothing in the field. This is a serious undertaking and only the Military and a few other organizations have the resources to accomplish this. Thorough requirements for this stage are given in ACCP 1 (1992). Typical numbers of subjects would be about 20–50 (Goldman, 2006).

The final stage is the operational use of the clothing by a large numbers of soldiers.

7.7   Selection of clothing for cold weather

7.7.1   Base-layer

In cold weather, standard cotton should be avoided. It is perfectly acceptable when sweat rates are very low, as in normal everyday use, but it becomes wet and clings to the body when sweat rates increase. In a recent survey of outdoor users in the UK, comprising professional field testers, outdoor instructors and shepherds, none of the 254 users used cotton (KLETS, 2007). There is a substantial body of practical experience and advice advocating this opinion. Many guided mountaineering expeditions forbid the use of cotton, as do outdoor training centres for cold weather activities. The most forthright opinion we have seen comes from the US Marine Corps Cold Weather Training Manual: 'cotton equal(s) death' (MCRP 3-35.1A, 2000). Although cotton and synthetic base-layers have been evaluated in chamber tests, the studies have failed to find a substantial benefit in favour of the synthetic materials. It is likely that the chamber test protocols are not simulating field use and that the purpose of the tests is not clear. In the tests we have seen so far, it is cotton which tends to emerge as the base-layer that aids heat loss, i.e. core temperature increases are less during exercise than when wearing synthetics (Rissanen *et al.*, 1994). This is to be expected, since materials which evaporate water closer to the body are more efficient. With synthetic base-layers that have been designed to keep sweat away from the skin, less efficient evaporation, and less cooling is expected (Uglene, 2000). The benefits of wearing a good synthetic base layer in the cold are explained next, and it is not to aid heat stress but to provide sensorial comfort. Recently, during one of our field tests, when commenting on wearing a cotton vest and comparing it to a good polyester base-layer, the testers remarked that, when wet, the cotton vest clung to the skin, took a long time to dry and felt cold. It is these properties that are important for cold weather base-layers. They can be easily measured and can serve to pre-select base-layers for more detailed tests. Wet cling can be measured on a tensile test machine adapted with a horizontal plate, as the force to pull a fabric covered boat over a wet surface. The feeling of cold can be measured as its thermal absorptivity

$$b = (k \cdot \rho \cdot C)^{0.5}$$

where b = thermal absorptivity
        k = thermal conductivity
        ρ = density
        C = specific heat
Units of b are J s$^{0.5}$ m^{-2} K^{-1}

A machine, the Alambeta, has been developed to do this (Hes, 2002). It consists of two plates, one heated to 10 °C higher than the other. A fabric is placed on the lower, unheated plate with the surface that would be next to the skin facing upwards. The upper plate then contacts the fabric with a pressure of 200 Pa. A heat flux sensor then monitors the peak heat flow, and subsequently the steady-state heat flux. As well as the thermal absorptivity, thermal conductivity is measured. Fabrics with a low thermal absorptivity are perceived as warm, and those with a larger absorptivity as cold. Depending on whether we are hot or cold will determine whether we perceive a fabric as pleasant or not. When we are hot, cotton, with its high thermal absorbtivity, is perceived as pleasant but the opposite is true when we are cold. When wet, cotton has a higher thermal absorptivity compared to good cold weather polyester base-layers. It should be pointed out that some polyester base-layers can have higher thermal absorptivities than cotton. But base-layer selection is not about natural versus synthetic. There are many good fine merino wool and wool/polyester mix base-layer fabrics that are available and very well suited for cold weather use. There are also many synthetic base-layers that are unsuitable for use in cold weather, including some made from polyester. It is not just the fibre that makes a good base-layer but the construction as well. Liquid should be attracted away from the surface by differential capillary pressure; whether it is moved away to the bulk of the fabric or to the outer face does not seem to be important, as long as it is removed from the skin. In cold weather it makes sense to use a heavier, thicker fabric.

The problems associated with chamber tests not agreeing with field tests, mentioned above, are nothing new. Here is a quote from Forbes (1949): 'There are other matters in which the measurements do not accord entirely with the subjective impressions of the users of the clothing. For example, the Brynje vest (a very coarsely knitted string undershirt) has been used by large numbers of people and for the most part their reports are very favorable, but measurements of insulation value, skin temperature and sweating show surprisingly little difference with and without it – a difference which, though it is in a favorable direction, is smaller than one would expect subjects to detect'. The outcomes of properly conducted field trials should take precedence over laboratory tests in cases like this.

7.7.2 Mid layers and insulations

Polyester fleece and pile fabrics are very popular in marine and outdoor sports occupations as well as for military clothing. Wool, which has been proven over many years in cold and extremely cold weather as a good insulating layer, is currently having a renaissance in the outdoor market. Down is still considered to be the best insulation but some polyester fillings can approach down in dry conditions whilst outperforming it in wet.

When selecting insulation, the first thing to decide is what insulation value is needed. Since the insulation depends on the immobilization of air within the textile structure, the fibre material itself plays a minor role. The insulation value depends, within a range of about ±10%, solely on the thickness of a textile layer. As a rule of thumb, an inch of insulation is equivalent to about 4 clo; in metric units, 1 cm of insulation equals $0.24 \text{ m}^2 \text{ C W}^{-1}$. The 10% variation is the influence of the fibre, especially the fibre diameter, on the suppression of radiative heat loss. Radiative heat loss, fibre to fibre, has been found to decrease with decreasing fibre diameter (Lee, 1984) and this is the main reason why finer fibres are better insulators. However, there is a minimum diameter beyond which the radiation suppression decreases again – this is when the fibre diameter is roughly the same as the wavelength of infra red radiation. An important property to measure is the sustainability of the insulation value during field use. Regarding sustainability, the most critical factor for insulation breakdown in the cold is moisture, since all insulating materials lose about two-thirds of their insulation when wet (Schneider and Hoschke, 1992). Because drying clothing in the field is not easy, what really counts when garments get wet, for example, by accidental immersion, is how much water you can squeeze out and then how quickly the residual moisture dries out so that the original insulation value and loft are obtained.

This definitely rules out cotton: normal cotton fabrics retain over three times more water than polyester fabrics. Wool does not retain much liquid water due to its natural hydrophobicity. What makes wool very interesting for military applications is that it is non-melting. Man-made fibres such as polyester (PES), polyamide (PA), and polypropylene (PP) have the advantage of taking up very little liquid water in the fibres. Even in the field, textiles made from these fibres that get wet can be wrung dry to remove most of the water.

Down is the best insulation in terms of warmth for weight – it also has good resilience to compression and retains its loft when unpacked. Weight and compression to a small size can be strong advantages for these layers.

There are two basic construction principles for the insulation layers: (i) textile-like, e.g. knits, fleeces and pile fabrics, and (ii) duvet-like, when more insulation is needed. The textile ones can reach excellent durability. During the Second World War 'The development of pile fabrics as clothing insulation for Arctic climates made possible for the first time in history the fabrication of washable combat garments which approached, in the protection they afforded, the fur parkas used by the Alaskan scouts before World War II' (Whayne, 1955).

With fleece and pile fabrics their thickness is limited, and if very thick packable and lightweight insulation for the extreme cold is needed, duvet type layers are the best choice. They can be filled with down or non-woven battings made from artificial fibres. Polyester fibre insulations are in general use in many military clothing and sleeping systems because they can approach the insulation and loft of down but perform much better when they get wet. As stated before, the vapour permeability of insulations should be considered in the form of the permeability index rather than the evaporative resistance. The water vapour permeability index for insulations can be classified into unsatisfactory when $i_{mt} < 0.35$, good when it is between 0.35 and 0.55, and very good above this latter value (Umbach, 1987).

Insulation depends on the immobilization of the air within its structure. A windproof shell, worn on top, is sufficient for this. Yet to make the insulation layer windproof on its own has a lot of advantages in practical use. It increases the versatility of that layer because it can be worn now as a shell in windy conditions. And, if in cold and windy conditions it is necessary to take off the outer shell for putting on or taking off a second insulation layer, a windproof insulation still protects in that situation. As the earlier example shows, in the extreme cold this moment can quickly become a life-threatening one.

The best way to get a textile layer windproof is to use a vapour-permeable membrane. Since the best membranes, e.g. those used in

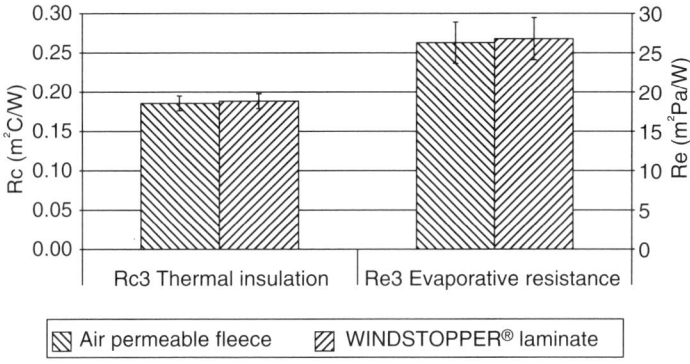

7.2 Comparison: WINDSTOPPER® fleece vs polyester fleece under GORE-TEX® rain suit.

WINDSTOPPER® fabrics, add hardly any resistance to the overall evaporative resistance of an insulation layer, an extra membrane in a clothing system has no measurable disadvantage – see Fig. 7.2. Also, we might predict that an additional membrane would hinder ventilation while active. Surprisingly this is not the case. In different measurements carried out on manikins and on humans, the ventilation rate with the windproof fleece mid-layer was either the same or slightly higher when walking compared to that of a standard fleece without a membrane – see Fig. 7.3. This figure also shows the effect of fit on ventilation rate, with a slimmer person having about 70% more than the well-built individual wearing the same size of garment.

In terms of the right insulation value, it is advisable when wearing or carrying two mid layers for colder weather, to make sure that they are of different thicknesses – this means that, when worn, the thermal insulation can be varied more than by having the two layers of equal insulative value.

7.7.3 Outer protective layer

For cold, wet weather this should be durably waterproof to protect the clothing worn underneath. A resistance to water penetration equivalent to a hydrostatic head pressure of about 28 m is needed in order to be classified as waterproof (Smedstad, 1995). Seams should be sealed to make it waterproof. The durability of waterproofness should be tested in the field. There are very few lab tests that are good indicators, but wet flex and abrasion as is experienced by a fabric in a washing machine is an exception. It cannot be emphasized too much – durable waterproofness is the essential attribute for this layer. A 'waterproof' layer that isn't waterproof is like a car without wheels. For dry cold, if waterproofness is not necessary, this layer should be

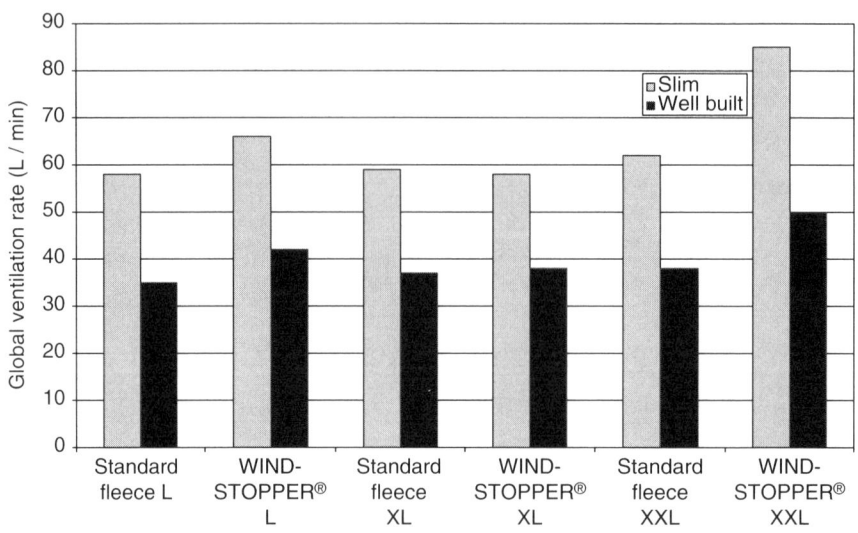

7.3 (a) Effect air permeable/non-air-permeable fleece on ventilation rate, (b) Effect of subject size on ventilation rate. L, XL and XXL Standard fleece and WINDSTOPPER® fleece garments under GORE-TEX® jacket (M for Slim, L for well built). GORE-TEX® trousers.

windproof with an air permeability < 5 L m^{-2} s^{-1}, measured according to ISO 9237 (1995) with a differential test pressure of 100 Pa.

Garments that have outer fabric air permeabilities in the range 0 to 5 l m^{-2} s^{-1} will show no difference in thermal insulation in the wind. These layers should also be water repellent to minimize water pick-up. To allow the evaporation of sweat, this layer should have a high water vapour permeability or a low evaporative resistance (Ret). Waterproof water vapour permeable fabrics can be classified into four groups (Umbach, 1986); see Table 7.3. The NATO standard for waterproof clothing (STANAG 4364, 2003) states that 'In order to provide a rain protective garment that may be worn in various climates, the water vapour permeability should be as high as possible. The water vapour resistance as defined in ACCP-1 should be less than 13 m^2 Pa/W (ISO 11092, 1993)'.

Water vapour permeability (WVP), in g m^{-2} Pa^{-1} s^{-1} and evaporative resistance (Ret) are related by the following relationship:

$$WVP = (Ret \times Latent\ heat\ evaporation\ water)^{-1}$$

Textiles can be made waterproof by either applying a coating or being laminated to a polymer membrane. There are other properties that may be needed to be built in to this layer, such as flame retardancy, camouflage and spectral reflectance.

Table 7.3 Classification of the evaporative resistance
of waterproof, moisture vapour permeable textiles

Evaporative resistance m^2 Pa W^{-1}	Classification
0–6	Very good
6–13	Good
13–20	Satisfactory
>20	Unsatisfactory

Lightweight white nylon over-garments are needed for snow-covered landscapes. For very cold, dry weather the need for waterproofness is not necessary, with some people in the Arctic and Antarctic managing perfectly well with windproof finely woven cotton fabrics or polyester microfibres, whilst others use water vapour permeable laminates. For clothing designed for use by US scientific staff in the Antarctic, a waterproof, water vapour-permeable laminate was chosen. One of the requirements was for an absence of cold crack at –80 °C, which is a very severe test for any product (Sawicki *et al.*, 1992). For military personnel, who need a minimum of garments that work well in a maximum of climates, it does not make sense to have two outer garments – one waterproof for cold, wet weather and one windproof for very cold, dry weather. A windproof, waterproof, water vapour-permeable garment can work in both environments. Recently, based on laboratory testing, the ability of hydrophilic laminates to work at low temperatures was questioned, with the claim that between 18 °C and –10 °C, the evaporative resistance was increased by a factor of 5 (Osczevski, 1996). However, human subject testing in a climatic chamber of these laminates at –20 °C has proved that there was no indication of a temperature dependency of the water vapour resistance and these laminates still provided a great benefit to wearers (Bartels and Umbach, 2002). Also tested in the same study were a water vapour-impermeable coating and a microfibre-woven fabric. In a comparison of the evaporated to produced sweat ratio, the waterproof, water vapour-permeable laminate had about 60% of the sweat evaporated, the woven microfibre about 75%, and the non-water vapour-permeable slightly less than 20%, which was presumably through ventilation through the garment openings.

Condensation in clothing – in cold weather, the best and frequently given advice is try not to sweat. The saturated water vapour pressure or concentration decreases sharply with temperature. The temperature gradient through a clothing system goes from skin temperature on the inside to the ambient temperature at the boundary air layer. For the same evaporative heat flux we are more likely to get condensation the colder it is.

For a given evaporative heat flux, as the evaporative resistance increases, the water vapour pressure difference has to increase to compensate and so we are also more likely to get condensation if the outer layer has a high evaporative resistance. This is shown in Fig 7.4 – here we have three examples of profiles of temperature, relative humidity, saturated and actual water vapour pressures through a garment. This consists of a base-layer, air gap, fleece insulation layer, air gap, outer fabric and a boundary air layer. The sweat rate is $100 \, g \, m^{-2} \, h^{-1}$ in all examples. The actual water vapour pressure is that needed to drive the sweat loss – in reality it cannot be greater than the saturated pressure and if it is, condensation occurs and the relative humidity is shown greater than 100%. If we compare the left and centre cross-sections, where the ambient temperature is 5 °C, we can see the effect that the evaporative resistance of the outer layer has. On the left we have a low Ret of $6 \, m^2 \, Pa \, W^{-1}$ whilst the Ret in the middle example is 20. With a Ret of 6 there is no condensation predicted whereas with the higher evaporative resistance we are predicting condensation from the fleece to the outer layer. On the right-hand example we have a Ret of 6 again, but here the ambient temperature has been reduced to 0 °C. Due to the 5 degree lower temperature we are now predicting condensation next to the outer shell. Damp clothing is to be avoided since it decreases the thermal insulation. When the temperature is less than 0 °C, ice may form in the clothing. This can be brushed off. Ideally if condensation occurs it should be on a surface, rather than inside an insulating layer.

7.7.4 Footwear, gloves and headgear

Footwear for cold, wet climatic conditions should be waterproof and water vapour permeable for the same reasons that clothing needs to be – to preserve the insulation from internal and external moisture. Leather boots with an inner waterproof lining are popular and can, in general, be worn between +20 °C down to about −10 °C. For colder temperature there are two further options – vapour barrier boots or liners and mukluks. In the vapour barrier boots, the insulating felts are encased in two layers of rubber. Socks need to be changed regularly. Vapour barriers can also be used in normal boots.

It is good practice to wear two socks – one next to the foot, which provides blister protection, and the other providing the insulation. The inner sock should also be non-absorbing. Boots should be fitted to be worn with enough insulation so as not to constrict the circulation. To prevent foot problems, soldiers are often advised to change their socks every day, with the spare socks kept next to the body in very cold weather. Protecting the hands is critical for survival. The major problem with gloves

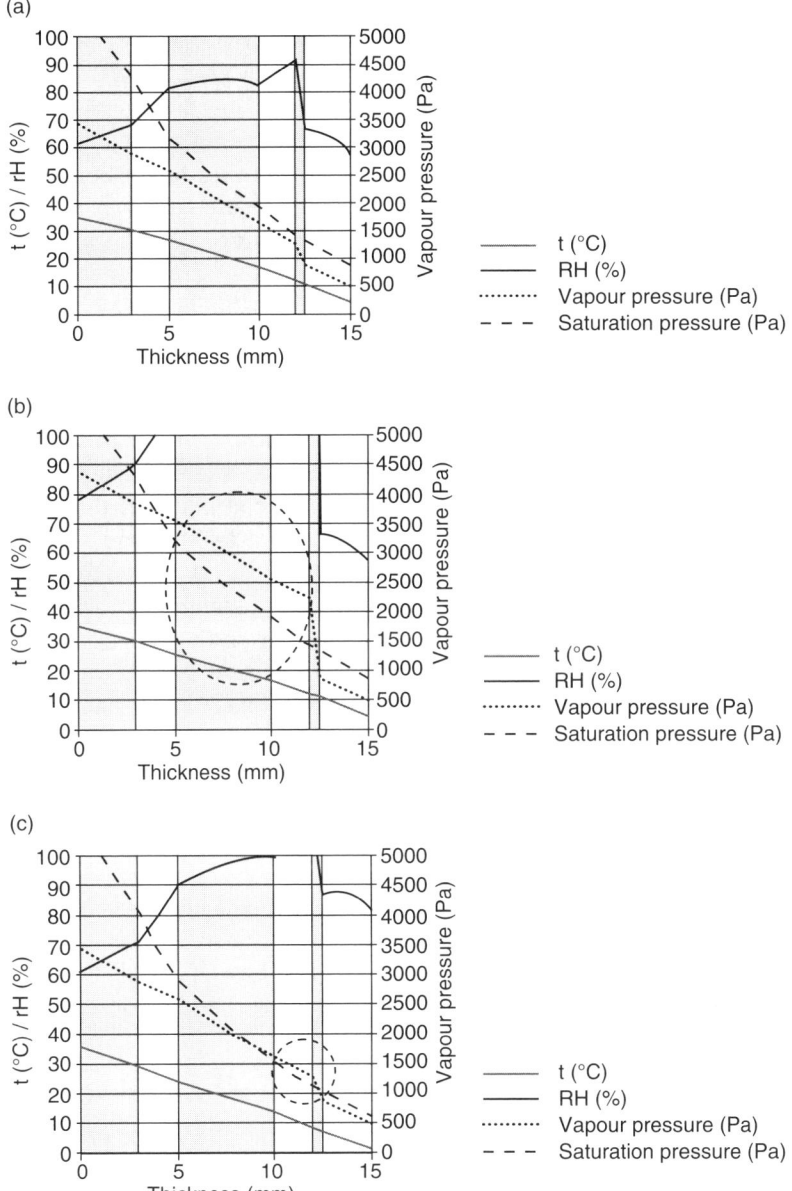

7.4 Cross-sections through a clothing ensemble consisting of a base-layer, a fleece and an outer waterproof layer. The lines show relative humidity, temperature, saturated water vapour pressure and the actual water vapour pressure. The ambient temperature for (a) and (b) is 5 °C and for (c) is 0 °C. The evaporative resistance for the outer is 6 for (a) and (c) and 20 m^2 Pa W^{-1} for (b). The sweat rate for all is 100 g m^{-2} h^{-1}. Skin temperature is 35°C. The dotted circles indicate condensation, when the actual vapour pressure required to drive the sweat rate evaporation is greater than the saturated vapour pressure. Condensation is also indicated when the relative humidity line is greater than 100%.

and mitts is how to maintain dexterity and keep the hands warm, or at least not incur an injury. In reality, it is very difficult to substantially insulate the fingers, as explained by van Dilla *et al.* (1949). Being small cylinders, the increase of thickness is opposed by the increase in surface area available for heat transfer. At very cold temperatures it is the cold-ness of the hands and feet which is the limiting factor for endurance. According to Candler and Freedman (2001), insulated gloves are adequate above −12 °C when reasonably active; trigger finger mittens are with wool inserts from −12 °C to −29 °C with arctic mittens are needed below this. Below −18 °C, thin synthetic contact gloves should always be worn to prevent flesh freezing. Materials are those suggested previously for cloth-ing insulation – wool and polyester pile fabrics with a windproof, water vapour permeable shell.

Heated footwear and gloves – auxiliary electrically heated gloves and footwear has been investigated, especially for aircrew, since the Second World War (Siple, 1949), and research has continued in the mean time (Goldman, 1964; Hickey *et al.*, 1993; House *et al.*, 2003). There is no insulated glove that can keep the hands warm when sitting at −40 °C for more than about 2 hours, and heating is the only viable solution to achieve longer duration. Although the power requirements are relatively low, in the order of 20 W in total for hands and feet, this would need a substantial weight of standard batteries to be carried. Although compact battery packs have been developed for mobile phones, etc., they require a charging unit which is not practical for general use in the field. Overheating the feet can also upset the thermoregulation of the rest of the body, with the very unpleasant feeling of simultaneous sweating and shivering being produced (Burton, 1963).

Hats – as it gets colder, hoods need to be built into jackets. Balaclavas offer the best protection since they can be worn with just the eyes exposed, on the neck or head to adjust heat loss. The best materials are those sug-gested for general clothing insulation – polyester and wool.

7.8   Sources of further information and advice

Two books, although published over 50 years ago, contain some of the most useful information about clothing in cold environments:

'The Physiology of Heat Regulation and the Science of Clothing' edited by Newburgh (1949) and 'Man in a Cold Environment' by Burton and Edholm (1955).
Another good source is 'Human Performance Physiology and Environmental Medicine at Terrestrial Extremes' edited by Pandolf *et al.* (1988).

Proceedings of Environmental Ergonomics Conferences, held approximately every 2 years since 1988, are a good up-to-date source of current research in this area.

GORE-TEX® and WINDSTOPPER® are registered trademarks of W L Gore and Associates.

7.9   References

ACCP 1 (1992), *Heat transfer and physiological evaluation of clothing*, NATO, Brussels.

ADOLPH E and MOLNAR G (1946), 'Exchanges of heat and tolerances to cold in men exposed to outdoor weather', *American Journal of Physiology* **146**, 507–537.

BARTELS V and UMBACH K (2002), 'Water vapour transport through protective textiles at low temperatures', *Textile Research Journal* **72** (10) 899–905.

BARTELS V and UMBACH K (2001), 'Skin Sensorial Wear Comfort of Sportswear', *40th International Man-made Fibres Congress*, Dornbirn, Austria, 19–21 September.

BURTON A C (1963), 'The Pattern of Response to Cold in Animals and the Evolution of Homeothermy' in Hardy J D, *Temperature – its Measurement and Control in Science and Industry* Vol. 3 Part 3, New York, Reinhold, 363–371.

BURTON A C and EDHOLM O G (1955), *Man in a Cold Environment*, Hafner, New York.

CANDLER W and IVEY H (1997), 'Cold weather injuries among US soldiers in Alaska: A five year review', *Military Medicine* **162** (12) 788–791.

CANDLER W and FREEDMAN M (2001), 'Military Medical Operations in Cold Environments', in *Medical Aspects of Harsh Environments* Volume 1, Eds. Pandolf K and Burr R, Office of the Surgeon General, U.S. Army, Virginia.

DEF STAN 0035, Part 4 (2006), *Environmental Handbook for Defence Matériel – Natural Environments*, MOD, UK.

DEGROOT D, CASTELLANI J, WILLIAMS J and AMOROSO P (2003), 'Epidemiology of U.S. Army cold weather injuries, 1980–1999', *Aviation Space and Environmental Medicine* **74** (5) 564–570.

DORMAN L (2007), *The effects of protective clothing and its properties on energy consumption during different activities*, Thesis (PhD), Loughborough University, UK.

EVANS M (2007), 'How fears of trench foot led Falklands leaders to take risks', *The Times*, London, August 27th.

FIALA D (1998), *Dynamic Simulation of Human Heat Transfer and Thermal Comfort*, Thesis (PhD), De Montfort University, Leicester, UK.

FIENNES R (1993), *Mind over Matter – The Epic Crossing of the Antarctic Continent*, Sinclair Stevenson, London.

FORBES W (1949), 'Laboratory and Field Studies: General Principles', in *Physiology of Heat Regulation and the Science of Clothing*, Ed Newburgh M, Saunders, Philadelphia.

FRIM J (1995), *Physiological Evaluation of the Improved Environmental Clothing System (IECS): The New Canadian Forces Cold Weather Clothing System*. Defence and Civil Institute of Environmental Medicine, Canada. DCIEM No 95–18.

FROESE G and BURTON A (1957), 'Heat losses from the human head', *Journal Applied Physiology* **10**, 235–241.

GIVONI B and GOLDMAN R (1972), 'Predicting rectal temperature response to work, environment and clothing', *Journal of Applied Physiology* **32** (6), 812–822.

GOLDMAN R F (1964), 'The Arctic soldier: Possible research solutions for his protection', in *Proceedings of the 15th Alaskan Science America Association for the Advancement of Science*, 114–135.

GOLDMAN R F (2006), 'Thermal Manikins, Their Origins and Role', *in Thermal Manikins and Modelling* (Ed Fan J), Hong Kong, Hong Kong Polytechnic Library.

HARDY J (1970), 'Thermal Comfort: Skin temperature and Physiological Thermoregulation', in *Physiological and Behavioural Temperature Regulation*, Eds. Hardy J, Gagge A and Stolwijk J, Charles C Thomas, Springfield, Illinois. 856–873.

HAWRYLUK O (1977), 'Why Johnny Can't March: Cold Injuries and Other Ills on Peacetime Manoeuvres', *Military Medicine* **142** (5), 377–379.

HAYES P (1989), 'A physiological basis of cold protection', in *Thermal Physiology*, 1989, Ed. Mercer J, Amsterdam, Excerpta Medica.

HES L (2002), 'Recent developments in the field of user friendly testing of mechanical and comfort properties of textile fabrics and garment', *World Congress of the Textile Institute*, Cairo.

HICKEY C A, WOODWARD A A and HANLON W E (1993), *A Pilot Study to Determine the Thermal Protective Capability of Electrically Heated Clothing and Boot Inserts.* U.S. Army Research Laboratory, Aberdeen Proving Ground, Report AD-A276 511.

HOUSE C M, LLOYD K and HOUSE J R (2003), 'Heated socks maintain toe temperature but not always skin blood flow as mean skin temperature falls', *Aviat. Space Environ Med* **74** (8), 891–893.

IREQalfa (2002), available at http://wwwold.eat.lth.se/Forskning/Termisk/Termisk_HP/Klimatfiler/IREQ2002alfa.htm Accessed December 2006.

ISO 8996 (1989), *Determination of metabolic Heat Production*, ISO, Geneva.

ISO 9237 (1995), *Determination of the Permeability of Fabrics to Air*, ISO, Geneva.

ISO/DIS 11079 (2005), *Determination and interpretation of cold stress when using required clothing insulation (IREQ) and local cooling effects*, ISO, Geneva.

ISO 11092 (1993), *Textiles – Physiological effects – Measurement of thermal and water vapour resistance under steady state conditions (sweating guarded hotplate test)*, ISO, Geneva.

KLETS (2007), *Natural vs. synthetic base layers*, accessed at http://www.klets.co.uk/test_reports.htm, January 2007.

LEE C K (1984), *Heat Transfer of Fibrous Insulation Batting*, Aero-Mechanical 1 Laboratory, NATICK, AD-A161 032.

LEVINE L, SAWKA M N, GONZALEZ R R (1995), *General procedures for clothing evaluations relative to heat stress*, USARIEM TN 95-5.

LLOYD E L (1994), 'ABC of Sports Medicine: Temperature and Performance 1: Cold', *BMJ* **309**, 531–534.

LOTENS W A (1993), *Heat transfer from humans wearing clothing*, Delft University, PhD thesis.

LOTENS W A (1988), 'Optimal Design Principles for Clothing Systems', Chapter 17 in *Handbook on Clothing: Biomedical Effects of Military Protective Clothing*, Ed. Vanggaard L, NATO, Brussels.

MCLEAN A D (2001), 'Burns and Military Clothing', *Journal Royal Army Medical Corps* **147**, 97–106.

MCRP 3-35.1A (2000), *Small Unit Leader's Guide to Cold Weather Operations*, U.S. Marine Corps.

MEKJAVIC J, KOCJAN N, VRHOVEC M, GOLJA P, HOUSE C and EIKEN O (2005), 'Foot Temperatures and Toe Blood Flow during a 12 km Winter Hike and Guard Duty', in *Prevention of Cold Injuries, Proceedings* RTO-MP-HFM-126.

NEWBURGH L H (1949), *The Physiology of Heat Regulation and the Science of Clothing*, Philadelphia, Saunders.

NOAKES T (2000), 'Exercise and the cold', *Ergonomics* **43** (10), 1461–1479.

OAKLEY E H (1984), 'The design and function of military footwear: A review following experiences in the South Atlantic', *Ergonomics* **27** (6), 631–637.

O'HEARN B, BENSEL C and POLCYN A (2005), *Biomechanical analyses of body movement and locomotion as affected by clothing and footwear for cold weather climates*, Technical Report NATICK/TR-05/013.

ORR K and and FAINER D (1952), 'Cold injuries in Korea during winter of 1950–51', *Medicine (Baltimore)* **39**, 177–220.

OSCZEVSKI R (1996), 'Water Vapour Transfer through a Hydrophilic Film at Subzero Temperatures', *Textile Research Journal* **66** (1) 24–29.

PANDOLF K, GIVONI B and GOLDMAN R (1977), 'Predicting energy expenditure with loads while standing or walking very slowly', *Journal Applied Physiology* **43** (4), 577–581.

PANDOLF K, HAISMAN M and GOLDMAN R (1976), 'Metabolic energy expenditure and terrain coefficients for walking on snow', *Ergonomics* **19** (6), 683–690.

PANDOLF K, SAWKA M and GONZALEZ R (1988), *Human Performance Physiology and Environmental Medicine at Terrestrial Extremes*, Benchmark Press, Indianapolis.

PIMENTAL N (1991), *Physiological Acceptance Criteria for Cold Weather Clothing*, Navy Clothing and Textile Research Facility, Natick. ADA 235670.

RINTAMAKI H, HASSI J, OKSA J and MAKINEN (1992), 'Rewarming of feet by lower and upper body exercise', *European Journal of Applied Physiology* **65**, 427–432.

RISSANEN S, HORI-YAMAGISHI M, TOKURA H, TOCHINARA Y, OHNAKA T and TSUZUKI H (1994), 'Thermal Responses Affected by Different Underwear Materials during Light Exercise and Rest in Cold', *Ann. Physiol. Anthrop.* **13** (3), 129–136.

SAWICKI J, GOLDMAN R and FINN E (1992), *Development of a Modular Antarctic Clothing System: Phase II*, Nokobetef IV, Kittila, Finland.

SCHISSEL D, KELLER R and BARNEY D (1998), 'Cold weather injuries in an Arctic environment', *Military Medicine* **163** (8), 568–571.

SCHNEIDER A and HOSCHKE B (1992), 'Heat transfer through Moist Fabrics', *Textile Research Journal* **63** (2), 61–66.

SIPLE A P (1949), 'Clothing and Climate' in *Physiology of Heat Regulation and the Science of Clothing* (Ed Newburgh). Philadelphia, Saunders, 389–442.

SMEDSTAD N (1995), 'Waterproof breathable fabrics for military clothing systems: An innovative approach to acquisition', in *Proceedings of Cold Weather Military Operations*, Ed Collins N, Burlington, Vermont, 243–264.

STANAG 4364 (2003), *Waterproof Clothing*, Edition 2, NATO, Brussels.

STEFANSSON V (1944), *Arctic Manual*, Macmillan, New York.

STOLWIJK J A J and HARDY J D (1966), 'Temperature regulation in man – a theoretical study', *Plugers Arch* **291**, 129–162.

STROUD M (1998), *Survival of the Fittest*, Jonathan Cape, London.

TAYLOR M (1992), 'Cold weather injuries during peacetime military training', *Military Medicine* **157** (11), 602–604.

TEITLEBAUM A and GOLDMAN R (1972), 'Increased energy cost with multiple clothing layers', *Journal of Applied Physiology*, (6), 743–744.

TEK D and MACKEY S (1993), 'Non freezing cold injury in a marine infantry battalion', *Journal of Wilderness Medicine* **4**, 353–357.

UGLENE W (2000), 'Effect of fabric composition on amount of energy required for drying moisture', *Proceedings of Environmental Ergonomics IX*, Eds. Werner J and Hexamer M, Aachen, Shaker Verlag.

UMBACH K (1984), *Universelle Beschreibung des Tragekomforts in Abehangigkeit von Kleidung und den Darin verwendeten Textilien Sowie variablen Randbedingungen von Klima und Arbeit*, AIF Report Number 4287, Hohenstein Institute, Germany.

UMBACH K (1986), 'Funktionelle Wettrschutzkleidung mit guten bekleidungphysiologischen Trageeigenschaften', *Melliand Textilberichte* **67**, 277–287.

UMBACH K (1987), 'Bekleidungsphysiologische Trageeigenschaften von wasserdichten, jedoch wasserdampfdurchlassigen Vliesstoff-Membran-Konstruktionen', *INDEX 87 Congress*, Geneva.

VAN DILLA M, DAY R and SIPLE P (1949), 'Special Problem of Hands', in *Physiology of Heat Regulation and the Science of Clothing*, Ed. Newburgh M, Saunders, Philadelphia.

VANGGAARD L (1975), 'Physiological Reactions to Wet–Cold', *Aviation Space Environmental Medicine* **46** (1), 33–36.

Weather Underground is available at http://www.wunderground.com/, accessed Feb 2007.

WERNSTEDT F (1972), *World Climatic Data*, Climatic Data Press, Lemont, PA.

WHAYNE T (1955), 'Clothing', Chapter 3 in *Preventive Medicine in World War II, Volume 3, Personal Health Measures and Immunization*, Office of the Surgeon General Department of the Army, Washington DC.

WHAYNE T and DEBAKEY M (1958), *Cold Injury Ground Type*, Office of the Surgeon General, Department of the Army, Washington DC.

WINSLOW C E and HERRINGTON L P (1949), *Temperature and Human Life*, Princeton University Press, Princeton, New Jersey.

XU X, GIESBRICHT G and GONZALEZ R (2003), *Real time thermoregulatory model for extreme cold stress: Applicable to Objective Force Warrior*, USARIEM, Natick.

8
Designing military uniforms with high-tech materials

C. A. GOMES, Foster-Miller, Inc., USA

8.1   Introduction

Materials of yesterday, today and tomorrow are used or will be used in military uniforms and in first responder uniforms to protect those who protect us. Materials of yesterday can be combined with a technology in a new way to create a high-tech material used today and tomorrow. For instance, cotton is not thought of as being a high-tech material; however, a reactive agent can be added to cotton to make it self-detoxifying for chemical and biological protection, hence, creating a high-tech cotton material.

Ideally, military uniforms will become lighter, better and cheaper as technology improves and high-tech materials are incorporated into uniforms. A military uniform needs to be lightweight and flexible to not hinder performance nor interfere with the execution of a mission. It needs to protect the soldier against known, as well as unknown, hazards; therefore, the uniform needs to be better than it is today. Unfortunately, protection may come at a cost. Take body armor as an example: side hard armor plates may increase protection of the soldier from a ballistic impact, but it may also hinder mobility or cause increased heat stress so a soldier cannot perform his/her duty.

Unfortunately, it is a reality that compromises take place. However, with new materials and uniform designs, those compromises can be minimized without sacrificing the safety of the soldier. Some high-tech materials and components that are being used or could be used to improve military uniforms, include: 'wear and forget' physiological status monitoring to provide remote health status for the soldier, flame-resistant fleece for flame resistance and thermal insulation, self-decontaminating fibers for chemical and biological protection, shear thickening fluids impregnated into fabrics to improve ballistic performance, and many more.

8.2   Design process

Understanding types of hazards or threats is the critical first step in developing systems for protection and enhancing safety[1] of the soldier, first responder, or even the individual. Table 8.1 lists examples of common hazards or threats, their effects, and possible protection mechanisms which could be designed into the soldier's uniform. A military uniform is a very complex system, so a systems approach is used for the design process, material selection, and verification testing. The approach starts with understanding the user needs (including hazard/threat identification), defining product subsystems, determining requirements, evaluation of alternatives for material selection, and ultimately verification of the design, as well as the individual materials. When asking a user what they need in a uniform, it is also important to ask them to prioritize their needs, as well as to identify the activities they perform. The inherent problem with asking opinions though, is that if five people are asked the same question, there can easily be ten opinions.

Military uniforms need to protect the user from many hazards as well as provide day-to-day comfort while working a desk job. Therefore, in

Table 8.1 Examples of hazards/threats, their effects, and protective mechanisms

Hazard or threat	Effect	Protection mechanism
High environmental temperature	Heat stress (e.g. heat stroke, heat exhaustion)	Passive or active thermal management
Cold environmental temperature	Cold stress (e.g. frostbite, hypothermia)	Passive or active thermal management
Identify location of soldier	Detection	Signature management, camouflage: visual thermal, olfactory, and auditory
Chemical (e.g. nerve, blister, and blood choking agents)	Damage through skin, eyes, and inhalation	Adsorption or barrier materials
Biological (e.g. pathogens and toxins)	Damage through skin and inhalation	Barrier materials
Flame resistance	Melting of material to skin or burns	Inherently flame resistant or no melt no drip fibers
Fragmentation	Ballistic impact, blunt force trauma	Ballistic plates and high modulus fibers
Ballistic	Ballistic impact, blunt force trauma	Ballistic plates and high modulus fibers

determining performance requirements for the uniform, it is also necessary to take into account normal use conditions, even care, such as laundering.

8.3   Features of military uniforms

This chapter will focus on high-tech materials used for protection versus day-to-day normal wear conditions. The features discussed in this chapter are:

- Physiological monitoring
- Thermal management
  - Hot and cold environments
  - Passive and active management
- Signature management
  - Visual
  - Infrared
  - Olfactory
  - Auditory
- Chemical and biological
- Flame resistance
- Environmental
- Body armor
  - Soft armor
  - Hard armor

8.4   Physiological monitoring

Is a soldier ready for battle? Are they hurt? In the future, an integral part of the soldier's uniform might be a 'wear-and-forget' physiological status monitoring (PSM) system to provide both medics and commanders with remote health status for improved medical treatment and planning for combat casualty care. Remote physiological monitoring could determine the readiness level of a soldier: whether they are well rested or not, are adequately hydrated or dehydrated, or which soldier needs to be rescued first on the battlefield.

An ambulatory PSM garment using electrotextile materials needs to be unobtrusive, lightweight, and comfortable. One PSM system being evaluated by the United States (US) Army is pictured in Figure 8.1 and was developed by Foster-Miller, Inc. (USA). This baseline system monitors heart rate, respiration rate, body posture, activity level, and skin temperature, then transmits the data wirelessly to a remote display. The electronic-textile shirt platform is scalable and can be upgraded with additional sensors

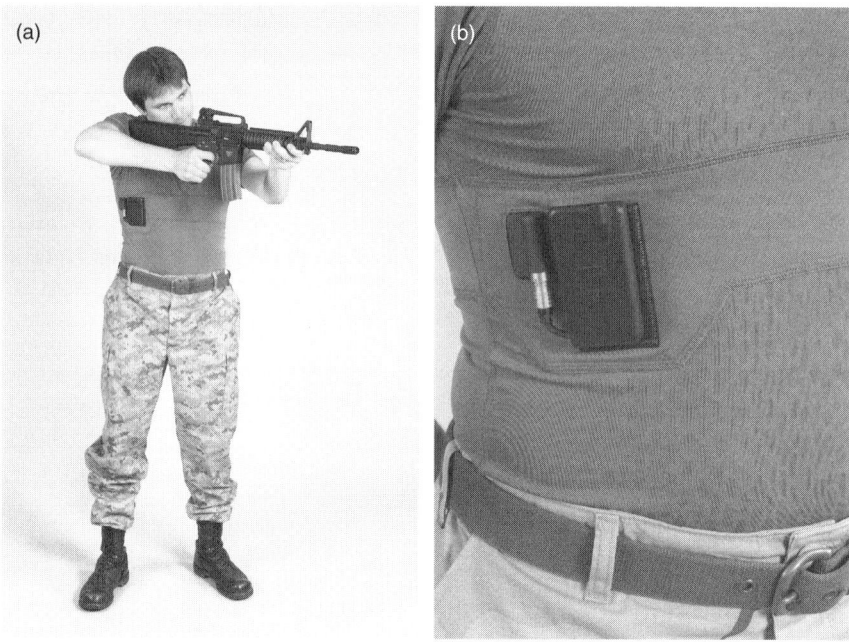

8.1 (a) 'Wear-and-Forget' physiological monitoring shirt with (b) detail. (Photograph from Foster-Miller, Inc.)

such as ballistic impact detection and full digital electrocardiogram. The shirt is also designed to be worn under body armor.[2]

The next-to-skin base fabric for the PSM shirt is a Polartec® fabric enhanced with Body Mapping™ which was engineered by Polartec, LLC (USA) to provide custom performance where it is needed. This material will be discussed more fully in the next section, on thermal management.

8.5   Thermal management

A thermal stress is categorized as acceptable when a person is able to compensate temperature changes without undue strain on their body. However, if their body is unable to compensate and incurs severe strain, either heat or cold injuries can occur. Heat transfer by radiation and convection between a person and their environment can result in a positive or negative heat balance; if the environment is warmer than the person, a positive heat balance (toward the subject) occurs and conversely if the person is warmer than the environment, a negative heat balance (toward the environment) occurs.[3] Therefore, a uniform needs to manage hot and cold environments without hindering performance or interfering with the execution of a mission. Designing the uniform to function within the wide spectrum of the operational environment, e.g. from desert to arctic conditions, is further complicated by donning body armor or the need for protection from chemical or biological hazards. Both of these protective clothing elements typically restrict air flow or moisture movement for evaporative cooling to occur.

Thermal management is typically divided into two methods, passive and active. Active thermal management usually requires power; therefore, passive thermal management is advantageous because no power source is required. Active thermal management could supplement passive, thereby minimizing weight, size, and power requirements on the system for both cooling and heating scenarios.

8.5.1 Passive thermal management

Hot environments

Evaporative cooling is a key in a hot environment. The ideal fabric worn next to the body allows sweat from the body to be wicked to the outside of the fabric where it spreads rapidly over the surface resulting in evaporative cooling. Historically, the US military used polypropylene fabrics as the next-to-skin layer. Currently the US military uses Polartec® Power Dry® as a next-to-skin material because it is manufactured using a bi-component construction that increases wicking of sweat from the skin and moves at least 30% more moisture away from the skin than single component fabrics.[4] The fiber content of the Polartec® Power Dry® varies depending on the style.

As previously noted, the PSM shirt is manufactured from a Polartec® fabric enhanced with Body Mapping™. This material offers multiple performance zones that are mapped to the body's needs. These zones perform the following functions:

- Increase comfort and mobility,
- Eliminate excess bulk,
- Provide cushioning and chafing protection, and
- Allow for breathability and cooling.

This technology was originally designed by the US military for integration with body armor so the combination of activity-specific pile requirements with the underlying principles of moisture management, offers state-of-the art fabric performance for any activity or environment. The key is also the seamless transition between functional zones, which reduces seam bulk.[2] The different pile heights offer:

- No pile: highest cooling efficiency, enhance breathability, and low bulk.
- Low density, low pile pillars and no pile channels: protection from chaffing, enhanced breathability, faster heat dissipation, and limited bulk.
- High density, low pile pillars and no pile channels: cushioning for carrying heavy loads (e.g. body armor, back packs), enhanced breathability, faster heat dissipation, and limited bulk.

Cold environments

Insulation is important for a protective clothing system in a cold environment. The objective is to trap warm air around the body. A fabric with low weight, low volume, and that is highly insulative would be ideal. Classic insulation is based upon fiber batting technology which creates a thick product, but it is not state-of-the-art.

The trend in military protective clothing is for a multi-layering system, such as the US Army's new Generation III Extended Cold Weather Clothing System (ECWCS) seven-layer, 12-component system. Based upon layering systems utilized by mountaineering professionals, the Generation III ECWCS system uses the latest textile science to keep soldiers comfortable, dry, and warm in the most inclement conditions. The seven layers of clothing include: light-weight and mid-weight Polartec® Power Dry® moisture wicking shirt and briefs; a Polartec® Thermal Pro® fleece jacket; a nylon/spandex wind jacket; a soft shell jacket and trousers using Nextec® fabric (further discussed below and in the Environmental section); an extended polytetra-fluoroethylene (e-PTFE) membrane, Gore-Tex® wet weather jacket and trousers; and a Primaloft® insulated loft parka and trousers for extreme cold weather conditions. Each piece fits and functions either alone or together as a system to maximize the options available to the soldier.[5] The system protects troops from temperature extremes, ranging from –51 °C (–60 °F) to 4 °C (40 °F).[6]

Another clothing system is the Multi-Climate Protection System (MCPS) developed by the Naval Air Warfare Center (USA). The flight-approved MCPS was developed to provide insulation, moisture management, flame resistance, and wind/water protection using state-of-the-art materials. The multiple insulating layers provide flexibility and versatility, allowing the user to customize the insulation to their needs by adding or removing individual layers to achieve a personal level of comfort over a broad range of conditions. The system is comprised of twelve garments, seven of which have been developed using three flame-resistant Polartec® fabrics, based upon the meta-aramid NOMEX® fiber, engineered specifically for aviator and aircrew protection.[2]

The MCPS components are:

- Silk-weight (light-weight) (NOMEX® blend) underwear, shirt and pants;
- Mid-weight insulation (Polartec® Power Stretch®-FR with NOMEX®) shirt and pants;
- Heavy-weight insulation (200 Polartec® Thermal-FR with NOMEX®) shirt, pants, and bib overalls;
- Fleece (300 Polartec® Wind Pro®-FR with NOMEX®) jacket and vest;
- Shell (GORE-TEX®) jacket and trousers;
- Face mask.

The Polartec® materials offer superior moisture management by wicking sweat away from the body, spreading it out over the exterior surface of the fabric, where it then evaporates. The fleece in the jacket and vest is two times more wind resistant than traditional fleece, and protects the inner layers of the system from wind and water, thus maintaining the system's efficacy. The fabric has a surface of tightly constructed yarns, providing a wind-resistant garment and, consequently, greatly reducing the harmful effects of wind chill. In high wind conditions, wind chill can greatly reduce the thermal properties of the insulating layers; however, the 300 Polartec® Wind Pro® fabric can help prevent this. In addition, this fabric surface has durable water repellency, allowing moisture to bead off it rather than be absorbed into the MCPS system which would reduce its insulation characteristics.[2] The outermost layers for the MCPS are a GORE-TEX® shell, jacket and trousers, worn over the entire system for complete protection. The shell fabric is made from a material composed of three layers: a NOMEX® N303 face, GORE-TEX® membrane, and lined with a NOMEX® jersey knit. This system provides flame protection and insulation while maintaining a breathable characteristic to prevent over-heating. In addition, it protects against penetration by chemicals such as JP-8 jet fuel.[2]

Insulation performance of a clothing system is based upon a unit called the CLO. The higher the value the more insulative the system. As a system (silk-weight NOMEX® underwear set, mid-weight insulation set, 27/P Nomex® flight coverall, heavy-weight fleece trousers, heavy-weight fleece jacket, and GORE-TEX® set), the CLO value is approximately 2.2 based upon an evaluation using a thermal manikin in a 15-knot wind: A CLO value of 0.5 is given by just the silk-weight NOMEX® underwear set and the 27/P NOMEX® flight coverall.[7]

Another uniform system worn by US Special Operations Forces (SOF) for warfare in cold weather is the Protective Combat Uniform (PCU) which is comprised of 16 garments, including pants, jackets, windshirts, and vests. The system is designed to withstand rain, snow and high winds in temperatures ranging from −43 °C (−45 °F) to 18 °C (65 °F).[8] The outer fabric of the PCU is EPIC from Nextec® by Nextec Applications, Inc. (USA). The fabric is water resistant, windproof, breathable, and washable. Unlike GORE-TEX®, the EPIC fabric is not a laminate nor is it coated. The fabric is finished using a unique encapsulation technology where a thin polymer layer surrounds the individual fibers creating a tough and durable barrier against harsh environments[8,9] (This is further discussed later in the Environmental section.)

8.5.2 Active thermal management

Hot environments

The US Army Natick Soldier Center developed the Air Warrior Microclimate Cooling Garment (MCG) used onboard certain rotary wing aircraft and armored vehicles. Figure 8.2 is a picture of the Air Warrior MCG and the microclimate cooling unit (MCU). The MCG is a lightweight, comfortable, breathable, tube-type undergarment worn against the skin. The tubing is between two layers of 100% cotton fabric and is machine launderable. The MCU circulates a coolant fluid to remove up to 180 watts of body heat from the torso.[10]

Although other areas of US military are evaluating active cooling, the Air Warrior MCG is the only flight-approved active cooling system.

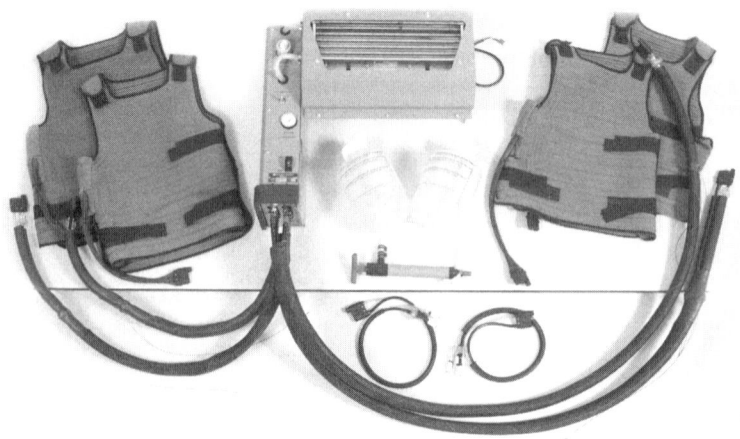

8.2 Combat Cool™ using technology from Air Warrior MCG. (Photograph from Foster-Miller, Inc.)

Cold environments

Electric resistive heating is one technology that can be used to provide heat in clothing. The mechanism is simple – voltage is applied to a series of conductive yarns or wires, which resist the flow of current. As current passes through them, heat is produced. The amount of heat produced can be directly and easily controlled by varying the voltage to the resistive network.[11] The US military has not yet adopted active heating in its uniforms to supplement passive thermal management; however, it is being evaluated for the future and numerous commercial manufacturers produce heated vests, jackets, gloves, and boots.

8.6   Signature management

Soldiers do not want to give their presence away to the enemy and therefore need to manage their 'signature'. They need to determine how the enemy 'sees' them, and then mask all the elements that make the soldier stand out. Signature management can typically be divided into several categories: visual, infrared (e.g. thermal), olfactory, and auditory.

8.6.1 Visual

Camouflage patterns are supposed to make the soldier blend into the background. Conventional camouflage fabric has two basic elements to help conceal a person: color and pattern. Since nature changes color on a regular basis, the soldier must match the environment as closely as possible, even resembling vegetation, buildings, etc. The environment in which the soldier needs to fight will dictate the predominate color of the uniform. It will be difficult to have the same camouflage coloring for all conditions.[11]

The ultimate goal is to provide the least amount of contrast between the soldiers and their background when viewed by the enemy, so color and pattern matching are utilized. Too large a pattern for a small background provides contrast, which is bad. On the other hand, too small a pattern for a large background also imparts contrast. For urban backgrounds, the pattern needs to have more straight, vertical and horizontal designs, so it blends with homes, buildings, and other urban structures.[11]

Numerous examples for camouflage in nature exist. A striped zebra blends in with tall grass, hence the camouflage is based upon pattern. On the other hand, all penguins are camouflaged using color – penguins have a white underside so they hide from predators in the water looking up from below, and a dark color on their backs to camouflage them from above.[12] A chameleon is a good example of camouflage that changes color with its environment.

For decades, US military branches wore the same type of camouflage uniforms until the Marines released a digital camouflage pattern, MARPAT, in 2001. Up close, the digital pattern resembles computer pixels; however, from a distance, it blends into the background faster than the current design.[13] The Canadian military is also using a version of a digital camouflage pattern called Canadian Disruptive Pattern for temperate woodland or CADPATTM (TW).[14]

The US Army released their universal camouflage pixelated pattern as part of their newly designed Army Combat Uniform (ACU) in 2004. Unlike the old camouflage, shown in Fig. 8.3(a), digital camouflage, shown in Fig. 8.3(b), suggests shapes and colors without actually being shapes and colors – like visual white noise.[15] There are two features that distinguish the Army's ACU from the Marines' MARPAT. First, the Marines use several camouflage schemes for different environments whereas the ACU is a universal pattern capable of blending into several environments. Second, the Army ACU does not use the color black as black is not found in nature.[15]

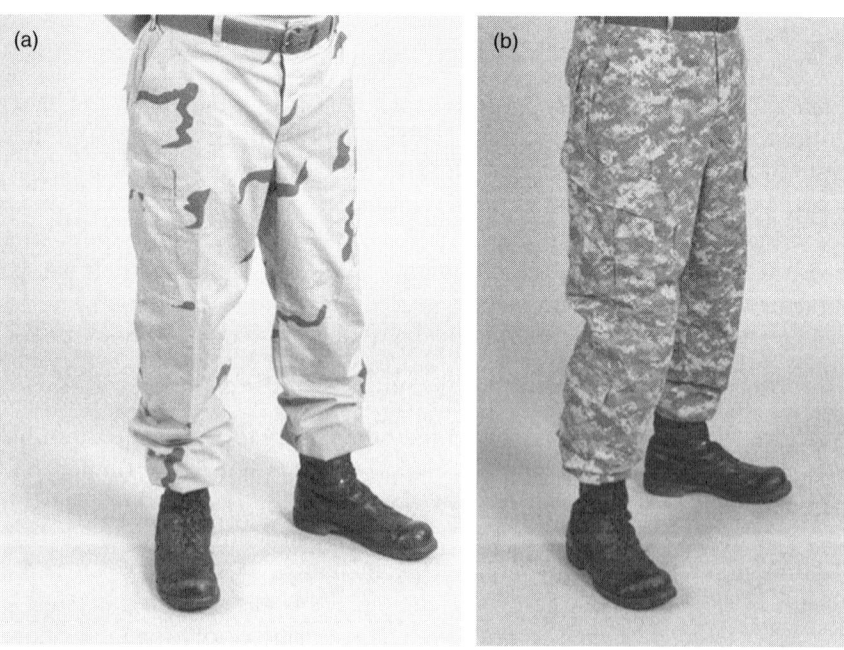

8.3 Old and new US Army camouflage uniform patterns. (a) old desert camouflage pattern and (b) new universal digital camouflage pattern. (Photograph from Foster-Miller, Inc.).

8.6.2 Infrared

Infrared (IR) imaging technology is used to measure the temperature of an object. All objects emit electromagnetic radiation, primarily in the IR wavelength, which cannot be seen by the naked eye. However, IR radiation can be felt as heat on one's skin. The hotter the object, the more IR radiation it emits.[8] IR can also penetrate smoke and fog better than visible light, revealing objects or people that are normally obscured.[16]

Thermal signature is one of the hardest signatures to manage because of the contrast of the human body against the environment; a hot body in a cold climate will be 'seen' as well as a cold body in a hot environment.[8] Special thermal imaging cameras can be used to detect hidden soldiers because of the contrast. However, Texplorer® has developed a special garment called Ghost® to reduce IR emissions and thus lower radiant temperature. The fabric incorporates metallized fibers and is available in several camouflage prints for visual concealment.[17]

Night vision devices also use the IR spectrum; however, they use near-IR (NIR), which is closest to visible light. The main difference is that thermal-IR is emitted by an object instead of reflected as is the case with NIR. The main mechanism to manage NIR is to use proper dyestuff selection for the fabric.[18] The color needs to mimic the spectral reflectance of various environments, like vegetation where the top of a leaf reflects light differently from the bottom of a leaf.

8.6.3 Olfactory

Another target indicator to locate the soldier is using the sense of smell. The olfactory target indicator will not identify exactly where the soldier is, only that they are in the area. Human odor originates from bacteria living on the skin. These bacteria then expel waste in the form of an odorous gas. Therefore, fabrics with antimicrobial properties can be used to reduce the body odor of the soldier, as well as to improve his/her health.

Commercial antimicrobial technologies reduce the presence of bacteria by different mechanisms. Some disrupt the bacterial membrane and others interfere with the metabolic bacteria function, destroying the organism.[19]

Metallic compounds based upon metals such as silver and copper have also been shown to exhibit antimicrobial behavior. X-Static® is a silver-coated textile fiber that provides permanent anti-odor and antimicrobial properties. X-Static® is a product of Noble Fiber Technologies, LLC (USA). X-Static® is currently used in US military socks and T-shirts. Using X-Static® offers more benefits than just odor elimination, it can be used for: temperature regulation so the wearer stays warmer in the winter and cooler in the summer; instantaneous static reduction to protect the wearer as well as electrical equipment; and for wound care.[20] X-Static® is also conductive and can be used to transfer data across a fabric matrix.

Other chemistries being evaluated for US military application as anti-microbials include: triclosan, silver zeolites (encapsulated), copper oxide, chitosan and quaternary ammonium silanes.[19]

8.6.4 Auditory

A soldier can be found by using the sense of hearing, so it is important to reduce the auditory (sound) signature of the soldier. Unnatural sounds will give away the presence of a soldier to the enemy. Metal-on-metal sounds (like a traditional snap) are unnatural so it might be preferable to use a plastic snap or a button. Hook and loop fasteners such as Velcro® also produce an unnatural sound; again, it might be preferable to use a button instead. Laminated fabrics have a crisp 'swishing' sound, so fabrics with a finish might be used instead.

8.7   Chemical and biological defense management

A chemical and biological defense (CBD) system needs to protect against chemical liquids, vapors, and bioaerosols. A breach in this system could result in serious injury or loss of life. As previously noted, there is a classic tradeoff between protection and function, and this is no different in a CBD system. Typically, the higher the protection level of the CBD system, the lower the comfort and breathability.

Classic CBD materials are based upon butyl rubber and activated carbon. Butyl rubber is a chemical-resistant impermeable elastomer; however, since butyl rubber is impermeable, it does not breathe and a soldier could overheat.

Because of extremely high adsorptive properties, activated carbon is widely used in CBD systems to adsorb chemical vapors as well as odors. Activated carbon has an extremely large surface area, in some cases, in excess of $2000\,m^2/g$. Activated carbon has a high adsorptive capacity for high boiling point gases (e.g. nerve agents), but a very low removal efficiency for low boiling point gases (e.g. blood gases such as cyanogen chloride and hydrogen cyanide) and choking agents such as phosgene and chlorine. Activated carbon impregnated with heavy metals and triethylenediamine can effectively remove both high and low boiling point gases. There are a few disadvantages of activated carbon; they are: (a) the large amount required, (b) the weight, (c) it is hot to wear, and (d) it has a limited life.

A CBD system is typically worn either as a next-to-skin garment or an overgarment. LANX Fabric Systems™ (USA) manufactures the Chemical Protective Undergarment (CPU), and it provides vapor and aerosol protection. The fabric stretches for user comfort, enhances flash fire protection, is good for ten launderings, and is made from a durable composite fabric containing polymerically encapsulated carbon for the adsorption of chemical warfare agents.[21] LANX also manufactures a Chemical Protective Overgarment (CPO) as does the Gentex Corporation (USA). The CPO is a two-piece ensemble consisting of a coat with integral hood plus separate trousers providing liquid and vapor protection. The Gentex tri-layer fabric is manufactured from an outer shell fabric, a sorptive layer, and an inner liner. The adsorbent spheres in the sorptive layer are made from an engineered polymer base.[22]

The Joint Service Lightweight Integrated Suit Technology (JSLIST) CBD protective clothing system is one of the most highly used CBD protective system for the US military. The JSLIST suit is a two-piece overgarment consisting of a coat with an integral hood plus separate trousers. The primary requirements for the JSLIST are that it provides 24 hours of continuous protection from $10 g/m^2$ of liquid chemical warfare agents after a 45-day wear period (720 cumulative hours) and six launderings.[23] The SARATOGA™ JSLIST suit was the only material combination to meet all the program requirements.[24] This suit is manufactured from a two-layer fabric system consisting of a liquid-repellent nylon/cotton outer shell with a SARATOGA™ carbon sphere liner (non-woven, laminated to activated carbon spheres bonded to a knit backing). JSLIST protects against chemical warfare vapors, liquids, and aerosols. It is launderable and also has a 24-hour wear-time after contamination.

The Joint Protective Air Crew Ensemble (JPACE) CBD protective clothing system is in development for all services of the US military. JPACE is a one-piece protective coverall. A key performance parameter for JPACE is that it provides equal or better CB protection than JSLIST.[25] Although JSLIST does not have a requirement for flame resistance, JPACE does.

Chemviron Carbon (Belgium) is the European operation of the Calgon Carbon Corporation (USA) and manufactures activated carbon cloth (ACC), ZORFLEX®-Defense, for a wide range of chemical, biological, and nuclear agent defense applications. The ZORFLEX®-Defense can also be impregnated with numerous chemicals to enhance the adsorption capacity for selected gases.[26]

As the US Department of Defense is searching for revolutionary solutions to CBD, a great deal of research is being conducted and will be discussed further within the Future trends section of this chapter.

8.8   Flame resistance

There is a significant burn risk associated with wearing some synthetic fibers such as polyester or nylon, as they melt and fuse to the skin when exposed to extreme heat and flames. As flame-resistant materials are expensive, a less costly alternative would be fibers that do not melt or drip;[2] examples are cotton, modacrylic (e.g. Protex M™), and wool. Some other commercial flame-resistant fibers are: carbon (polyacrylonitrile, e.g. CarbonX™), melamine (e.g. Basofil®), oxidized acrylic (e.g. Panox®), novoloid (e.g. Kynol®), and polybenzimidazole (PBI). These, as well as other fibers and fiber blends, have been and are being evaluated by the US military. Also included are flame-retardant treated (FRT) cotton, FRT cotton/nylon, FRT lyocell (Tenecel®), and FR rayon.[27]

It is important for these fabrics to block heat transfer in order to reduce burn injury potential. In addition, the integrity of the garment is important as damage can occur from seam splitting, hole formation, shrinkage, etc.[28] Therefore, the key in any protective uniform design is to test the uniform as a system as well as the individual materials. A system test uses an instrumented manikin, as in ASTM F 1930, Standard Test Method for Evaluation of Flame-Resistant Clothing for Protection Against Flash Fire Simulations Using an Instrumented Manikin.

US Army tankers and aviators of all military services are required to wear flame-resistant clothing made from NOMEX® and Kevlar® fiber blends, while the infantry wear a nylon/cotton blend fabric. (The infantry desire flame-resistant materials, but Nomex® and Kevlar® blend fabrics are too expensive to issue to every soldier.[27,29])

8.9   Environmental defense

Soldiers need to be protected from the wind, rain, and snow when they are outdoors. Therefore, ideally they need a garment that is windproof, waterproof, and breathable. Typical technologies for imparting resistance to water and wind include: finish, coating, laminated film, and fiber encapsulation. A durable, water-repellent finish is added to the surface of a fabric and does not typically penetrate the yarn. A solution-coated fabric covers the interstices of the yarns, but it also does not typically penetrate the yarn. A film is laminated on the surface of a fabric, generally using an adhesive with heat. The latest technology to impart environmental protection is a micro-thin coating that encapsulates the fibers of the fabric.

As previously noted, for environmental protection the ECWCS III is a tri-laminate manufactured with an e-PTFE membrane while the PCU utilizes encapsulation technology from Nextec. The EPIC™ fabrics are more supple than the laminated fabrics and typically do not have a 'swishing' sound.

8.10   Body armor

Body armor is arguably the most critical protective clothing item that the soldier has to wear. Armor needs to protect from bullets, fragmentation, knives, armor piercing threats, and more. No armor is bullet proof, it is bullet resistant. The soldier also needs protection from impacts and blunt force trauma. Armor needs to be flexible enough to enable the soldiers to be mobile and to fire their weapons; they must also be comfortable. Modularity is another desired feature for body armor. Typically, when one thinks of body armor, one thinks of a vest that protects the upper torso; however, additional protection is afforded by attachable subcomponents such as throat and groin protectors.

Body armor is typically classified as soft or hard. Depending upon requirements for particular body armor or the mission, different types of body armor are worn.

8.10.1 Soft body armor

When a bullet hits a woven ballistic fabric, thousands of individual fibers engage the projectile to 'catch' it and disperse its energy throughout the area. The points where the fibers cross over each other in the weave absorb the most energy.[11,30] A round penetrator like an awl, ice pick, or nail has an extremely small tip so it does not have to break fibers – it simply pushes them apart and slips through to penetrate the body. A knife poses a unique problem. Once its tip penetrates, the cutting edge slices through loosely woven ballistic fibers, reducing drag, to penetrate the body.[11,30] To reduce or eliminate the impact of such threats, ideal fibers for ballistic protective equipment should be based upon the following: high tensile and compressive modulus, high tensile and compressive strength, high damage tolerance, and low specific weight.[11]

Soft body armor usually has a shell covering the ballistic filler. Depending upon the requirements, some military shells, such as the US Army Air Warrior flexible body armor vest shell, are made using meta-aramid (e.g. NOMEX®) while others may use a rugged nylon shell. Ballistic filler in soft body armor is typically manufactured from fibers such as para-aramid (e.g. Kevlar®, Twaron®) and ultrahigh molecular weight polyethylene (e.g. Spectra®, Dyneema®). Para-aramids such as Kevlar® are five times stronger than steel while Spectra® 2000 is ten times stronger than steel.[31] For soft body armor to be effective, the ballistic filler is manufactured using numerous layers to protect from threats such as fragmentation and 9 mm rounds. Some vest constructions require 28 layers of ballistic filler to withstand the ballistic event.

The US Army Air Warrior flexible body armor vest utilizes a layered ballistic para-aramid system. Interceptor Body Armor (IBA) is also used by the US military and offers increased protection and comfort compared with the Personnel Armor System Ground Troops (PASGT) body armor system. The IBA soft body armor is manufactured with Cordura® as the shell covering, and Kevlar® and/or Twaron® as the ballistic filler. The IBA is modular and, besides a soft body armor vest, the IBA offers arm protection via deltoid (upper arm) and axillary (side arm) protectors, as well as attachable throat and groin protectors.[32]

8.10.2 Hard body armor

To protect the soldier from armor piercing threats, hard body armor is necessary. Lightweight hard armor plates typically consist of a ceramic face supported by a fiber-reinforced composite. Upon impact, the projectile core is fractured by the ceramic, and the ceramic also absorbs a significant portion of the kinetic energy. The lightest ceramic component is manufactured from boron carbide, B_4C. Composite backing materials are characteristically manufactured using para-aramid, ultrahigh molecular weight polyethylene, and fiberglass.[11] The composite backing is needed to contain spalled fragments and absorb energy that may result from a ballistic hit.[32]

As the weaving process can damage high strength yarns, Honeywell (USA) developed Spectra Shield® by lining up the fibers parallel to one another, then binding them together with a resin. A second layer is created in the same manner; however, the pattern is rotated 90 degrees so the fibers lie crosswise. These two layers are fused to form a composite that lines the back of a hard ceramic plate, thus reducing its weight and increasing performance compared with a plate with a para-aramid fiber composite backing.[31] The modular IBA includes front and rear enhanced small arms protective inserts as well as enhanced side ballistic inserts. These hard armor plates provide additional protection to the soft body armor that can withstand multiple small arms hits.[31]

8.11   Future trends

Research into new materials and technologies for use in military uniforms is crucial to the improved safety and success of the soldier. Some critical areas of research are summarized below.

8.11.1 Chemical and biological

Although the JSLIST suit is the current state-of-the-art for the US military CBD system, there is still room for improvement. Areas of particular interest include:

- Protection against a broader range of known and unknown chemical agents, bioaerosols, toxic industrial chemicals/toxic industrial materials (TICs/TIMs);
- Further reduction in weight and heat stress;
- Increased durability; and
- Self-detoxifying materials.[33]

Although being researched, self-detoxifying materials are not currently used in US military uniforms.

An advanced CBD self-decontaminating suit is being jointly developed by Foster-Miller, Inc. (USA) and GE Energy (USA), with support from the Natick Soldier Center, Research, Development & Engineering Command. The general architecture of the suit material is an outer textile layer, an inner reactive membrane, an absorptive layer, and an inner comfort layer. The outer shell is woven from very high surface area reactive Capillary-channeled Polymer (C-CP™) fibers from Specialty & Custom Fibers, LLC (USA); see Fig. 8.4. The fibers are embedded with a catalytic system to endure multiple chemical warfare challenges.

An ideal catalytic system for chemical decontamination should provide full protection against nerve and sulfur mustard agents. For that purpose, organophosphorous hydrolase (OPH), organophosphorous acid anhydrolase (OPAA), and haloalkane dehalogenase (HD) are being used. The inner e-PTFE membrane may be infused with nano-metal oxides and the microporous membrane acts like a sieve for biological agents.[33] The activated carbon is embedded with heavy metals to combat nerve and choking agents.

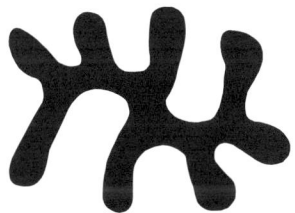

8.4 Very high surface area reactive Capillary-Channeled Polymer (C-CP™) fibers from Specialty & Custom Fibers, LLC (USA).

8.11.2 Flame resistance

TenCate Protective Fabrics, USA, has recently developed Defender™ M, a new inherently flame-resistant fabric that self-extinguishes and will not melt or drip. The composition of the fabric is 65% FR rayon/25% para-aramid/10% nylon printed with universal camouflage or MARPAT designs. The protection offered by Defender™ M is equivalent to more expensive flame-resistant fabrics.[34]

8.11.3 Body armor

One new material on the horizon for body armor is M5®, produced by Magellan Systems International (USA). The M5 fiber is based upon poly {diimidazo pyridinylene (dihydroxy) phenylene} and is estimated to provide a 42% weight reduction for soft body armor over Kevlar KM2®. Also, in the future, M5® will most likely be used as a reinforcing fiber in the hard armor backing composite.[11]

The Army Research Laboratory (USA) and the University of Delaware (USA) are introducing advances in soft body armor with fabrics utilizing shear thickening fluids. The objective is to impregnate shear thickening fluid (STF) into fabrics to improve their protective properties. STF becomes rigid during high speed deformation, such as a ballistic impact. Conversely, the fabric is drapable like an ordinary fabric with no applied deformation. STF addition significantly improves the puncture resistance of Kevlar KM2®. At the same areal density, STF-Kevlar KM2® has dramatically higher spike (e.g. ice pick) protection than untreated Kevlar KM2®. At the same areal density, STF-Kevlar KM2® has comparable knife protection versus untreated Kevlar KM2®; however, the STF-Kevlar KM2® has significantly fewer fabric layers (i.e. could result in a thinner cross-section) than the untreated Kevlar KM2®. At higher speeds, during archery testing, the STF-Kevlar KM2® offers significantly more puncture resistance than untreated Kevlar KM2®.[35]

8.11.4 Wearable power

Power will be one of the big differences of military uniforms in the future compared with those of yesteryear. As previously noted, electronic textiles are used for physiological monitoring and resistive heating. Various conductive components such as wires, metallic yarns, and composite yarns have been woven and knitted into narrow and broad width fabrics to transmit power and data. A wearable computer using these low profile, light, and flexible electronic textiles is revolutionary because traditionally power and data are transferred through stiff and bulky wires. The challenge, however,

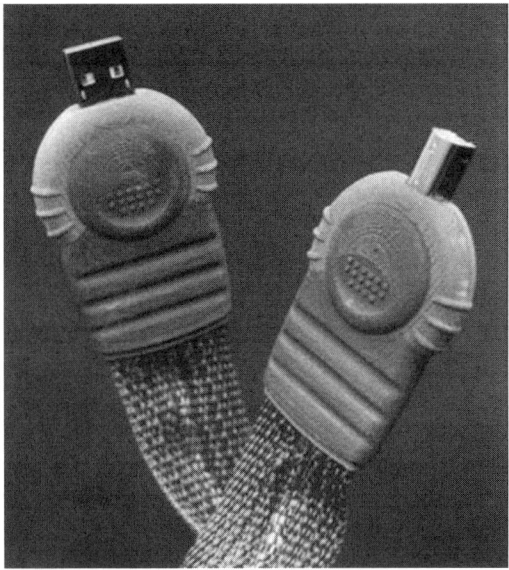

8.5 USB connection.

is not necessarily in the weaving or knitting process, but in the connections of those conductive yarns that make the product robust to withstand flexing and laundering. Figure 8.5 is an example of a USB connection with conductive yarns in a flexible webbing developed by Foster-Miller, Inc., Offray Specialty Narrow Fabrics (USA) and the US Army Natick Soldier Center.

8.12   Sources of further information and advice

From what has been discussed, it is easy to envision future military and first responder uniforms which are light weight, flexible, very versatile and able to perform a variety of functions. Certainly, as technologies and materials improve, they will be mass produced and become cheaper and available to servicemen throughout the military. Moreover, as material and technology advances are made, new uses and combinations will be discovered, further enhancing uniform functionality and the performance of military personnel and first responders.

High-tech materials can be found when one is not necessarily looking for them. It is important to attend trade shows and conferences to find state-of-the-art technologies and materials. These trade shows and conferences do not need to necessarily promote military clothing and equipment because the technologies and materials cross over, for instance, truck and tarp cover fabrics are used in military tents. Some examples of non-military-oriented trade shows and conferences are: Industrial Fabrics Association International (IFAI) annual exposition (www.ifai.com), Techtextil international trade fairs (www.techtextil.com), American Association of Textile Chemists & Colorists (www.aatcc.org), Outdoor Retailer winter and summer markets (www.outdoorretailer.com), International Trade Fairs for Sports Equipment and Fashion (ISPO) (www.ispo.de), The Fiber Society (www.thefibersociety.org), ASTM International (www.astm.org), and many more.

8.13   References

1 PARK, S. and JAYARAMAN, S., 'Intelligent Textiles for Personal Protection and Safety: The Emerging Discipline', *Intelligent Textiles for Personal Protection and Safety*, S. Jayaraman *et al.* (eds.), IOS Press, 2006.
2 GOMES, C., 'Fabrics in the military: Material demands', *Industrial Fabrics Products Review*, June 2006.
3 Naval Aerospace Medical Institute, *Hyperthermia, United States Naval Flight Surgeon's Manual*: Third Edition 1991: Chapter 20: 'Thermal Stresses and Injuries'.
4 *Extended Cold Weather Clothing System (ECWCS) Generation III*, www.polartec.com/military.
5 ADS, Inc. press release, *ADS, Inc. Awarded $220 Million Contract to Manage and Procure the U.S. Army's New Generation III Extended Cold Weather Clothing System*, January 2007.
6 KENNEDY, H., 'Army Selects New Winter Gear to Give Troops Edge in Combat', *National Defense Magazine*, January 2007.
7 W.L. GORE, & Associates, Inc. draft *Multi-Climate Protection System (MCPS) Technical Users Guide*, February 2005.
8 *U.S. Military's Special Operations Introduces New High-tech Combat Uniforms for Extreme Winter Conditions*, Press Release, Vista, CA, February 2003.
9 http://www.epicfabrics.com.
10 *Air Warrior Microclimate Cooling Garment (MCG)*, US Army Natick Soldier Center, Rev 04-27-06, OPSEC 03-237.
11 GOMES, C., 'High-tech military uniforms. Designing for the future', *Industrial Fabrics Products Review*, May 2003.
12 Wikipedia.
13 STONE, A., 'New Marine uniforms blend in but stand out', *USA Today*, June 2001.
14 'The hidden side of the matter . . . and more! Camouflage', *Canadian Textile Journal*, March/April 2002.
15 VANDERBILT, T., 'The U.S. Army's New Clothes, Why has the Army redesigned its uniforms'? *Slate*, September 2004.
16 http://www. indigosystems.com.
17 http://texplorer.de
18 DUGAS, A. *et al.*, 'Universal Camouflage for the Future Warrior', *4th International Conference on Safety and Protective Fabrics*, October 2004.
19 *Antimicrobial Technologies for the Warfighter*, US Army Natick Soldier Center Fact Sheet, March 2006.

20 JOWERS, K., 'Silver fibers help fabric fight disease, odor', *Army Times*, September 2006.

21 http://www.geodatasys.com/suit.htm

22 http://www.gentexcorp.com

23 MIL-DTL-32102, *Detailed Specification Joint Service Lightweight Integrated Suit Technology (JSLIST) Coat and Trouser, Chemical Protective*, April 2002.

24 *SARATOGA^{TM} JSLIST Chemical Warfare Protective Overgarment*, Tex-Shield, Inc. 2004.

25 'JPACE – MOPP 4 in the Aviation Environment', *Chemical/Biological Individual Protection Conference*, March 2006.

26 http://www.chemvironcarbon.com

27 WINTERHALTER, C., LOMBA, R., TUCKER, D., MARTIN, D., 'Novel Approach to Soldier Flame Protection', *Journal of ASTM International*, February 2005.

28 RAHEEL, M., PERENICH, T., KIM, C., 'Heat- and Fire-resistant Fibers for Protective Clothing', *Protective Clothing Systems and Materials*, (M. Raheel, ed.) Marcel Dekker, Inc., 1994.

29 *Flame Resistant Combat Uniform Fabrics and Clothing Systems*, US Army Natick Soldier Center Fact Sheet, July 2003.

30 CUNNIFF, P., AUERBACH, M., VETTER, E., SIKKEMA, D., *High Performance 'M5' Fiber for Ballistics/Structural Composites*. http://web.mit.edu/course/3/3.91/www/slides/cunniff.pdf

31 MAYNARD, B., 'Extreme Textiles', American Chemical Society, *Chemistry*, Autumn 2006.

32 www.peosoldier.army.mil, *PM Soldier Equipment* 2006.

33 LEE, Y., CHADHA, S., RIECKER, A., MENDUM, T., PUGLIA, J., 'Dynamic Nanocomposite Self-deactivating Fabrics for the Individual and Collective Protection', *Army Science Conference*, December 2006.

34 *TenCate Protective Fabrics brochure* #940.2351, July 2007.

35 WETZEL, E., WAGNER, N., 'Protective Fabrics Utilizing Shear Thickening Fluids (STFs)', *4th International Conference on Safety and Protective Fabrics*, October 2004.

Part II

Protection

9

High-performance ballistic fibers

T. TAM and A. BHATNAGAR,
Honeywell International Inc., USA

9.1   Introduction

High-performance fibers (HPF) are engineered for specific end uses that require exceptional strength, heat resistance and/or chemical resistance. They are generally niche products, such as lightweight composite materials for aircraft, ballistic fibers and bullet-resistant vests or body armor, protective gear for fire officers, and cut or stab resistant articles. On the lighter side, examples are fishing line, bowstring, and marine rope and sail cloths such as those used in the Americas Cup race.

9.2   Classical high-performance fibers

9.2.1   Glass fibers

The oldest, and most familiar, high-performance fiber is glass. Glass fibers are relatively inflexible and not suitable for many textile applications. However, they can be found in a wide range of end uses, such as insulation, fire resistant fabrics, and reinforcing materials for plastic composites. In recent years optical quality fiberglass has revolutionized the communications industry.

9.2.2   Carbon fibers

The next classic HPF is carbon fiber which can be engineered for strength and stiffness to reinforce composites; or can, in various forms, improve the electrical conductivity, thermal and chemical resistance of textile materials. The primary factors governing its physical properties are degree of carbonization and orientation of the layered carbon planes. Carbon fibers are made from specially purified rayon or top quality acrylics (PAN), or pitch fibers from liquid crystal (for reinforcement and other applications). The almost perfect carbon fiber is graphite.

9.3   Rigid chain aromatic high-performance fibers

The best-known high-performance, synthetic, organic fibers are aramids, which like nylons are polyamides derived from organic acids and amines. Figures 9.1 and 9.2 show nylon 6, and nylon 66, which have a flexible chain between the amide group whereas Fig. 9.3 shows Nomex® which has aromatic rings between the amide groups that give its unique properties.

Because of the stability of the aromatic rings and the added strength of the amide linkages, due to conjugation with the aromatic structures, aramids exhibit higher tensile strength and thermal resistance than the aliphatic polyamides (nylons). The para-aramids (trade names Kevlar® and Twaron®) based on terephthalic acid and *p*-phenylene diamine, or *p*-aminobenzoic acid, exhibit higher strength and thermal resistance than that with the linkages in meta positions on the benzene rings (trade name Nomex®). The greater degree of conjugation and more linear geometry of the *para*-linkages, combined with the greater chain orientation derived from this linearity, are primarily responsible for the increased strength. The high

9.1 Structure of nylon 6.

9.2 Structure of nylon 66.

9.3 Nomex® structure.

9.4 Structure of aramid fiber.

9.5 The benzimidazole link within PBI polymer.

impact resistance of the *para*-aramids makes them popular for first-generation bullet-resistant body armor. Aramid fiber (Fig. 9.4) can be chopped into staple form to make felt. Applications such as chain saw protective garments may be made with blends with other fibers for other end uses. Aramid fiber is lyotropic. It is solution spun and it melts at a lower temperature than a thermotropic liquid crystal fiber.

9.4   High-temperature performance fibers

9.4.1 PBI fiber

PBI (polybenzimidazole) (Fig. 9.5) is another fiber that takes advantage of the high stability of conjugated aromatic structures to produce high thermal resistance.

The ladder-like structure of the polymer further increases the thermal stability. PBI® was first synthesized in the 1950s. In the 1960s, Celanese developed a dry spinning and polymerization process for a high-temperature resistant PBI® polymer. Following a fire in an Apollo spaceship in 1967, NASA cooperated with Celanese to develop PBI® textiles. The fibers were launched in 1983. PBI is noted for its high cost, due both to high raw material costs and a demanding manufacturing process. The PBI fiber has a yellow color (PBIgold) but with high moisture regain (7–8%). When converted into fabric, it yields a soft hand and feels comfortable (due to high moisture regain). Blending with other high-temperature resistant fibers such as aramid to reduce cost and/or increase fabric strength may optimize the utilization of PBI.

9.4.2 PBO fiber

PBO (polyphenylenebenzobisoxazole) is another high-temperature fiber based on repeating aromatic structures, which is a recent addition to the

9.6 PBO fiber structure.

9.7 Structure of liquid crystal fiber.

market (see Fig. 9.6). PBO exhibits very good tensile strength and high modulus, which are useful in reinforcing applications. Currently, Toyobo's commercial rigid-rod chain molecules of poly (*p*-phenylene-2, 6-benzobisoxazole) (PBO) is called Zylon.

9.5   High-performance thermoplastic fibers

9.5.1 Liquid crystal fiber·

Liquid crystal fiber (Fig. 9.7) is a melt spun fiber made by high-temperature melting and spinning liquid crystal polymer. Vectran® is the only commercially available melt spun LCP fiber in the world. The lightweight Vectran® reinforcement fibers and matrix fibers have exceptional strength and rigidity, which make them a very good alternative to steel: pound for pound, Vectran® is five times stronger than steel. Its cross-section shape and distribution make it ideal for high-temperature filtration applications. It is sometimes blended with aramid or other performance fibers to increase final fabric strength.[1]

9.5.2 HMPE

HMPE (ultra-high molecular weight polyethylene) can be extruded using special gel spun technology to produce very high molecular orientation. The resulting fiber combines high strength, chemical resistance and good wear properties with light weight, making it highly desirable for applications ranging from cut-proof protective gear to marine ropes. Since it is lighter than water, ropes made of HMPE float. Pound for pound, gel spun HMPE fiber (Spectra®) is ten times stronger than steel. Its primary drawback is its low softening and melting temperature, as well as its tendency to creep under high load.

9.6   Physical properties comparison

Graphical comparisons of representative high-performance fibers are illustrated in Fig. 9.8.

9.7   Requirements for high-performance fibers

In order to achieve high-performance fiber with exceptional tenacity and modulus properties, there are at least three necessary requirements.

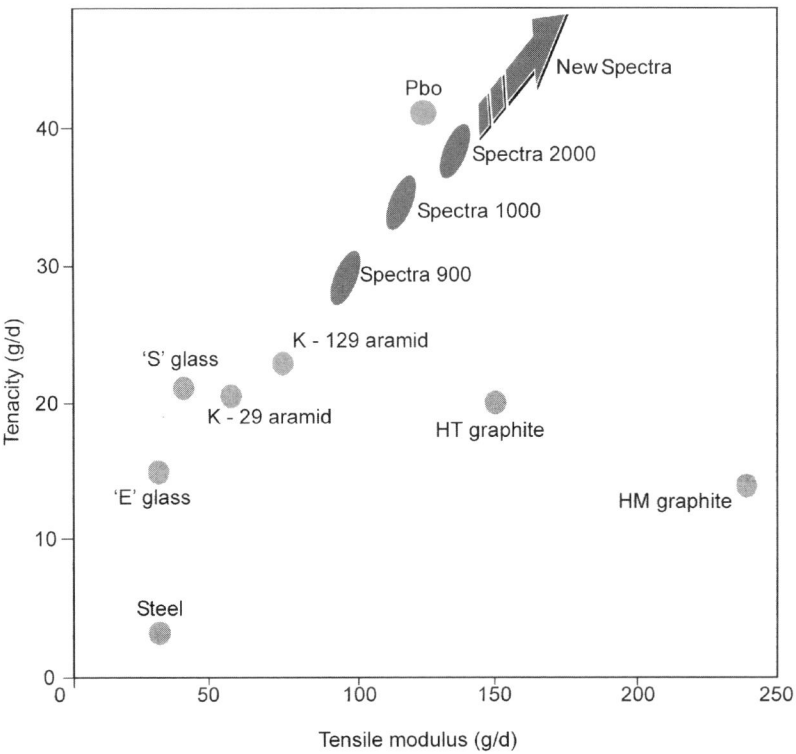

9.8 Modulus versus tenacity of commercial high-performance fibers.[5]

1. The molecule must be highly oriented in the fiber axis direction.
2. The molecular weight or the molecular chain length must be very high.
3. The fiber must be highly crystalline with few defects.

There are generally two approaches in manufacturing high-performance fibers to meet the above criteria. One can start with a highly oriented but relatively low molecular weight, rigid chain and rod-like polymer (Fig. 9.9), such as an aramid (lyotropic) or liquid crystal (thermotropic) polymer.[2,3] This can then be spun into fiber and given a high molecular weight by drawing and/or annealing processes. Aramid spinning will be used as an example for this approach.

On the other hand, one can start with an ultra-high molecular weight, flexible long chain randomly coiled polymer like ultra-high molecular weight polyethylene (HMPE)[2,3] (see Fig. 9.10). Since the ultra-high molecular weight polymer cannot be melt spun (polymer will decompose before it will flow at the melting temperature), one must spin this polymer with a dilute solution in the range from 2 to 30% concentration. In this dilute solution, the ultra-high molecular weight polymeric chain will 'uncoil' and form

9.9 Random rods of polymers.

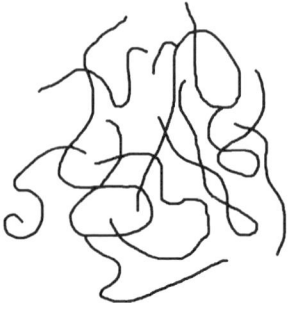

9.10 Random coils of polymers.

a network called a gel. By this 'gel spinning' method, a long molecular chain xerogel fiber with a loosely connected network can be made. The xerogel fiber can be drawn to a highly oriented, highly crystalline high-performance fiber via specially developed drawing techniques. High-performance HMPE fibers like Spectra® or Dyneema® will be used to illustrate these processes. A more in-depth discussion of these fibers will follow.

9.8   Aramid fibers

An aramid fiber is based on poly (p-phenylene terephthalamide) (PPD-T) polymer: a classical polycondensation of PPD (p-phenylene diamine) and terephthaloyl chloride (TCI) in amide solvent. The condensation polymerization is described below[2,3] (see Fig. 9.11).

While the PPD-T polymer is not soluble in conventional solvent like most of the *para*-oriented aromatic polyamides, the rod-like aramid fiber can be dissolved in strong sulfuric acid[2,3] (see Fig. 9.12).

The degree of molecular order of aramid in solution depends on the concentration as in Fig. 9.13.[2,3] As the polymer concentration increases from 5 to about 12%, the solution viscosity increases as expected. The rod-like molecule will take a form as in Fig. 9.14.

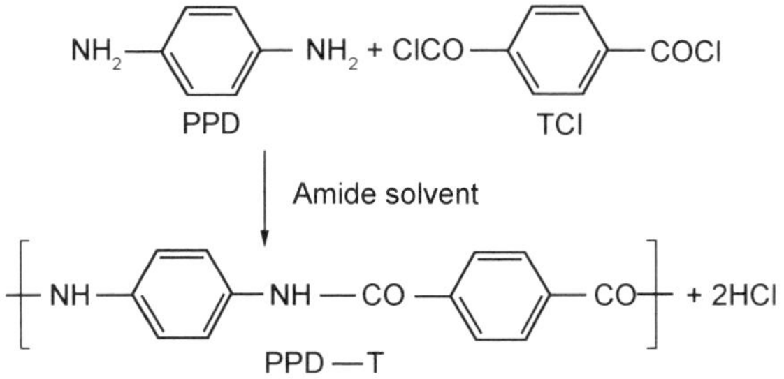

9.11 Condensation polymerization.

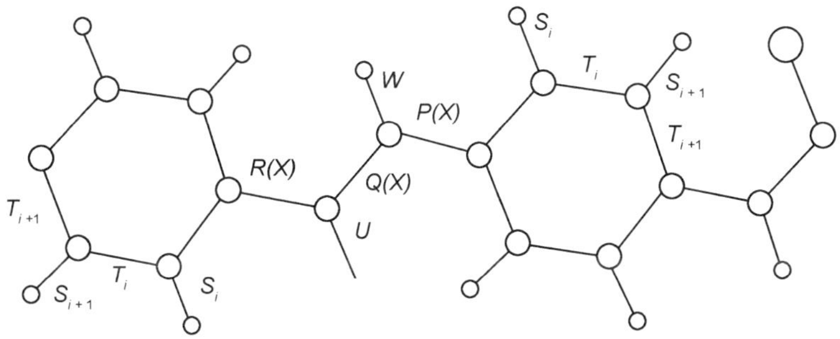

9.12 Aramid in sulfuric acid.

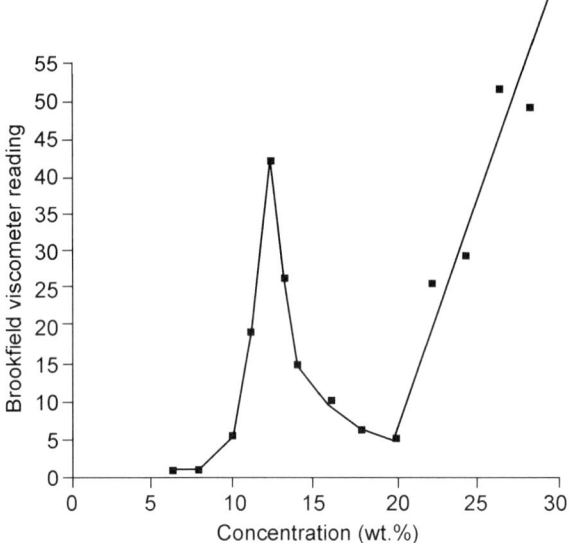

9.13 Viscosity versus polymer concentration in sulfuric acid solution.

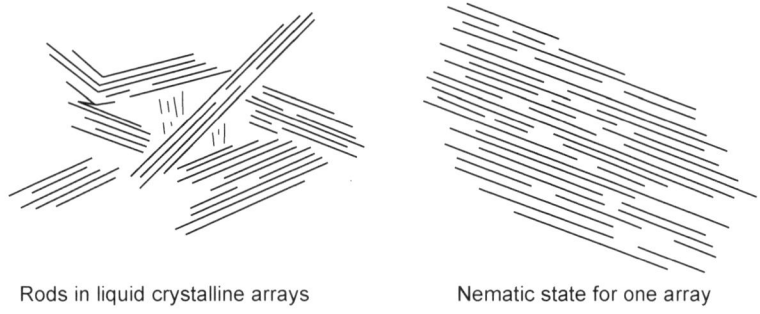

Rods in liquid crystalline arrays   Nematic state for one array

9.14 Distribution of rod-like structures in diluted solvent.

However, as the concentration increases further, the rod-like polymer will form a nematic state with high degree of orientation. As a result, the solution viscosity will drop instead of increase as shown in Fig. 9.13. When this highly anisotropic solution is under shear, or elongation flow like fiber spinning process, the molecule of the extrudate will further align with the fiber axis to give the resulting fiber its orientation.

9.8.1 Dry-jet wet aramid fiber spinning

The aramid solution is spun by a process called the dry-jet wet spinning (Fig. 9.15). In this process, an anisotropic solution of PPD-T is extruded through the air gap into a coagulated bath as shown in Fig. 9.15. The resultant yarn after coagulation is washed and dried.[2,3]

The keys for the dry-jet wet spinning method to orient the anisotropy molecule are both shear orientation and elongation flow, through the spinneret's capillary, and this is represented graphically in Fig. 9.16. In addition, the 'relaxation' of the molecule after the exit of the capillary is kept to a minimum by filament tension or attenuation in the air gap and through the coagulate bath as the filament precipitates into the highly oriented crystalline fiber. This fiber is also heat treated under tension to increase its modulus. Various properties of the Kevlar fibers are listed in Table 9.1.[2,3]

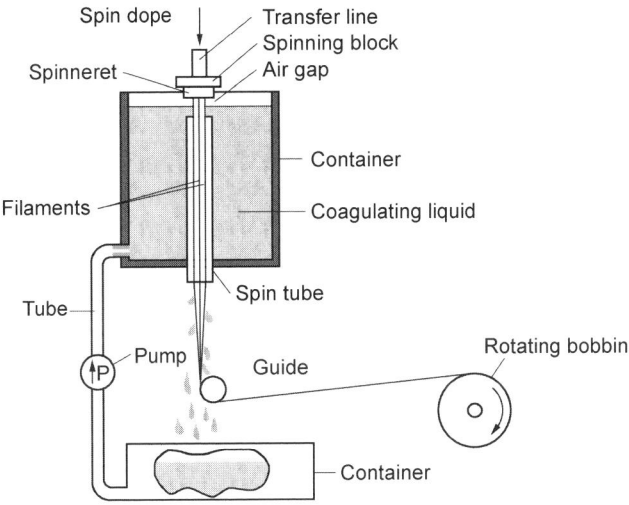

9.15 Schematic diagram of the dry-jet wet spinning process for aramids.

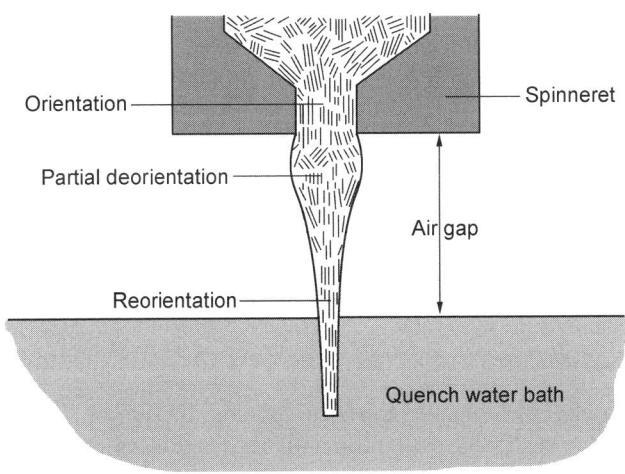

9.16 Orientation through the capillary die: elongation and shear flow.

Table 9.1 Typical properties of kevlar aramid yarns

Yarn property	Ballistic fiber	High modulus fibers
Tensile strength		
gpd	23.0–26.5	18.0–26.5
Kpsi	420–485	340–420
Initial modulus		
gpd	550–750	950–1100
Mpsi	10.3–14	17.4–21
Elongation, %	3.6–4.4	1.5–2.8
Density g/cm^3	1.44	1.44
Moisture regain, % 25 °C, 65% RH	6	1.5–4.3

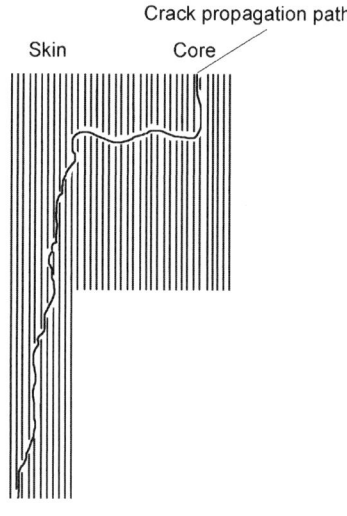

9.17 Crack in fiber.

9.8.2   Aramid fiber structure and morphology

Aramid fibers contain several levels of microscopic and macroscopic morphology. A brief discussion of each is described below using the individual fiber as a starting point.

Skin core fibril structure

When aramid fiber is subjected to tensile testing, its typical fracture modes are generally a fibrillated type failure. This fracture mode represents a highly ordered lateral fiber structure (see Figs 9.17 and 9.18).

Fiber fibrillar structure

Aramid fiber fibrillates easily upon abrasion especially in the perpendicular direction to the fiber axis. In fact, almost all highly oriented fibers like UHMWPE (such as Spectra® fiber) are easily fibrillated. It is because the macro-molecules were only held together by the van der Waals force, and/or the hydrogen bond force. Figure 9.19 is a proposed model of the fibrillar structures for most of the highly oriented performance fibers. The individual fibrils are the load-bearing elements for the fiber whereas the tie molecule

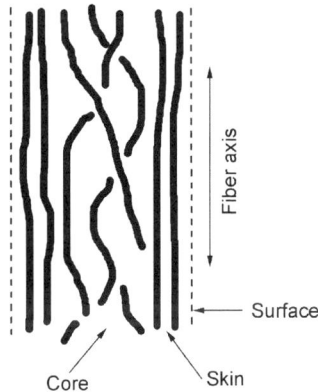

9.18 Skin and core of fiber.

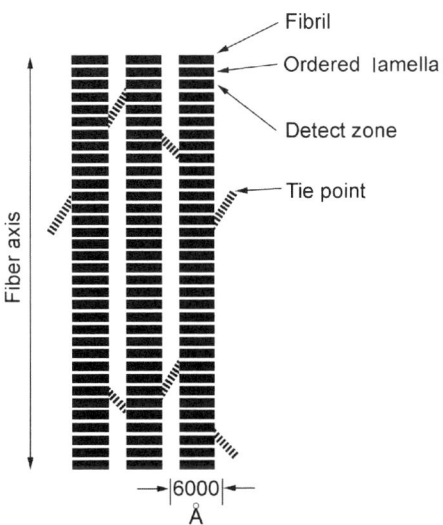

9.19 Fibrillar structure model of aramid fiber.

is the load-bearing element for the conventional fibers. The widths of the fibril are about 600 nm and the lengths up to several cms.[2,3]

9.8.3 Aramid fiber morphology and orientation

Figure 9.20 illustrates a fibril. On each fibril, the straight line represents the PPD-T molecular chain. Some of these chains contain breaks or bends. These defects or amorphous layers are the weak links in the fiber structure. However, some of the PPD-T chain can be oriented and extended to bridge several 'amorphous' or defect layers. This unique 'extended chain tie molecule' should give satisfactory fiber strength as shown in Fig. 9.20.

9.8.4 Pleat structure

Aramid fiber has a unique feature when observed under a cross-polarized microscope light field, in that it displays transverse bands. However, these transverse bands diminish when the filament is under tension.[2,3] This leads to the hypothesis that aramid fiber has a pleated structure as in Fig. 9.21. The occurrence of a pleat sheet structure in aramid is not well understood.

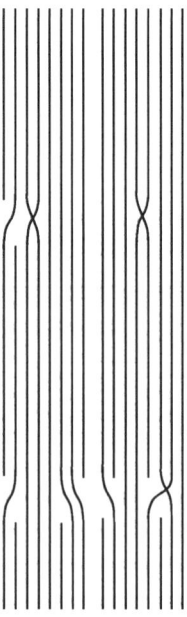

9.20 Crystalline structure model of aramid fiber.

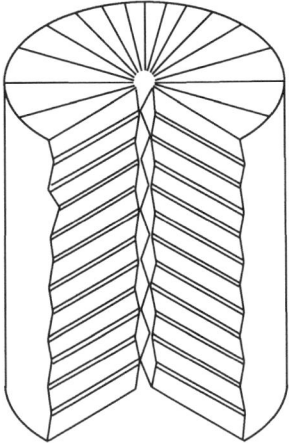

9.21 The pleat structure model of aramid fiber.

For the formation of the pleated structure it has been hypothesized that during the coagulation of the aramid fiber, the skin is first formed and is subjected to attenuation stress on a spinning filament. This allows the 'core' fiber to relax and form pleats at a uniform[2,3] periodicity. The formation of the pleat structure gives the fiber an inherent elongation or elasticity. That may be the reason why, when Kevlar fiber is under stress, the transverse bands diminish as observed under the microscope.

9.8.5 Crystalline structure

Aramid fiber has a highly crystalline, highly ordered molecular structure. Wide angle X-ray diffraction shows no amorphous halo indicating a highly crystalline fiber. There is a pair of sharp rings in the equatorial scan, indicating that the fiber may contain a few percent of unoriented crystals.

9.9   Gel spinning of ultra-high molecular weight polyethylene (HMPE) fiber

The process of making the high-performance HMPE fiber, based on the simplest and flexible polyethylene, is another extreme spectrum of processing methods for high-performance fibers. While the chemical structure of the HMPE is identical to the normal high or low density polyethylene (HDPE, LDPE) such as those found in engineering plastics, the HMPE is not melt spinnable due to its extreme high melt viscosity. In addition, because of the very high degree of entanglement in the flexible molecular chain, the drawing for high tenacity yarn HMPE is almost impossible even at a slow drawing rate.

The key to achieve high strength, high modulus properties of the HMPE is by the gel spinning process. In this process, the long, flexible and entangled molecules are dissolved in a solvent from 2–15% concentrations (depending on the molecular weight) and mixed thoroughly via an extruder, helicon mixer or other mixing means as shown Fig. 9.22.

In the solution, the molecules become disentangled and form a loosely connected network called gel. The gel is then spun through a spinneret just like a conventional melt spinning process. After quenching or cooling of the gel fiber, the loosely entangled molecule fiber can be drawn at a very high draw ratio to a highly oriented, long chain crystalline high-performance fiber. The solvent to dissolve or disentangle the HMPE can be volatile or non-volatile but the principle of the gel spinning will be the same. The schematics shown in Figs 9.22, 9.23 and 9.24 were proposed by Pennings and colleagues from spinning of the gel to drawing into high-performance fiber.[4]

9.9.1 The morphology of the HMPE fiber

Similar to aramid fibers, ultra-high strength HMPE fiber also contains microscopic and macroscopic fiber morphology. The SEM picture shows regular micro- and macro-structures. Figure 9.25 is a representation of the current model consisting of micro and macro fibrils. The longitudinal structure consists of micro-fibrils which have a proposed structure in which nearly perfect crystals are covalently linked through a relatively small amorphous domain (see Fig. 9.26). This micro-fibril structure is far from the perfect uniaxial fiber structure and thus the strength of the HMPE fiber, while ten times stronger than steel, is still far from the theoretical strength of the covalent C–C bond (see Fig. 9.27).

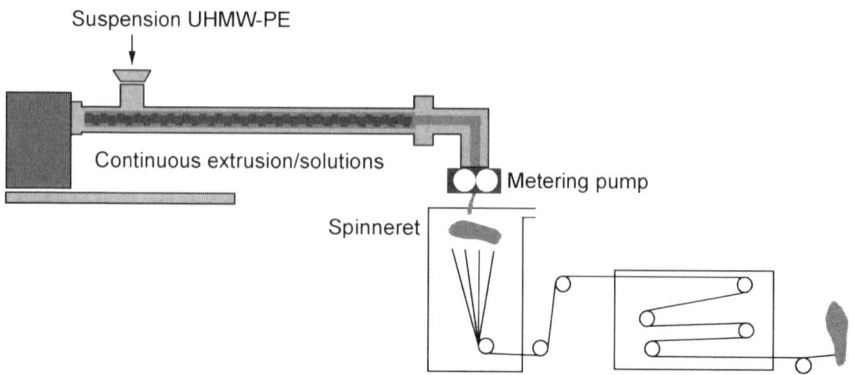

9.22 Schematic of gel spinning process.[6]

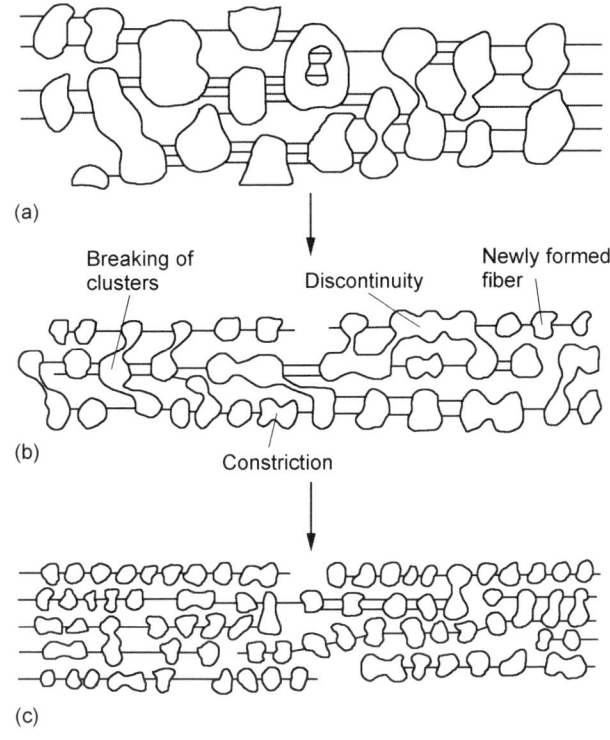

(a)

Breaking of
clusters
Discontinuity
Newly formed
fiber

(b)

Constriction

(c)

9.23 Deformation stages of gel fiber with solvent.

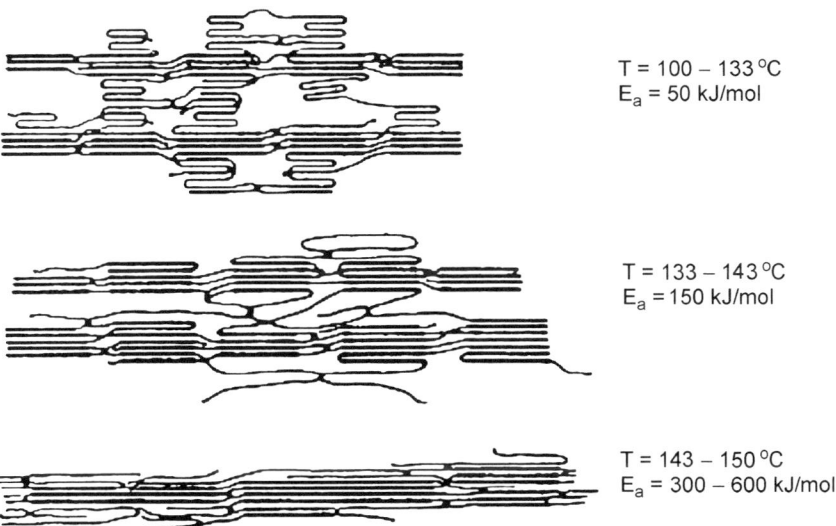

T = 100 – 133 °C
E_a = 50 kJ/mol

T = 133 – 143 °C
E_a = 150 kJ/mol

T = 143 – 150 °C
E_a = 300 – 600 kJ/mol

9.24 Deformation mechanism during hot drawing of HMPE.

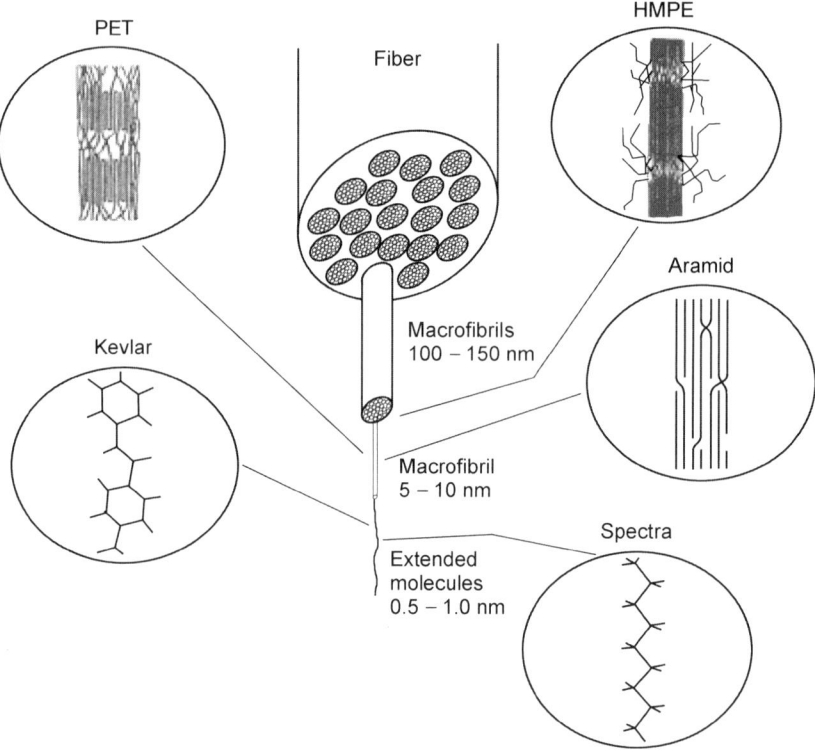

9.25 Micro- and macro-fibrillar structure of PET, aramid and HMPE fibers.

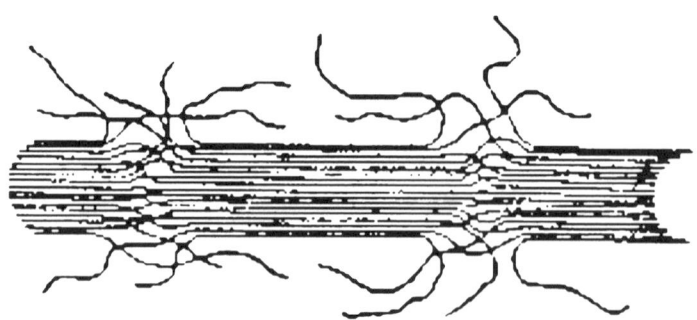

9.26 Micro-fibrillars of HMPE fiber.

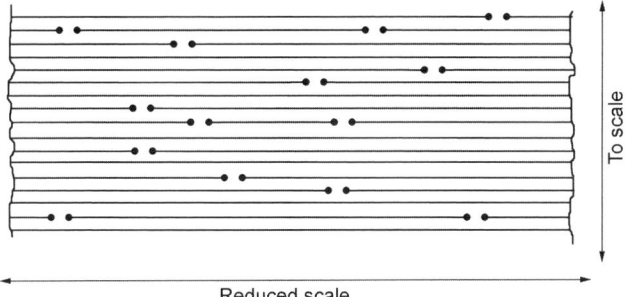

9.27 Proposed longitudinal structure of HMPE micro-fibrils.

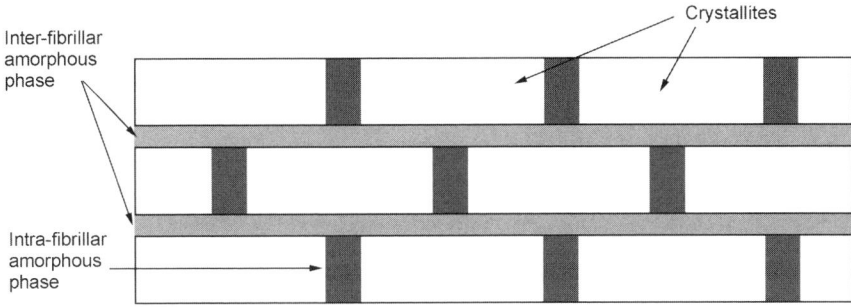

9.28 Characterisitcs and properties of high-performance HMPE fiber showing macro- and micro-fibrillar.

It is speculated that an increase in the number of 'extended chain' molecules that span the amorphous domain would increase both strength and modulus. The potential is certainly there to further advance the properties of the HMPE fibers (see Fig. 9.27).

Figure 9.28 represents a proposed model for the macro-fibrils. Because amphorous matter also exists between the micro-fibrils, the structure appears to be a composite of near perfect oriental crystalline micro fibrils imbedded in an amorphous matrix. This model appears to be similar to the aramid model discussed earlier. However, the aramid model suggests that a strong inter-macrofil linkage exists because of hydrogen bonding of the polyamide molecules. Figure 9.28 shows a 'clear-cut' amorphous and crystalline region.[5] However, there are extended chain molecules that can bridge through several layers of 'amorphous' region. It is speculated that the more of this type of 'bridging' molecule, or a new term called the extended chain tie molecule, the higher strength and more dimensionally stable the HMPE fiber will be.

The typical HMPE fiber's properties are listed in Table 9.2.[6] As the gel spinning and drawing technology mature, fiber properties improve to meet

Table 9.2 Properties of HMPE fibers

Yarn property	Standard fiber	High strength fibers
Tensile strength		
gpd	25.5–30.5	37.5–41.0
Gpa	420–485	3.21–3.61
Initial modulus		
gpd	775–920	1320–1450
Gpa	66–79	113–124
Elongation, %	3.6–4.4	1.9–3.6
Density g/cm^3	0.97	0.97

different end uses. As a result, there are different grades of Spectra fibers such as S-900, S-1000 and S-2000 or in case of DSM, SK 75 and SK 76. In short, the new generation product tends to be in lower denier per filament (dpf), higher tenacity and higher modulus.

9.10   Poly(*p*-phenylenebenzobisoxazole) (PBO) fiber

Synthetic fibers for ballistic applications have been getting stronger and more effective to defeat ballistic threats since the first development of nylon fiber, followed by aramid fiber, HMPE fiber, and PBO fiber is the latest commercial fiber in this field.

High-performance properties of PBO are originated from the rod-like nature of the polymer chain which also makes the processing of fiber from polymer fairly difficult. The development of production technology on PBO fiber spinning took a long time due to the difficult nature of the rod-like structure. In 1991 Dow Chemical decided to work with Toyobo. Their joint development resulted in a unique spinning technology, opening the way to the industrial production of PBO fiber.

Toyobo received a license from Dow Chemical and has worked on further development. The pilot plant for PBO fiber production was completed in early 1995. The commercial production started in 1998.

9.10.1 Polymerization and spinning

PBO is polymerized from diaminorescocinol dihydrochloride (DAR 2HCl) and terephthalic acid (TA) in polyphosphoric acid (PPA). Current PBO fiber is spun from spinning dope with phosphoric acid solution using air-gap wet spinning technology. On a coagulation process, fiber structure formation through phase separation should occur. The first filaments extruded from a spinneret transform to a swollen micro-fibrillar network when the nematic rigid-rod solution touches a coagulant. Passing through the coagulation process, the network loses its open spaces and forms dense fibrillar structure. The coagulated fiber is subsequently washed and dried.

As-spun PBO shows the tenacity of 42 g/d (5.7 GPa) or more and the modulus of 1300 g/d (175 GPa) or more. By heat treatment at around 600 °C, the as-spun fiber achieves the increase of modulus up to 2000 g/d (275 GPa) without tenacity loss.

9.10.2 Micro fibril and void

Scanning electron micrographs taken on a fractured surface of high modulus PBO fiber show that the fiber is formed from assembly of fibrils, the diameter of which varies from 10 to 50 nanometers. On such fractography, however, careful analysis should be performed to elucidate structural entities, because there may exist some artificial structures generated in the fracturing process.

In PBO fiber, streak-like scattering patterns, which would come from elongated micro voids to the fiber direction, appear on the equator. During the heat treatment process this streak disappears and the four-point pattern, similar to the shape of a butterfly, appears. This kind of striation was reported on PPTA fibers. It is interesting that current high modulus fiber, even stronger than former fibers in tenacity, gives us the same pattern. In the case of high-strength polyethylene fiber, this periodic density fluctuation acts as a weak point on tensile strength.

To estimate the cross-sectional diameter of micro-voids of PBO fiber intensity profile along the equator was taken from a two-dimensional small-angle X-ray scattering (SAXS) pattern. The logarithm of the intensity after background correction is plotted against the square of the scattering vector. The data exhibits linearity and the slope gives the average diameter of the micro voids which is measured as 24 A.

9.10.3 Fiber structure and physical properties

Structure

Structure of PBO fiber formation is through coagulation, washing and drying. Since 86% of PPA is extracted from the dope, the structure of as-spun filaments has a fibrillar nature with a capillary void of diameter of around 20 Å which is determined from the plot of small-angle X-ray scattering SAXS. As-spun fiber has an extended chain structure which is confirmed by the lattice image of electron micrograph and its inverse FT image.

The crystal size of the as-spun fiber is about 100 Å and increases up to 200 Å by heat treatment. SAXS pattern of as-spun fiber shows a four-point pattern. This four-point pattern disappears with heat treatment.

The standard PBO fiber is formed from micro-fibrils (preliminary 10–50 nm in diameter) and contains many capillary-like micro-voids, which exist between micro-fibrils before drying. These micro-voids are connected with each other through cracks or openings between micro-fibrils. There is a void-free region in the very surface of the fiber. The micro-fibril is made of extended PBO molecules, highly oriented to the fiber axis. The Hermann's orientation function measured by WAXS is estimated to be over 0.95. The preferential orientation exists and the a-axis of the PBO crystal is aligned radically in the cross-section of the fiber. In the case of higher modulus PBO fiber, the Hermann's orientation function value becomes 0.99 or higher.

Properties of PBO fibers

Tenacity, modulus, heat resistance and flame resistance are the four main physical attributes of the PBO fiber. PBO is the first organic fiber which exceeds steel and even carbon fiber in strength per cross-sectional area. The theoretical modulus of polymers can be easily calculated due to the recent remarkable progress in computer chemistry. The PBO-HM from the Toyobo pilot plant shows only the 60% of crystalline modulus of PBO. Fiber modulus of many super fibers achieved crystalline modulus. When PBO fiber achieves the crystalline modulus value, no other fiber from linear polymer will exceed PBO, which is the ultimate fiber in terms of modulus.

The heat-resistant property of PBO is about 100 °C higher than p-Aramids. Flame resistance (limiting oxygen index (LOI)) is surprisingly higher than other FR organic fibers such as PBI (LOI 41), which is the former record holder, and p-Aramid (LOI 29).

Thermal stability

PBO fiber shows very high heat resistance. Temperature dependence of physical properties is also very small as compared to other organic fibers. The temperature dependence of crystalline modulus does not change up to 400 °C. Fiber modulus also does not show significant loss even at high temperature. Only 20% loss of modulus is observed at 400 °C. Tenacity at high temperature is also superior to p-Aramid. 15 g/d of tenacity of fiber still remains at 500 °C.

Table 9.3 Properties of PBO fibers

		PBO-AS	PBO-HM
Filament denier		1.5	1.5
Density	g/cm^3	1.54	1.56
Tensile strength	g/d	42	42
	GPa	5.8	5.8
Tensile modulus	g/d	1300	2000
	GPa	180	280
Elongation break	%	3.5	2.5
Moisture gain	%	2.0	0.6
Decomposition temp.	°C	650	650

Other properties

Moisture regain is very low, 0.6 wt% for PBO-HM and 2.0 wt% for PBO-AS at 25 °C and 65RH condition. Dimensional stability against moisture and temperature is excellent. Creep rate is about half of that for *p*-Aramid in the same stress ratio to breaking stress. Chemical resistance against organic solvents and alkaline conditions is excellent and no loss of strength is observed. As for bleach, PBO is superior to other organic super fibers. In acidic conditions, PBO is not as strong as in alkaline, but still is stronger than *p*-Aramids (Table 9.3).

9.11   Sources of further information and advice

AFMA website Fiber Source, High-Performance Fiber. Chinese patent CN 2392788Y.

HEARLE, J. W. S. (ed.) *High-performance Fibres*, Woodhead Publishing Limited, 2000.

KTAGAWA, TOORU, MURASE, HIROKI and YABUKI, KAZUYUKI, 'Morphological Study on Poly-p-phenylenbenzobisoxazole (PBO) Fiber', *J. of Polymer Science: Part B: Polymer Physics*, **36**, 39–48 (1996).

US patent 4536536, Karesh and Prevorsek Assigned to Allied Corporation, 20 August 1985.

VAN DINGENEN, JAN L. J. Gel-spun high-performance polyethylene fibers. In Hearle, J. W. S. (ed.) *High-performance Fibres*, Woodhead Publishing Limited, 2000, pp. 62–92.

YABUKI, K. Poly(*p*-phenylenebenobisoxazole) fiber, *The Twelfth Annual Meeting, the Polymer Processing Society*, Sorrento, Italy, May 27–31, 1996, 279–281.

9.12   References

1 Engineering Brochure. *www.Vectran.net.*
2 YANG, H. H. *Kevlar Aramid Fiber*, John Wiley & Son, 1993.

3 YANG, H. H. *Aromatic High Strength Fibers*, SPE Mongraph, John Wiley & Son, 1989.
4 SMOOK, J. and PENNINGS, J. *Journal of Material Science*, **19**, 31 (1984).
5 PREVORSEK, D. 'Spectra: The latest entry in the field of high-performance fibers. In Lewin, M. and Preston, J. (eds) *Handbook of Fiber Science and Technology, Vol. 3 High Technology Fibers Part D*, Marcel Dekker, 1996, p. 17.
6 Spectra® fiber technical information.

10
Ballistics testing of textile materials

D. R. DUNN, H. P. White Laboratory, Inc., USA

10.1 Introduction

Armor may be defined as a covering intended to defend against an exterior threat to the covered entity. It may be categorized by the entity being protected (vehicles, structures, human body, etc.), the nature of the perceived threat to that entity (pointed and edged weaponry, radiological, biological, chemical, or ballistic) and the physical configuration of the armor (monolithic rigid or flexible materials and composite rigid or flexible materials). The scope of this chapter covers the protection of the full range of protected entities from ballistic threats with rigid or flexible armor whose makeup includes a textile.

10.2 Military usage of textiles

Dictionaries define a textile as 'any material that is woven' and the product of weaving (fabric) as a 'cloth made by weaving'. Military usage of textiles includes multiple plies of cloth woven from fibers of high tensile and shear strength, and the resultant multi-ply material, fashioned into ballistic resistant materials.

Synthetic fibers have been developed which, when woven into fabrics, provide lightweight, flexible protection from low-level fragmentation and bulleted ballistic threats as well. While several forms of these synthetic materials are not woven, but are flexible sheets of 'film' and not therefore textiles, their weight, flexibility, resistance to puncture, and use in personal protection have promoted their inclusion, along with their woven counterparts, in the broadened military definition of ballistic resistance textiles.

Flexible armor is constructed from a multiplicity of plies of cloth, or film, of ballistic resistant textiles. Some configurations of *rigid* armor imbed multiple plies of this cloth in a matrix, which renders the protection more suitable to vehicular and/or structural applications.

Among the primary design concerns for personal armoring materials is personal discomfort, restricted mobility, and intra-body movements imposed by the protection (weight, bulk, heat retention, rigidity, etc.). These factors are far less significant in vehicle and structural armoring materials, and may even be desirable in rigid armor applications. For these reasons, flexible textiles are used for most personal armor, while impregnated/laminated rigid textiles are often found in vehicle and structural armoring materials. Exceptions to this general rule are the use of rigid, laminated textiles in helmets, and as limited area inserts in otherwise flexible textile torso armor.

Military personal armor has always provided state-of-the-art protection from perceived hand and hand-delivered threats of the period – clubs, spears, arrows, etc. Until recent times, the threat from firearms has been beyond the state-of-the-art of body armor. As recently as World War II, the only protection provided to the soldier was from light fragmentation (helmets and flak vests), conceding mortality to threats of military personal weaponry. This approach remained unchanged until the introduction of law enforcement protection from handgun threats in the late 1950s and 1960s, reacting to the increased role of the military in counter terrorist activities, and capitalizing on advances in the protection from hand and shoulder-fired firearms.

Currently, fielded armor includes helmets, which provide protection from a variety of handgun threats and fragmenting munitions, torso armor, which provides all-around handgun and fragmentation protection and limited chest, back and side protection from rifle levels of threat, and arm and leg protection from handgun and light fragmentation.

Military helmets are most often monolithic, woven aramid fiber in a rigid matrix, which provides protection from light fragmentation and handgun ammunition. Basic handgun and fragmentation torso protection is provided by flexible woven vests with back, front, and side pouches to accommodate rigid ceramic panels and provide limited area (chest, back and side) protection from rifle levels of ball ammunition and armor piercing ammunition.

All materials that provide ballistic protection (armoring materials) are designed not only to prevent entry of the ballistic threat into the protected area, but also to deny entry to fragments of the ballistic threat and fragmentation of the armor itself (spall). Materials that provide this protection when worn on the body must resist lethal levels of injury from deflection of the armoring materials into the protected area from a non-penetrating, ballistic impact. This secondary threat is frequently referred to as 'blunt-trauma'.

The successful penetration of a ballistic threat is a function of its striking energy, its resistance to deformation upon impact, its obliquity, and the overmatching of the threat material with the material of the armor. The striking energy of the ballistic threat is a function of its mass and impact velocity. The level to which the ballistic threat is deformed by the impact with the armor is directly related to its hardness relative to the hardness of the armor. The level to which the target projectile is able to resist deformation and concentrate that striking energy on as small an area of armor as possible enhances the likelihood of it completely penetrating the armor. Conversely, the armor must cause the ballistic threat to deform and spread its energy over as large an area of the armor as possible, thereby reducing the energy delivered per unit of armor area to levels below the failure point of the armoring material. In short, given the mass, striking velocity, and shape of the threat are constant, a threat constructed of harder material will more readily maintain its undistorted profile, and more readily penetrate an armoring material, than its softer material counterpart. For that reason, the more-deformable lead of military ball ammunition is far less likely to completely penetrate armoring material than the less-deformable steel or tungsten cores of armor piercing ammunition.

The extent to which the backface of the armor is linearly deformed by a non-penetrating ballistic threat is generally more indicative of the lethality level of the deformation than is the volumetric deformation of armor. In other words, a small area of extreme linear deformation is more likely to be lethal than a larger area of lesser linear deformation. Armor is critical to the reduction of the lethal effect of blunt-trauma.

The characteristic of the material that determines its usefulness as armoring material is the ratio of its weight to its shear and tensile failure points. That characteristic, and the flexibility of armoring textiles, makes them extremely useful in personal and many rigid, vehicular/structural armoring applications.

A single-ply of any textile materials is an under match for virtually any ballistic threat, but by layering multiple plies of textile material the threat is required to sequentially stress each ply to its failure point. A portion of kinetic energy is thus dissipated with each ply the projectile penetrates. The energy necessary to penetrate each ply is a function of the area of the ply being penetrated, as well as the progressively increasing cross-sectional diameter as the projectile distorts. Each successive ply requires more energy for penetration than the proceeding ply.

10.3   Armor testing

10.3.1 General

Ballistic threats may be categorized as either kinetic energy or chemical energy (explosive) threats. The effects on armoring materials of fragmentation from exploding munitions are included in the scope of this chapter, but the explosive effects are not.

Kinetic energy projectiles (bullets) from fielded ammunition are used for ballistic testing, but, due to the virtually limitless size and weight of fragments from fragmenting munitions encountered in the field, resistance to fragmentation is determined with fragment simulator projectiles (FSPs). MIL-P-46593A puts forth the physical characteristics for each of four FSP's – caliber .22 (17 grain), caliber .30 (44 grain), caliber .50 (207 grain), and 20 mm (830 grain). Recently, residual velocity testing of personal armor has been used in support of casualty reduction studies. This testing is usually conducted with right circular cylinders (RCC's) weighing either 2, 4, 16, or 64 grains. Configuration control documentation for RCCs has yet to be developed.

The type and scope of ballistic testing of military armor will be determined by the objective of the testing, which may be research and development (R&D), performance certification or confirmation of continuing performance (quality control).

The scope of performance certification, frequently referred to as First Article Testing (FAT), normally involves testing with the anticipated threat(s) over a broad range of environmental conditions (hot, cold, water and oil immersion, etc.) and usage (impact, etc.). Once the performance of a specific armor configuration has been approved through FAT testing, random samples of production armor are tested to ensure the continuing quality. This is frequently referred to as Lot Acceptance Testing (LAT), the scope of which is normally less than that of the FAT.

The objective of R&D Testing, while infinite in scope, generally consists of variations of the procedures used in Performance and Quality Control Testing.

Procurement of military body armor is a multifaceted process, Each of the elements, e.g. flexible vests, rigid chest and back inserts, rigid side inserts, etc. is separately procured. There is no single vendor of rigid and flexible elements. Each of the elements is procured from a host of vendors. For each of these vendors, contracting is dependent on the vendor successfully demonstrating compliance with the full range of physical characteristics and ballistic resistance characteristics for the element of the system being procured. Once a contract is awarded, the vendor must demonstrate continued compliance with the full range of requirements. The pre-award compliance demonstration is known as First Article Testing (FAT), while the continuing compliance demonstration is through a random sampling and testing process of production lots known as Lot Acceptance Testing (LAT).

FAT testing requirements are intended to demonstrate consistently acceptable ballistic performance over a broad range of environmental conditions, while the ambient condition LAT testing is intended to demonstrate consistent configuration and continued ballistic performance of production units of a successful FAT configuration.

The ballistic compliance of helmets of both FAT and LAT is conducted with completed, ready-to-wear assemblies. The ballistic compliance of a candidate flexible armor is conducted with rectangular, 15 × 15 inch panels of ballistic elements known as 'shoot packs'. The LAT ballistic compliance testing of the flexible armor is conducted with complete production assemblies randomly selected from each production lot. Both FAT and LAT ballistic resistance testing of rigid elements is conducted with finished, rigid samples backed with 15 × 15 inch shoot packs provided by approved vendors of the flexible armor.

Ballistic protection testing of all armor, including flexible and rigid textiles, is of two types. One type is intended to determine, in effect, the failure point of the armor by establishing the velocity at which a specific ballistic threat has a 50 percent chance of penetrating the armor. This methodology is known as Ballistic Limit, Protection (V50BL[P]) testing, and is frequently referred as V50 Testing. The other type of testing is intended to confirm or deny the penetration resistance of the armor to a specific ballistic threat at a specified range of fair impact velocities. This methodology is known as ballistic resistance testing or penetration testing, and is frequently, yet erroneously, referred to as V0 Testing.

Since materials with greater resistance to penetration will produce higher V50 values than less resistant armor, V50 testing is most useful in evaluating the relative performance of an armor material with respect to other candidate armors. It is also used to indicate changes in performance with respect to age and/or environmental stress, although it is most often used to confirm or deny compliance with contract requirements (quality assurance testing).

Penetration testing is frequently used to confirm or deny resistance to penetration of a ballistic threat at a specified velocity. The result of this form of testing will support a pass/fail determination only, and is of little value in assessing the relative performance of environmental stress, although it is also frequently used to confirm or deny compliance with contract requirements (quality assurance testing).

Materials intended to protect the torso and head are frequently tested for backface deformation of modeling clay from a non-penetrating impact. This testing, known as blunt trauma testing, is usually conducted in conjunction with ballistic resistance testing of armor materials to ensure non-penetrations do not produce lethal levels of blunt trauma injuries.

Ballistic resistance testing of military textiles often includes a method which determines the loss in velocity of a projectile penetrating armor material. This method is frequently referred to a Vs − Vr (velocity, striking minus velocity, residual) testing, or as Casualty Reduction Testing. By subtracting the residual velocity from the striking velocity, the energy absorbed by the armor material can be estimated.

Procedures, general – The performance of ballistic resistance materials and commodities is tested by either, or both, of two test methodologies (i) Ballistic Limit (V50) testing with, or without, Casualty Reduction determinations and (ii) Ballistic resistance testing, with or without supplemental backface deformation (blunt trauma) testing. Prior to initiation of any testing, there should be a clear understanding of a variety of procedural variations, which will affect the validity and repeatability of the test.

Mounting, rectangular samples – Rectangular samples are most often mounted in rigid frames, and the framing is rigidly supported.

Mounting irregular shapes – Irregularly shaped flexible armor panels, such as body armor elements, are frequently mounted by clamping weighted metallic bar stock (top, bottom and sides) which are independent of one another, leaving as large an impact area as possible but ensuring against slippage between the sample and clamping. The clamped sample is then suspended from its horizontal, upper edge clamping, and is free to swing in the line-of-fire. The bottom and vertical edges are free to move independently of their adjacent edges.

Irregularly shaped rigid armor panels are generally rigidly fixtured, which may involve drilling and bolting, or clamping, of the armor sample to an articulating test stand which provides for impacting of all areas of the sample, in the full range of obliquity requirements.

In addition to drilling or clamping, helmet testing procedures may require mounting of the helmet on a head form using only the strapping of the helmet. The head form of this mounting method will include slotting, front-to-back and side-to-side, to accommodate a witness panel for penetration determinations. Blunt-trauma testing of helmets will require filling those slots with clay for determinations of backface deformations.

Mounting, backed – Many body armor and body armor material tests are conducted with the test sample strapped to clay backing material, which provides post-shot evidence of the extent to which the backface of the armor is deformed.

Shot spacing – Regardless of the mounting configuration of the sample, the minimum distance between impact locations, the minimum distance from any framing used in the mounting, and the minimum distance from the edge of the sample itself must be specified. Each impact will result not only in the disruption of the sample at the point of impact, but will produce a visually discernable distortion in the area surrounding the point of impact. This area, frequently referred to as the 'disturbed area', will vary in size, depending on the make-up of the test sample (cracking, delamination, etc.). It may also be used as a shot spacing requirement. Shot spacing requirements often include the provisions that no disturbed area will extend to the edge of the sample, or interlock, or overlap, the disturbed area of any other impact. Testing procedures of flexible textiles often include a requirement for staggering of the shots to avoid more than one impact on any single horizontal or vertical strand of the weave.

Testing of finished commodities, or assemblies, will include, within the impact spacing requirements, distances of impacts from features of the commodity, such as seams, weldments, curvatures, holes, mounting hardware, etc., or any other feature which differs from the base materials.

Remounting of the test sample between shots may be acceptable or prohibited. When flexible materials are tested, and remounting between shots is permissible, it is usually acceptable to smooth the bunching between plies from previous shots.

Maximum impacts allowable on any single sample should be specified, and this requirement may be dependent on the remaining undisturbed area, which, based on shot spacing, might otherwise support additional impacts.

10.4 Ballistic limit (V50) testing

Ballistic Limit, Protection (V50BL[P]) testing is intended to evaluate armor material by testing either a single armor sample or multiple ostensibly identical armor samples. The test is based on the Bruceton method of evaluating a non-quantifiable reaction to a variable stimulus, e.g. a match ignites or does not, a light bulb flashes or does not, etc. These 'go/no-go' results are used to establish a 50 percent probability of penetration only. The full penetration curve is not explored.

V50 values are based on the average of equal numbers of velocities associated with complete penetration (witness panel penetrated) and partial penetration (witness panel not penetrated). The number of velocities required to determine the V50 value is required to be specified. In order to prevent skewing of the results, the lowest velocity associated with a complete penetration, and the highest velocity associated with a partial penetration, must be included in the V50 calculation.

The following parameters must be specified:

- The maximum permissible range of velocities used to determine a V50 value.
- The minimum distance between impacts, and from fixturing.
- The projectile obliquity, i.e. the angle between the projectile line of flight and the surface of the armor sample.
- In unbacked armor testing, the witness panel from which penetrations are determined, e.g. material, thickness, and distance behind the armor.

V50 testing of armor samples whose area is insufficient to permit the minimum shot spacing and minimum number of impacts can be conducted using two or more ostensibly identical armor samples. This method is termed constructed V50 testing, where one V50 value is constructed from testing multiple ostensibly-identical armor samples.

V50 values cannot be determined if the greatest velocity associated with partial penetration and the lowest velocity associated with complete penetration are not within the maximum permissible range of velocities. Such testing is deemed to be inconclusive, and requires further testing on undisturbed, ostensibly identical samples.

If the lowest velocity associated with complete penetration is lower than the greatest velocity associated with partial penetration, the arithmetic difference is reported as the range-of-mixed results.

V50 testing procedures for helmets most often require that the helmet be divided in five sections – crown, front, back, left, and right sides – and that the fair impacts used to calculate the V50 value be equally distributed among those sections. These procedures usually specify that one of the crown impacts be on, or near, the centroid of the crown area.

V50 values should be reported with other data that are an index of the reliability of the test itself. Such data include:

- V50 – an evaluation of the ballistic limit of the material.
- Qualified impacts – the larger the number of qualified impacts used to determine the V50 value, the greater is the confidence level in the V50 value, i.e. a V50 value based on the velocities associated with five complete penetrations and five partial penetrations is more reliable than a V50 value based on the velocities associated with one complete penetration and one partial penetration.
- Range-of-results – the smaller the range-of-results, the greater is the confidence level in the V50, i.e. the 2000 fps average of one complete penetration of 2001 fps and one partial penetration (1999 fps) is far more reliable than the 2000 fps average of one complete penetration of 3000 fps and one partial penetration of 1000 fps. Except for excesses related to material inconsistencies, excessive ranges-of-results are generally an indication of inexperienced technicians unable to load test cartridges to produce the intended velocities.
- Range-of-mixed results – velocities associated with complete penetrations that are lower in value than velocities associated with partial penetrations represent an inversion in logic that can be explained by inconsistencies in the armor material only. The greater the range-of-mixed results, the less homogeneous the armor material, and the less is the reliability of the V50 value.

10.5   Residual velocity testing

Residual velocity testing is not intended to determine the resistance of an armor sample to penetration, but rather the potential lethality of a projectile after it has penetrated the armor. This form of testing generally does not have an associated pass/fail criterion, as the results are, generally, used to support research studies.

Residual velocity (Vs – Vr) testing is usually conducted in conjunction with V50 testing, provided the V50 testing is conducted with an unbacked armor sample. Velocity computing instrumentation, positioned behind the armor sample, is added to the V50 test and the residual velocity (Vr) is compared with the striking velocity (Vs) for each penetrating shot. The difference (Vs – Vr) is used to assess the level of lethality to be expected from ballistic threats whose velocities exceed the level of protection of the material. In support of this testing, the striking velocity of the initial V50 test shot is often 200 percent, or more, of the expected V50 value of the sample. The velocities of subsequent shots are reduced by a specified amount until a non-penetration is produced. The normal, up-and-down method of V50 testing is then used until the required number of shots to produce a V50 is obtained.

10.6   Ballistic resistance testing

Ballistic resistance testing is intended to confirm, or deny, compliance with minimum performance requirements of ballistic resistant materials. This testing may be conducted with the same, or different, ballistic threat(s) used to conduct V50 testing of the same material. Unlike V50 testing, however, ballistic resistance testing is conducted using a fixed number of impacts fired within a specified range of fair velocities. It may be conducted on backed, or unbacked, flexible, or rigid, material. The armor material may, or may not, have had pre-test exposure to some form of environmental conditioning, e.g. extremes of temperature and humidity, immersion in fresh water or salt water or diesel oil, drop or impact, etc. Since textiles are usually the material used in flexible as well as in rigid body armor, ballistic resistance testing of textiles is usually conducted with the armor sample backed with a deformable material, such as non-hardening modeling clay, to more nearly simulate actual usage and to support blunt-trauma evaluations. Clay-backed testing is conducted with the test sample strapped to a clay block using elastic straps found on body armor.

Ballistic resistance testing is of limited value in determining the relative performance of two or more candidate armors, or the effect of changes in the armor. Specifically, if two armor samples pass a ballistic resistance test, one does not know which of the two armor samples is the more resistant to ballistic penetration.

Experience has shown that multiple-ply, flexible armor can be penetrated by oblique impacts of ballistic threats, even though the armor has success-fully resisted zero degree obliquity impacts. For that reason, ballistic resis-tance testing can include requirements for oblique impacts, e.g. 30 and 45 degrees.

Ballistic resistance test procedures often include requirements for impacts intended to evaluate features of a commodity, or assembly, whose construction varies from that of the base material; e.g. seams, weld-ments, mounting hardware, sub-assemblies of rigid armor, closures of body armor, etc.

Ballistic resistance testing of helmets presents a specialized form of rigid armor testing due to the shape and standoff between the helmet and the skull. Ballistic resistance testing of helmets is usually conducted on a head form using only the mounting straps and suspension system of the assembly. The head form used for this testing has been slotted – front-to-back and side-to-side – to accommodate a penetration witness panel at a specified distance from the inside surface of the helmet. The helmets are pre-marked to establish five areas – crown, front, back, left and right sides. The test procedures will include impacting of each of these locations and require at least one impacting of the suspension system mounting hardware, and the centroid of the crown, which are locations known to be susceptible to pen-etration. Testing may include backface deformation as well as penetration testing. When this applies, the slots of the head form are filled with clay, as required for deformation testing, and penetrations confirmed or denied by the evidence of the projectile, fragments of the projectile, or fragments of the helmet in the clay filler.

10.7   Blunt trauma (back-face deformation) testing

Blunt trauma testing, often referred to as back-face deformation testing, is usually conducted in conjunction with ballistic resistance testing, and may be performed on rigid material, as well as flexible material. The test samples are backed with a substance (usually non-hardening modeling clay) that will be readily deformed by the back-face deformation of the armoring material, from which the extent of that deformation may be determined. The method of determining the level of deformation is dependent on the characteristics of the backing material. Clay backing readily supports post-test, linear and/or volumetric measurements. Other more resilient materials, such as ballistic gelatin, do not retain the deformation and must be instrumented with photographic equipment, force gauges, etc. to examine the dynamics of deformation.

Until the middle of the 20th century, little was known of the lethality of blunt trauma injuries from ballistic impacts. Based on research conducted by the US Army, the US National Institute of Justice included blunt trauma requirements in their standards for body armor intended for use by law enforcement officers. These standards have been accepted worldwide. There are efforts to establish alternative methods for evaluating the lethality of blunt trauma. However, there remains no universally accepted procedure for that purpose, other than procedures based on NIJ standards. Helmet blunt trauma test requirements may provide for testing to be conducted in conjunction with ballistic resistance testing, and, procedurally, requires no change in the penetration testing procedures.

It is possible for well-intended, but often illogical, contractual requirements to result in unintended legal circumvention of otherwise indisputable test results. For example, temperature and relative humidity conditions are frequently included in test standards and imposed contractually. Failure of the testing facility to comply with these environmental requirements has been used by armor manufacturers to void failing test results. It is illogical to assert that such test range temperature or humidity values are more extreme than the temperature or humidity values encountered on the battlefield. Ballistic limit requirements of armor materials occasionally specify a maximum acceptable V50 value. This is, again, illogical, for its inclusion can be used to eliminate a superior performing armor for an inferior one.

Ballistic resistance testing requirements occasionally require additional impacts to be made on an armor sample when a previous impact exceeded, or did not meet, the range of fair velocities. Frequently, these requirements do not specify if an otherwise failing result of this extra impact, or impacts, is to be declared unfair, to the extent it exceeds the number of required impacts, or if this result is to be used to fail the armor sample. Such ballistic resistance test requirements should include provisions for interpretation of the failing results of such an impact when it is preceded by an excessive velocity impact on the same sample. The assumption is that the higher velocity impact has unfairly stressed the sample, thereby rendering any subsequent failure unfair. This matter should be clarified to preclude the possibility of contractual disputes. This claim of unfairly stressing an armor sample is not without merit. However, it does permit a scenario whereby the first impact on every sample is intentionally made at a slightly excessive velocity, thereby invalidating any subsequent, failing impact. A fair compromise may well be to establish a limited range of high velocities within which no claim of unfair stressing is permissible, but beyond which such a claim is valid.

Appendix 10.1 US military standards for armoring materials and commodities

The following is a brief, representative listing of applicable US Military Standards for armoring materials and commodities:

MIL-STD-662F – Department of Defense Test Method Standard (V50 Ballistic Test for Armor)

NIJ-STD-0101.04 – US Department of Justice (Ballistic Resistance of Personal Body Armor)

MIL-P-46593A – Military Specification (Projectile, Calibers .22, .30, .50 and 20 mm Fragment-simulating)

MIL-C-12369F – Military Specification (Cloth, Ballistic, Nylon)

MIL-C-44050A – Military Specification (Cloth, Ballistic, Aramid)

MIL-G-25871B – Military Specification (Glass, Laminated, Aircraft Glazing)

MIL-L-62474B – Military Specification (Laminate: Aramid-fabric-reinforced, Plastic)

MIL-A-12560G – Military Specification (Armor, Steel, Wrought)

MIL-V-43511C – Military Specification (Visor, Flyer's Helmet, Polycarbonate)

Appendix 10.2 Glossary

The following definitions are offered for clarity and to avoid any ambiguities in this chapter:

Areal density – the weight per unit area of armoring material; expressed in units of pounds per square foot (grams per square centimeter); useful in the relative evaluation of two or more candidate materials which otherwise satisfy the same ballistic resistance requirements.

Backface deformation – deformation of the rear surface of the armor material resulting from a non-penetrating impact; usually measured on clay backing.

Ballistic threat – includes all facets of the projectile impact, e.g. projectile caliber, projectile configuration, projectile weight, projectile velocity, and impact obliquity.

Blunt trauma – damage or injury to the protected area of the body from the backface deformation of the armor by a non-penetrating impact.

Environmental stress – including, but not necessarily limited to, extremes of temperature, controlled impacting, and exposure to a variety of liquids; e.g. saltwater, lube oil.

Impact energy – function of the mass and velocity of impact of a projectile; mathematically derived from Kinetic Energy $= \frac{1}{2}\, mv^2$.

Obliquity – angle between the line-of-flight of the projectile and the armor surface at the point of impact.

Penetration: Complete – perforation of a witness panel of thin material behind an unbacked armor sample; caused by projectile, projectile fragments, armor spall, or any combination the three. Partial – any result other than a complete penetration.

Projectile: Ball – lead core with or without thin copper or steel jacket; may include mild steel core ammunition. Armor Piercing – hardened steel or tungsten core with lead filler for weight; thin copper or steel jacket. Fragment Simulator – laboratory projectile of closely controlled configuration used to simulate fragments from exploding munitions, or behind-armor spall.

Spall – fragments of the ballistic threat and/or the armor, which, upon impact, are projected into the protected area.

Textile – woven fabric, or flexible film, used for ballistic protection.

Ballistic Element – that portion of an armor assembly which imparts ballistic resistance protection; usually multiple fabric plies.

Velocity, projectile:– mean velocity of a bullet expressed in feet per second (fps) or meters per second (mps); derived value using instrumentation that determines elapsed time for a projectile to traverse a known, linear distance.

Velocity, striking (Vs) – instrumental mean projectile velocity minus velocity loss from the point at which the instrumental velocity was determined to the armor target.

Velocity, residual (Vr) – instrumental after-armor mean projectile velocity plus velocity loss from the rear of the armor target to the point at which the after-armor instrumental velocity was determined.

Velocity loss (VL) – loss in projectile velocity as it traverses a known distance; derived from tables or from instrumentation used to determine velocity at two or more locations of known distances.

V50, Ballistic limit – mean of velocities associated with an even number of penetrating and non-penetrating impacts within a range of permissible velocities; V50 values usually consist of two, four, six, eight, or ten such velocities.

V50, Range-of-results – the range of velocities associated with impacts used to compute a V50 value.

V50, Range-of-mixed-results – the range of velocities associated with impacts, wherein a complete penetration occurs at a lower velocity than the velocity of a partial penetration.

V50, Inconclusive – a V50 test is deemed to be inconclusive when (i) the range-ofresults exceeds the maximum allowable, or (ii) the range-of-mixed results exceeds the maximum allowable.

Witness panel – a panel of easily penetrated material positioned closely behind, and parallel to, the rear surface of the armor; used to confirm, or deny, penetrating impacts.

11
Chemical and biological protection

Q. TRUONG and E. WILUSZ, US Army Natick Soldier
Research, Development and Engineering Center, USA

11.1 Introduction

In many aspects of everyday living it is necessary to provide protection
against hazardous chemical/biological (CB) materials. Proper protective
clothing is needed during everyday household chores; in industrial, agricultural, and medical work; during military operations, and in response to terrorism incidents. This clothing generally involves a respirator or dust mask,
hooded jacket and trousers or one-piece coverall, gloves, and overboots,
individually or together in an ensemble. Choices must be made as to which
items of protective clothing to select for a given situation or environment.
A number of variables to be considered include weight, comfort, level of
protection, and the duration of protection required. In addition, the types
of challenge to be encountered is of significant consideration. Due to the
large number of variables involved, a spectrum of CB protective materials
and clothing systems have been developed. Fully encapsulating ensembles
made from air impermeable materials with proper closures provide the
highest levels of protection. These ensembles are recommended for protection in situations where exposure to hazardous chemicals or biologicals
would pose an immediate danger to life and health (IDLH).

11.1.1 History of CB warfare and current threats

Chemical warfare agents (CWAs) such as chlorine, phosgene, and mustard
gas (also known as blistering agents) were used in World War I (WWI).
There were over a million casualties with approximately 90000 deaths.[1]
CWAs were not used during World War II (WWII) presumably for fear of
retaliation, and because CB agents are not easy to control on the battlefield.
A sudden shift of wind direction could potentially harm or slow down the
advancing army that deployed the agents.[2] Many countries still resorted to
the use of CWAs over the years. Some examples include the following historical records: Italy sprayed mustard gas from aircraft against Ethiopia in

242

1935; Japan used CWAs when they invaded China in 1936; Egypt used phosgene and mustard gas bombs in the 1960s in the Yemeni Civil War; during the Iran–Iraq war between 1980 and 1988, Iraq used sulfur mustard and nerve agents on their own Kurdish civilians in northern Iraq and in the city of Sardasht.[3]

The Aum Shinrikyo cult used Sarin nerve agent to terrorize Matsumoto city in 1994, and attacked the Tokyo subway system in 1995. The Tokyo subway incident injured about 5000 and killed 12 people. On 23 October 2002, 115 people died as a result of the Russian government's use of the BZ chemical, a 'knockout gas', to subdue about 50 Chechen armed guerillas holding about 800 Russians in a Moscow theater.

The use of biological warfare agents (BWAs) has been recorded as early as the 6th century BC when the Assyrians poisoned enemy wells with rye ergot. More recently, BWAs were used by Germany in WWI. In 1937, Japan used aerosolized anthrax in experiments on its prisoners. There were several suspected 'yellow rain' incidents in Southeast Asia, and the suspected use of trichothecene toxins (T2 mycotoxin) in Laos and Cambodia. In 1979, there was the accidental release of anthrax at Sverdlovsk. In 1978, ricin was used as an assassination weapon in London. In 1991, Iraq admitted its research, development, and BWA weapon productions of anthrax, botulinum toxin, Clostridium perfringens, aflatoxins, wheat cover smut, and ricin to the UN. In 1993, a Russian BW program manager, who had defected, revealed that Russia had a robust biological warfare program including active research into genetic engineering and binary biologicals. Table 11.1 provides a short summary of CWA users and the locations where they were used since WWI.

11.1.2 CB warfare agents and their effects

To design and to fabricate effective CB protective clothing, it is necessary to have an understanding of the hazardous threats that must be prevented from reaching the wearer. The threats comprise the entire spectrum of hazardous CBs. Since the specific characteristics of the threats vary so greatly, effort to develop materials to defeat these threats is an ongoing technological challenge. The military has been concerned since WWI about traditional CWAs, including mustard and organophosphorous nerve agents such as those listed in Table 11.2 and their derivatives, and other toxic chemical groups such as cyanide (causing loss of consciousness, convulsions, and temporary cessation of respiration), pulmonary agents (causing shortness of breath and coughing), and riot control agents (causing burning, stinging of eyes, nose, airways, and skin).[4]

CWAs are defined as natural or synthesized chemical substances, whether gaseous, liquid or solid, which might be employed because of their direct

Table 11.1 CWAs used since WWI

(a)

Year	User	Location
1936	Italy	Ethiopia
1939–40	Japan	Yemen
1960	Egypt	Eritrean rebels
1970	Ethiopia	Laos
1975–81	Soviets	Kampuchea
1979	Soviets	China
1979	Vietnam	Afghanistan
1979–80	Soviets	Iran
1983–87	Iraq	Iran
1988	Iraq	Iraq (Halabja)
1994	Japan	Japan (Matsumoto)
1995	Japan*	Tokyo
2002	Russia**	Moscow

Source: US Army Chemical School.
* Use by Aum Shirinkyo Terrorists, http://www.emergency.com/wter0896.htm.
** Use by the Russian government to solve hostage crisis, http://english.pravda.ru/main/2002/10/28/38794.html.

(b) Yellow rain attacks between 1975 and 1981

Location	Number of attacks	Deaths
Laos	261	6504
Kampuchea	124	981
Afghanistan	47	3042

Table 11.2 Common war chemicals

Nerve chemical warfare agents (CWAs):

VX: O-Ethyl S-2-diisopropylaminoethyl methyl phosphonothiolate
GB: (Sarin) O-isopropyl methylphosphonofluoridate
GD: (Soman) O-pinacolyl methylphosphonofluoridate
GA: (Tabun) O-ethyl *N*, *N*-dimethyl phosphoramidocyanidate

Blister CWAs:

L: (Lewisite) 1:2-chlorovinyldichloroarsine
HD: (Mustard) Bis (2-chloroethyl) sulfide

toxic effects on man, animals, and plants. They are used to produce death, or incapacitation in humans, animals, or plants. Typical effects of selected CWAs[5] are listed in Table 11.3, and Appendix 11.1 shows the effects of other CWAs and their characteristics.[6] Examples of selected volatile CWAs are: hydrogen cyanide, sarin, soman, tabun, and tear gases; and persistent CWAs include VX, thickened soman, mustard agent, and thickened mustard agent. CWAs are usually classified according to their effects on the organisms (Fig. 11.1).

BWAs are microorganisms (viruses and bacteria) or toxins derived from living organisms. They are used to produce death, diseases, or incapacitation in humans, animals, or plants. Typical effects of selected BWAs[7] are also listed in Table 11.3, and Appendix 11.2 shows the effects of other BWAs and their characteristics in more detail.[8] Toxins are chemical substances extracted from plants, animals, or microorganisms, which have a poisonous effect on other living organisms (some can be synthesized in a laboratory.) The sizes of bacterial and viral agents, and toxins which range from sub-micron to the micron-sized carrier aerosol particulates, make the development of protective materials challenging.

Table 11.3 Typical effects of toxic chemicals, microorganisms, and toxins

Toxic chemicals	
Nerve	affect nervous system, skin, eyes
Blood	prevent oxygen reaching body tissues
Blister	affect eyes, lungs, and skin
Choking	affect nose, throat, and especially lungs
Psycho-chemical	cause sleepiness
Irritant	cause eye, lung, and skin irritations
Vomiting	cause severe headache, nausea, vomiting
Tear	affect eyes and irriate skin
Microorganisms	
Anthrax	cause pulmonary complications
Plague	cause pneumonic problems (inflammation of the lung)
Tularemia	cause irregular fever lasting several weeks
Viral encephalitis	affect nervous system (inflammation of the brain)
Toxins	
Saxiloxin (STX)	cause shellfish poisoning – highly lethal
Botulinum A (BTA)	cause food poisoning – extremely lethal
Staphenterotoxin B	cause incapacitating effects

Source: *Jane's NBC Protection Equipment, 1990–91.*

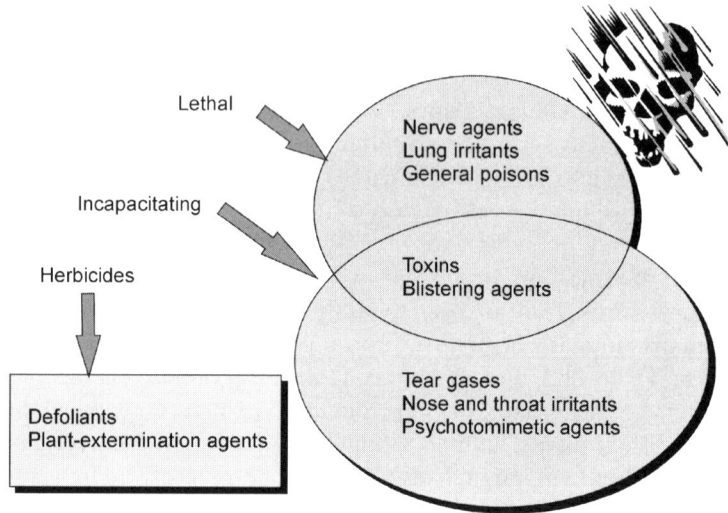

11.1 CWA classification.

11.1.3 Emerging threats

Recent military concerns include toxic industrial chemicals (TICs) as well as any novel CWAs which may be developed. TICs include chemicals such as common acids, alkalis, and organic solvents. It should be noted that TICs have long been a concern for civilian emergency responders and industrial chemical handlers. The military have become concerned more recently as they have encountered industrial chemicals during various deployments. Protection from TICs needs to be assured.

11.2   Current chemical/biological (CB) protective clothing and individual equipment standards

There are many different types and designs of CB protective clothing that are available both for military and civilian use in different environments and threat scenarios. Some of these clothing systems are mentioned below.

11.2.1 Military

Currently, the US army uses chemical protective (CP) combat clothing such as the Joint Service Lightweight Integrated Suit Technology (JSLIST) over-garment[9] over the Battle Dress Uniform (BDU) during missions where there is a chemical threat. The JSLIST, when worn with gloves and boots as shown in Appendix 11.3, provides protection against CWAs for 24 hours. Pictures of the JSLIST overgarment and other chemical protective clothing systems are shown in Appendix 11.4. It is designed for extended use, can be laundered every seven days, and is disposable after 45 days wear even if not contaminated. It contains a cleaner and more breathable sorptive liner material than its predecessor, the Battledress Overgarment (BDO).[10,11] The JSLIST overgarment has an integrated hood and raglan sleeve design which allows more freedom of movement. Its integrated suspenders (braces) allow individualized fitting for soldiers of different sizes. Its wraparound hook and loop leg closures allow easy donning and doffing. Other similar carbon-based air-permeable CP overgarments include the French Paul Boyé's NRBC Protective Suit,[12] and the United Kingdom Mark IV NBC Suit. Another U.S. CP clothing system is the chemical protective undergarment (CPU).[13]

The CPU is a two-piece, snug-fitting undergarment worn under any standard-duty uniform. It is a stretchable fabric that is designed to provide up to twelve hours of vapor protection, and it has a 15-day service life. Recent US Army R&D efforts in individual CB protection have been on the development of advanced CB protective clothing systems that are launderable, lighter, more comfortable, waterproof, and offer equal or better protection to warfare agents, TICs, and emerging CB agent threats as compared to current clothing systems. The aim is to enable the future soldier to operate longer in a CB contaminated environment comfortably, safely, and effectively. One of the current emphases is the use of selectively permeable membranes (SPM) as a component in future military systems. SPM-based CB protective clothing systems are about a third to half the weight of the standard CP clothing systems, depending on the clothing systems designed for different environments.[14]

The CB protective field duty uniform (CBDU) is a concept of an advanced clothing system that is based on SPM technology. Its development has demonstrated that it is possible to limit or eliminate the need for activated carbon, the use of chemical protective overgarments, the use of chemical protective undergarments, the use of butyl gloves, and the use of overboots. The elimination of any or all of these clothing items would represent a significant weight reduction, reduce logistics concerns, and costs, as well as provide an increase in protection and comfort. Other technologies that are in their early R&D stages are being investigated by the US Army and include electro-spun membranes and reactive membranes.

11.2.2 Civilian

Civilians as well as soldiers use special-purpose clothing such as the Improved Toxicological Agent Protective Ensemble (ITAP) and the Self

Contained, Toxic Environment Protective Outfit (STEPO) during domestic emergency operations for chemical spills and toxic chemical maintenance and cleanups in environments with higher threat concentrations. Pictures of the ITAP, STEPO, and other selected civilian emergency response clothing systems are included in Appendix 11.5. The ITAP is used in IDLH toxic chemical environments for up to one hour. It is used in emergency and incident response, routine chemical activity, and initial entry monitoring. ITAP is a CP suit that offers splash and vapor protection, and dissipates static electricity. It can be decontaminated for reuse after five vapor exposures, and it has a five-year minimum shelf life. The US Air Force fire-fighters use the ITAP with a self-contained breathing apparatus (SCBA), a personal ice cooling system (PICS), and standard TAP gloves and boots. The STEPO is a totally encapsulating protective ensemble that provides four hours of protection against all known CB agents, missile/rocket fuels, petroleum oil and lubricants (POL)s and industrial chemicals. The Explosive Ordinance Disposal (EOD) and Chemical Facility (Depot) munitions personnel engaged in special operations in IDLH environments use the STEPO. It can be decontaminated for reuse after five vapor exposures. Since complete encapsulation is very cumbersome, the work duration in the suit is strictly limited because of the limited air supply, and microclimate cooling is necessary for comfort. The STEPO has four hours of self-contained breathing and cooling capabilities. It has a tether/emergency breathing apparatus option. It also has a built-in hands-free communication system.

If the major concern is only liquid splash protection, then full encapsulation may not be necessary. The use of a coverall or apron may be more appropriate. Such items are typically fabricated from the same type of materials as are used in fully encapsulating suits. Vapor protection is then sacrificed for increased comfort and mobility. For lower threat environments, the suit, contamination avoidance liquid protection (SCALP), and Toxicological Agent Protective (TAP) suits are used. The SCALP is made of polyethylene coated Tyvek and it is worn over the BDO. It is designed to protect the users from gross liquid contamination during short-term operations for up to one hour. Decontamination personnel also use it. The TAP suit is issued to personnel (civilian and military) engaged in monitoring and routine clean up at U.S. CB agent stockpile sites, i.e., chemical activities. The TAP suit offers liquid splash protection. It has an adjustable collar, double sleeves, trouser cuffs and adjustable belt. Its hood has a semi-permanent mounting for an M40 mask. The lower portion of hood is a two-layer shawl. The TAP footwear covers protect its butyl TAP boots from gross contamination. It is used with filtered air or SCBA (for up to one hour).

Similar special-purpose clothing to that being discussed is available commercially. As with combat clothing, special-purpose clothing also has limited wear time. They can be constantly cleaned and reused, and repaired if not contaminated. There are various commercially available suits that are actively being marketed for use in events or incidents involving the use of CB agents. Examples of these suits include the air-permeable Rampart suit[15] and the Saratoga Hammer suit,[16] various air-impermeable Tychem[17] suits, and Kappler's Commander Brigade suit.[18] The weights of these uniforms typically range from 4.10 lb (Level B Dupont Tychem BR) to 7.05 lb (Level A Kappler Responder System CPF) or more. Other examples of protective ensembles include suit technologies from Sweden,[19] Germany,[20] and Russia.[21] These suits are different in design and protective capabilities; therefore the potential users must understand their capabilities in order to use them efficiently in different operational environments for specific durations of use. Ongoing efforts by the emergency responders are aimed at developing and field-testing better, lighter, and less costly suits, and SPM-based clothing is being considered and tested for domestic use.

11.3   Different types of protective materials

There are basically four different types of CB Protective Materials.[22] Figure 11.2 illustrates the differences in their protective capabilities.

11.3.1 Air-permeable materials

Permeable fabrics usually consist of a woven shell fabric, a layer of sorptive material such as activated carbon impregnated foam or a carbon-loaded non-woven felt, and a liner fabric. Since the woven shell fabric is not only permeable to air, liquids, and aerosols, but also vapors, a sorptive material is required to adsorb toxic chemical vapors. Liquids can easily penetrate permeable materials at low hydrostatic pressures; therefore, functional finishes such as Quarpel and other fluoro-polymer coatings are usually applied

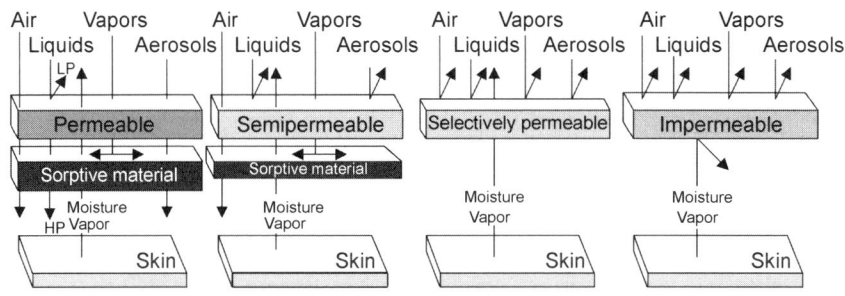

LP: Low hydrostatic pressure   HP: High hydrostatic pressure

11.2 Different types of protective materials.

to the outer-shell fabric to provide liquid repellency. Additionally, a liquid and/or an aerosol-proof overgarment such as non-perforated Tyvek protective clothing must be used in addition to permeable clothing in a contaminated environment to provide liquid and aerosol protection. Many users like to use permeable clothing because convective flow of air is possible through the clothing and open closures. This evaporative action cools the body. Examples of air-permeable protective clothing that contain activated carbon include the US, British, and Canadian current CP clothing.

11.3.2 Semipermeable materials

There are two different types of semipermeable membranes: porous and solution–diffusion membranes.[23,24] Porous membranes include macroporous, microporous, and ultraporous membrane structures. A macroporous membrane allows a convective flow of air, aerosols, vapors, etc., through their large pores. No separation occurs. A microporous membrane follows Knudsen diffusion through pores with diameters less than the mean free path of the gas molecules allowing lighter molecules to preferentially diffuse through its pores. An ultraporous membrane has also been referred to as a molecular sieving membrane where large molecules are excluded from the pores by virtue of their size. A solution-diffusion membrane has also been called a nonporous or a monolithic membrane. This membrane follows Fickian permeation through the nonporous membrane where gas dissolves into the membrane, diffuses across it, and desorbs on the other side based on concentration gradient, time, and membrane thickness. Examples of some semipermeable materials include W.L. Gore & Associates, Inc.'s Gore-Tex[®],[25] polytetrafluoroethylene (PTFE) micro-porous membrane, Mitsubishi's Diaplex[TM,26] polyurethane nonporous membrane, and Akzo's Sympatex[TM,27] copolyester ether nonporous membrane.

11.3.3 Impermeable materials

Impermeable materials such as butyl, halogenated butyl rubber, neoprene, and other elastomers have been commonly used over the years to provide CB agent protection.[28] These types of materials, while providing excellent barriers to penetration of CB agents in liquid, vapor, and aerosol forms, impede the transmission of moisture vapor (sweat) from the body to the environment. Prolonged use of impermeable materials in protective clothing in the warm/hot climates of tropical areas, significantly increases the danger of heat stress. Likewise, hypothermia will likely occur if impermeable materials are used in the colder climates. Based on these limitations, a microclimate cooling/heating system is an integral part of the impermeable protective clothing system to compensate for its inability to allow moisture permeation. ITAP, STEPO, and other OSHA approved Level A suits are examples of impermeable clothing systems. They have been effectively used for protection from CB warfare agents and TICs, but they are costly, heavy, bulky, and incur heat stress very quickly without an expensive and/or heavy microclimate cooling system after donning.

11.3.4 Selectively permeable materials (SPMs)

An SPM is an extremely thin, lightweight, and flexible protective barrier material to CB agents and selected TICs listed in Appendix 11.3, but without the requirement for a thick, heavy, and bulky sorptive material such as the activated carbon material layer being used in current CP protective systems that are discussed above. It allows elective permeation of moisture vapor from the body to escape through the protective clothing layers so that the body of a soldier is continuously evaporatively cooled during missions while being protected from the passage of common vesicant chemical agents in liquid, vapor, and aerosol forms.[29] SPMs have the combined properties of impermeable and semipermeable materials. The protection mechanism of selectively permeable fabrics relies on a selective solution/diffusion process, whereas carbon-based fabrics rely on the adsorption process of activated carbon materials, which has limited aerosol protection, and activated carbon based clothing provides insufficient cooling due to its inherent bulk/insulative properties.

SPMs represent the US Army's pioneering advanced technology.[30] Figure 11.3 shows its material concept. SPMs have been widely used throughout

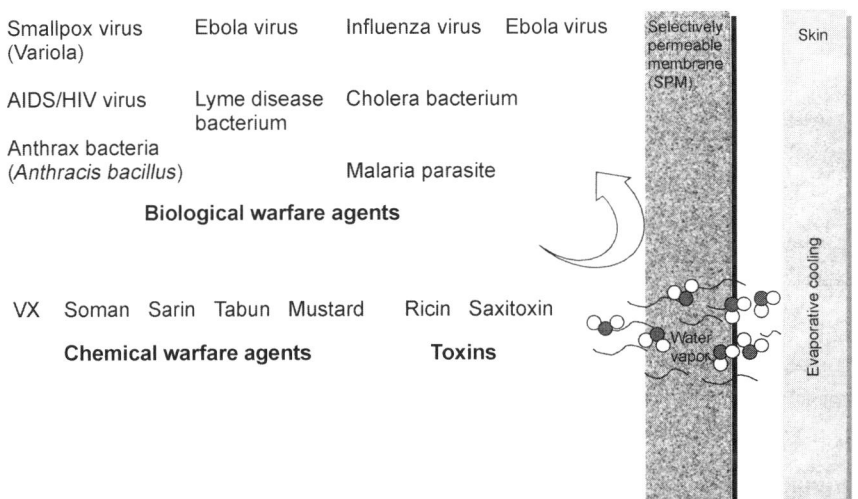

11.3 SPM material concept.

the chemical industry in gas separation, water purification, and in medical/metabolic waste filtration.[31] There have been many different material technologies co-developed by industry and the US Army Natick Soldier Center (NSC). SPMs consist of multi-layer composite polymer systems produced using various different base polymers such as cellulose, polytetrafluoroethylene (PTFE), polyallylamine, polyvinyl alcohol, among other gas or liquid molecular separation membranes. W.L. Gore & Associates, Inc., Texplorer GmbH, Dupont, and Innovative Chemical and Environmental Technologies (ICET), Inc. are among the leading companies that have been pursuing SPM developments with NSC.

Self-detoxification

Catalysts are under development, which are intended to cause the chemical transformation of warfare agents into less hazardous chemicals. These agent-reactive catalysts, when developed and incorporated into fabric systems, will serve to reduce the hazard from chemical contamination, particularly while doffing the contaminated clothing. The addition of catalysts to CB protective clothing systems is not a trivial matter. The US Army has several R&D efforts to incorporate catalysts (reactive materials) such as OPAA-C18 Organophosphorous acid anhydrolase to neutralize G and VX, and polyoxometalate to neutralize mustard agents (Fig. 11.4). N-halamine and quarternary ammonium salts with alkyl chains are also being investigated as biocides, for use in protective clothing via electro-spun fiber pro-

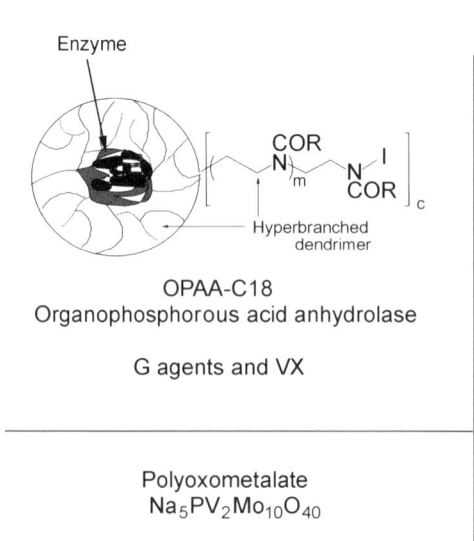

OPAA-C18
Organophosphorous acid anhydrolase

G agents and VX

Polyoxometalate
$Na_5PV_2Mo_{10}O_{40}$

Mustard

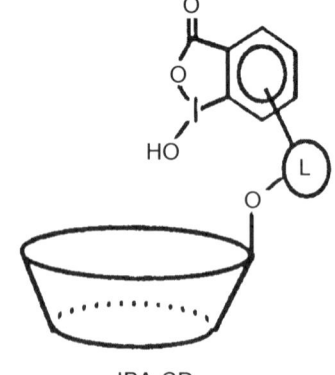

IBA-CD
Iodosobenzoic acid-substituted
β-cyclodextrin

G-agents

11.4 Agent reactive catalysts.

cesses.[32] In addition, the effectiveness of proprietary copper and silver-based nanoparticulates is also being investigated as biocide additives for development of self-decontaminable/biocidal SPMs.

Catalytic reactions, by their very nature, are specific to particular types of chemical. Since there are several different types of chemical warfare agents, one catalyst will likely not be sufficient to do the job against the spectrum of possible challenges.

11.4   Proper protective material designs

Material design is critical in the development of a desirable chemical/biological (CB) protective garment. Users often seek material/clothing that is lightweight, comfortable, durable, low cost, easy care, requires little maintenance, and is compatible with existing individual equipment. In order to develop such material and clothing systems, optimal design work involves contributions from multi-disciplinary engineers and scientists from government, academia, and private industry, in addition to clothing designers, coaters/laminators, fabricators, and the ultimate end-users for final wear assessments. Several different aspects of the importance of proper material designs must be considered. These aspects include: (i) the different types of protective fabrics which are discussed in Section 11.3 and trade-off between protection from toxic chemicals and hazardous microorganisms, and user comfort; (ii) material performance, garment durability, designs of garments and their closure interfaces; (iii) the intended use, environment, productivity, and cost. These aspects are discussed in this section.

CB protective clothing systems have been continually developed and improved over the years. These clothing systems differ in their protective materials, shell and liner fabrics, garment designs, and interfaces/closures (e.g., between gloves and jacket) based on the different levels of protection required and the operating environment. Their common purpose, however, is to provide the user with appropriate protection from hazardous chemicals, toxins, and deadly microscopic organisms. Therefore, the material designers as well as garment designers must understand the protective mechanism(s) in current garments, their components, and functions in order to develop effective CB protective clothing. The users' needs and the intended operational environment are also important. Information on these needs is frequently obtained through interviews, surveys, and/or questionnaires. However, since the protective material is the main component in the design of a CB protective clothing uniform against harmful chemicals and microorganisms, a basic understanding of the following areas is necessary: (i) protection capabilities of current materials; (ii) different concepts of protective materials; (iii) compatibility and integration of protective materials; (iv) wear comfort and material durability; (v) affordability.

Table 11.4 Typical fabric structures and their performance

Structure	Fabric systems	Aerosol penetration (%/10 min.)	Hydrostatic resistance (psi)	Moisture vapor transmission rate (g m/24 h)$^{-2}$
Permeable	Carbon loaded foam	–	0	1087
	7 oz/yd^{2}Nylon/ Cotton	36	0	915
Semipermeable	Plastolon membrane/ 5 oz/yd^{2}Nylon/ Cotton	0	200	1035
	Gore-Tex® II membrane/ 5 oz/yd^{2}Nylon/ Cotton	0	239	713
Sorptive semipermeable	3M Empore membrane	0	52	815
	Soreq NRC membrane	–	290	674
Selectively permeable	'ChemPk Lt-Green'	–	240	764
	'Dehydration' fabric	–	78	824
Impermeable	Parka & Trouser, Wet weather	–	250	<100

11.4.1 Protection capabilities of current materials

The protection capabilities of existing fabrics can be represented in the four different types of materials that were discussed in Section 11.3: permeable, semipermeable, impermeable, and SPMs. These material groups are represented in Fig. 11.2, and their typical performances are summarized in Table 11.4.

11.4.2 Different concepts of protective materials

Differences in protective materials are discussed in Sections 11.3.1–11.3.4.

11.4.3 Compatibility and integration of protective materials

A CB protective ensemble includes three main components: a textile outer layer material (non-woven or woven fabric), an inner layer of CB protective material, and a textile liner fabric. These components must be designed to work synergistically with each other. Individual equipment is also an important part of the ensemble which could include gas masks, breathing filters/devices, micro-climate cooling/heating system, and CB agent detection devices. These individual pieces of equipment and others that are not listed must also be considered for their compatibility and ease of integration into a total CB protective ensemble/garment. Depending on the specific mission or uses, the textile layer could be designed with different functional finishes (e.g., flame protection, water repellency, waterproofing, anti-static), garment designs, boots, gloves, hood, and closures/interfaces (between clothing and boots, gloves, and hood).

Protective materials can be different in their protective capabilities. Adsorption, reaction, and barrier are three different mechanisms that have been identified. Examples of adsorptive materials are activated carbon, zeolites, and aerogels. These materials work by adsorbing chemical vapors in nano-pore structures. The larger the pore surface area, the more desirable the adsorptive material will be. Reactive materials such as Chitosan, chitin, Amberlyst, etc. have been used at Natick to utilize their chemical reactivity in attempts to neutralize chemical agents. The use of reactive materials has been limited because they are reaction-specific to certain chemicals; however, they have great promise for future material development, especially in the area of self-detoxifying and decontaminating clothing.

Barrier materials such as perm-selective membranes and coatings have been gaining acceptance in the user communities as more research and development are carried out by government, industry, and academic institutions. The obvious benefit of a selectively permeable material is that it is extremely light compared with conventional activated carbon-based protective materials. Films, skin creams, aerosol sprays, vaccines, etc., are other materials that have been investigated by other US government agencies for different user scenarios. Combinations of adsorptive, reactive, and/or barrier materials have also been used.

11.4.4 Wear comfort and material durability

Comfort is perhaps the next most important concern after CB. Therefore, the development and application of moisture vapor permeable membranes, water- and oil-resistant coatings, waterproof materials, flexible, elastic, thin, and lightweight materials, as well as those that have high tensile strengths and resistance to tear and puncture damage are being conducted to provide comfortable clothing for the user. Semipermeable (commonly referred to as 'breathable' or moisture vapor permeable) materials are preferred over impermeable materials because they reduce heat stress in warm climates and minimize hypothermia in cold climates. Water-resistant coatings/

functional finishes are used to minimize weight gain by water or other liquid adsorption. Waterproofing is to keep the individual dry when navigating in wet environments. If waterproof fabrics are used together with waterproof closures, the users will be kept dry when crossing streams and rivers. Materials that are thin, flexible, and lightweight offer textile comforts and ease of garment fabrication. Materials with high tensile strength, resistance to tears and puncture or that possess elasticity offer fabric durability.

11.4.5 Affordability

Life cycle cost, or the total government cost to acquire and own an item or a system over its useful life, is very important in the material design stage. Designers must consider the projected cost of development, acquisition, support, and disposal of the item or system for which the material will be used. Affordability involves life cycle costs in the concept exploration/ definition phase, concept demonstration/validation phase, during full-scale development, and finally in operation and support of an end-item. It is a concern in all phases. This report will only emphasize the fact that an increase in competition would result in lower cost materials. Therefore, Natick has been encouraging new material development that would afford comparable CB agent protection, but with lower costs. Affordability is an important consideration in material design because, even if a very expensive protective garment provides the best CB protection, it would probably be prohibitively so for common users such as the infantry. In special military or laboratory operations, an expensive, protect-all garment may be purchased and used; however, this would make the CB protective clothing market an unattractive R&D investment by industry's material developers.

11.5 Clothing system designs

The use of excellent protective materials, effective closures, and ergonomic survival equipment for an individual soldier will be meaningless and unproductive without proper garment designs. Therefore, in designing materials, a designer should be familiar with garment design and fabrication. Material designers should understand that garments are designed differently, based on the characteristics of the protective materials, different applications, and environment to protect, and to maximize the time that a user can operate while wearing the protective garment. There are different garment designs with one-piece garments, two-piece garments, over-garments, under-garments, multi-layer garments, and last but not least important, closures and interfaces.

11.5.1 Coverall or one-piece garments

A one-piece garment eliminates agent penetration through the opening between jacket and trouser/pants. It allows for quick donning and doffing. Another advantage is that it has simplified seaming and sewing in joining fabric pieces during garment fabrication. However, there is no option to open jacket and/or pants for quick release of heat stress/body chill, or exchange of torn/defective jacket or pants. The whole garment must be replaced when it becomes defective and loses its protection.

11.5.2 Two-piece garments

A two-piece garment needs a closure system to seal the opening between jacket and pants. It also requires more seam sealing, sewing, and stitching in joining fabric pieces during garment fabrication. However, it allows donning and doffing for quick release of heat stress/body chill, and exchange of torn/defective jacket or pants. This also allows greater flexibility in sizing users with different dimensions.

11.5.3 Undergarments

Undergarments include underwear and other liner fabrics. They provide protection from the inside and must be worn before the mission. Protection materials/fabric is concealed. They are best used in situations where concealment of protective clothing is required such as in special security operations.

11.5.4 Multilayered garments

Multilayered garments seem to be most popular since they can be oriented toward specific mission(s) in different environments. Clothing layers with specific functions can be donned and doffed for various protection levels (e.g. environmental, chemical, thermal insulation, and/or ballistic protection). They also provide the option for quick release of heat stress or to alleviate hypothermia, or to exchange torn/defective jacket or pants. However, the users must be conscious of heat stress, as more layers are added, in order to prevent heat stress injuries. When using multilayer garments, a microclimate cooling system may be necessary. Users must be educated in the protective capabilities of all available layers for maximum protection and environmental adaptability. It will also be time consuming for donning and doffing of clothing, and compatibility between different layers may be an issue since they may not have been developed synergistically.

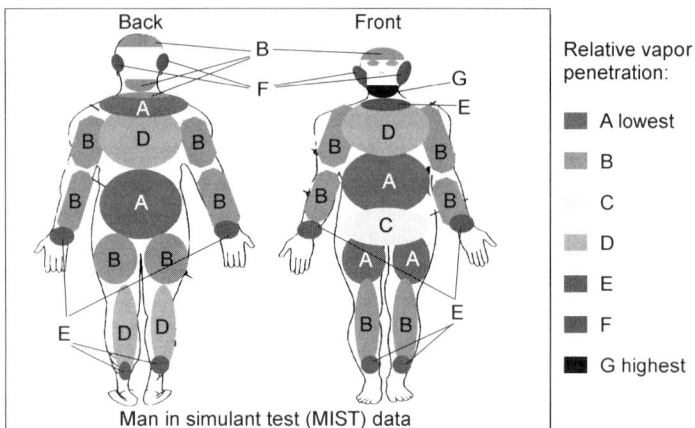

11.5 Vapor penetration resistance of a candidate ensemble.

11.5.5 Closure system, components, and systems

Closure interfaces between hood and gas mask, jacket and gloves, jacket and trousers, and trousers and boots are very important in a CB protective garment. Closure systems are very important because protection is a function of fabric, closure/interfaces, activity level, and the motions of the user. To assess these systems, the US Army has developed a test called Man-In-Simulant-Test, commonly referred to as the MIST test. Figure 11.5 confirms that closure systems are necessary for all CB protective fabric systems to improve protection. Natick has begun to develop a closure system for use with a selectively permeable fabric system. However, there remain technical barriers to overcome.

Current work at NSC addresses concerns for closure system weight, add-on cost to current CP uniforms, and time factor for donning and doffing optimized closures vs. soldiers' comfort and performance. Soldiers' comfort perception of being encapsulated is being studied. Redesign of current gas mask(s) and uniform(s) may be needed and therefore is being addressed by the US Army Edgewood Chemical and Biological Command (ECBC) and the NSC, respectively.

11.6   Testing and evaluation of chemical/biological (CB) protective materials and clothing systems

11.6.1 Material level testing

Table 11.5 shows the performance goals used for research and development of CB protective materials.

Table 11.5 Performance goals for CB protective materials

Chemical protection
Blister (HD, L) agent ≤4 µg/cm^2
Nerve (GB, GD, VX) agents
≤10 µg/cm^2
TOP 8-2-501 (AVLAG test). Preferred.
CRDC-SP-84010 (Mary Jo Waters test used in the past)

Chemical agent deactivation
Diisopropylfluorophosphate (DFP):
Must exhibit significant chemical reactivity
(>50 wt% of CWA neutralized)
ECBC CWA TM or NSC TM (will find TM#)

Moisture vapor transmission rate
≥700 g.m^{-2}/24 h
ASTM E96-95, Procedure B.

Water vapor flux @ 32 °C
≥1800 g.m^{-2}/24 h
ASTM F2298, Procedure B. (DMPC)

Hydrostatic resistance ≥35 lb/in^2
(water → liner)
ASTM D3393-75 or FTMS191A TM 5512

Bonding strength ≥10 lb/in^2
FTMS191A TM 5512 (water → shell fabric)

Weight ≤7 oz/yd^2 (3-layer fabric laminate)
FTMS191A TM 5041

Torsional flexibility Pass
FTMS101A TM 2017

Water permeability after flexing
@ 70 °F and −25 °F: Pass
FTMS191A TM 5516

Biological protection Zero penetration of all microorganisms (10 to 0.001 µm)
USARDEC/NSC's Aerosol Penetration Test Method.

Biocidal activities Must exhibit biostatic (retard attraction of bacteria and viruses on surface) and sporicidal activities (bacteria and virus kill ability)
ASTM TM or AATCC TM (will find TM#)

Tensile strength @ break Warp: >200 lb; Fill: >125 lb.

Elongation @ break >35%
FTMS191A TM 5034

Abrasion resistance >5000 cycles
FTMS191A TM 3884

Delamination Pass
FTMS191A TM2724

Stiffness ≤0.01 lb
FTMS191A TM5202

Thickness ≤18 mils (3-layer fabric laminate)
FTMS191A TM5030

Dimensional stability (Unidirectional shrinkage <3%)
FTMS191A TM2646

Laundering Pass >5 times without delamination
FTMS191A TM2724

Chemical warfare agent simulation permeation ≤25 g/m^2/24 h
USARDEC Inhouse Test Method

USARDEC/NSC: US Army Research, Development, and Engineering Command/Natick Soldier Center; ECBC: Edgewood Chemical/Biological Center/SBCCOM; FTMS: Federal Test of Material Standard; TM: Test Method; ASTM: American Standard of Testing Materials.

11.6.2 Chemical barrier properties

The chemical surety test (known as the live chemical agent test) has been evolved from older methods such as the US Army ECBC's EATM 311-3[33] and the CRDC-SP-84010[34] to the current TOP 8-2-501[35] test method which is used by the US Army to qualify clothing prior to its formal acceptance and classification. These barrier tests include a flooded surface test and laid drop test where the surface of the test sample mounted in the test cells is either saturated with liquid CWAs over the entire surface of the test sample, or the surface is gently laid with droplets of live agent simulants, and the agent permeation is measured over time. An agent contamination density of 10 g/m^2 is often selected in a 24 h test. MINICAMS is used to monitor agent vapor permeation. Vapor permeation (cumulative) will be reported as nanograms/cm^2 versus time. Agent simulant tests[36] with simulants such as trichloroethylene (TCE), methyl salicylate (MeS), dimethyl methyl phosphonate (DMMP), dichloropentane (DCP), dichlorohexane (DCH) and triethyl phosphate (TEP) are often used as 'quick checks' or guides during the material development phase.

NFPA 1994: this is a performance standard released in August 2001 for testing protective ensembles for CB terrorism incidents.[37] This standard defines three classes of ensembles based on the perceived threat at the emergency scene. Differences between the three classes are based on: (i) the ability of the ensemble design to resist the inward leakage of chemical and biological contaminants; (ii) the resistance of the materials used in the construction of the ensembles to chemical warfare agents and toxic industrial chemicals; (iii) the strength and durability of these materials. All NFPA 1994 ensembles are designed for a single exposure (use). Ensembles must consist of garments, gloves, and footwear. Table 11.6 shows the differences between the three classes of NFPA 1994 approved materials.

Toxic industrial chemical testing includes testing by the American Society for Testing and Materials (ASTM) F739/1000, NFPA 1994, and ITF 25 test procedures. These tests measure the permeation of toxic chemicals that are being used by the industry. Although TIC testing is as stringent as the safety protocols of warfare chemicals (nerve and blistering agents) testing, but similar test precautions are taken because in sufficient dosage, TICs can be as deadly as that of CWAs. Appendix 11.6 lists these chemicals. NFPA 1994-test procedure is briefly described in Section 6.1.1, and its full text could be requested from the National Fire and Protection Agency or reviewed online.[37] The ASTM Test Method F739 measures the permeation of chemicals through protective materials, and the ASTM 1001-89 lists these chemicals. This method evaluates the materials' chemical resistance to liquids or gases where their breakthrough time and permeation rate are measured. The test results are reported as belonging to indices 0 to 3. Index 0 is the

Table 11.6 National Fire Protection Agency (NFPA) 1994 Standard

Class	Challenge	Skin contact	Vapor threat	Liquid threat	Condition of victims
1	Vapors Aerosols Pathogens	Not permitted	Unknown or not verified	High	Unconscious, not symptomatic and not ambulatory
2	Limited vapors Liquid splash Aerosols Pathogens	Not probable	IDLH	Moderate	Mostly alive, but not ambulatory
3	Liquid drops Pathogens	Not likely	STEL	Low to none	Self-ambulatory

best and most resistant material and is recommended. Index 1 indicates a highly resistant material and may often be accepted by an industrial hygienist for harmful chemicals. Index 2 requires a greater degree of judgement by an industrial hygienist before it will be accepted. Index 3 materials are not usually sufficiently protective to be recommended by industrial hygienists unless there is no other choice or unless the work involves protection only against occasional splashes or compounds that are not very harmful.

11.6.3 Chemical reactive properties

Chemical reactivity testing, catalytically and non-catalytically, measures the performance of reactive materials to the challenging CWAs or simulants. Although these chemically reactive materials have been in existence for a long time, recent efforts focus on incorporating them into clothing for potential development of self-decontaminable CB clothing systems.

11.6.4 Biological barrier properties

The barrier properties of CB protective fabrics are tested using NSC's in-house test method to measure the aerosolized penetration of MS2 viral and *Bacillus globigii* bacterial spores.

11.6.5 Biocidal activity/properties

The US Army Edgewood Chemical & Biological Command (ECBC) test protocol is used to test the biocide-containing materials and a fabric system's ability to kill BWAs such as anthrax. The kill rate of BWAs and simulants such as spores of non-virulent *Bacillus anthracis* are measured. The two-test protocols to assess the test material and fabric's sporicidal/bactericidal effects include Protocol A, which involves spore plating on nutrient (DIFCO) plates, while the test material is in close contact with the spores. Testing was done to determine the biocidal activity of the test membrane/fabric to anthrax. The anthrax spore dilutions were prepared ranging from 10^{-1} to 10^{-5}. The stock spore titer was $\sim 1.5 \times 10^8$/ml. Spore counts (survival or colonies formed) in the presence of the test material are documented. Protocol B involved growth of spores in nutrient broth media (DIFCO) in the presence of test materials. The procedure in this protocol used 2-ml nutrient broth in 12×75 mm tubes. The 1 inch diameter test sample was put at the bottom of the tube. The test samples were completely submerged in the broth. An aliquot of $50\,\mu l$ from 10^{-2} dilution ($\sim 30\,000$ spores) was added to each tube, and the tubes were shaken in a 'New Brunswick' shaker at 180 rpm at 30 °C for 36 hours. The absorbance was read at 600 nm.

11.6.6 Physical properties

Thickness[38]

The thickness of the membranes, fabrics and fabric systems were measured at 4.1 KPa pressure head using FTMS 191A TM 5030.

Weight[39]

The test samples' weights were measured using FTMS 191A TM 5041.

Aerosol penetration resistance[40]

Figure 11.6 displays the diagram of NSC's in-house aerosol penetration testing apparatus. The apparatus contains two important parts, namely, an aerosol generator and a detector. A potassium iodide salt–water solution is used to generate salt aerosols. The solid particle sizes are in the range of 2 to 10 μm with a 4.5 μm mean size. With 0.5 weight percent, the particle sizes shift to a range of 1 to 10 μm with a 3.5 μm mean size. An AEROSIZERO®, Amherst Instruments Inc. (software version 6.10.09), is used to analyze counts and the size of particles that can range from 0.5 μm to 200 μm.

Hydrostatic resistance

The water penetration resistance of the membrane-fabric was measured by Federal Test Method Standard (FTMS) 191-A, Test Method (TM) 5512.[41]

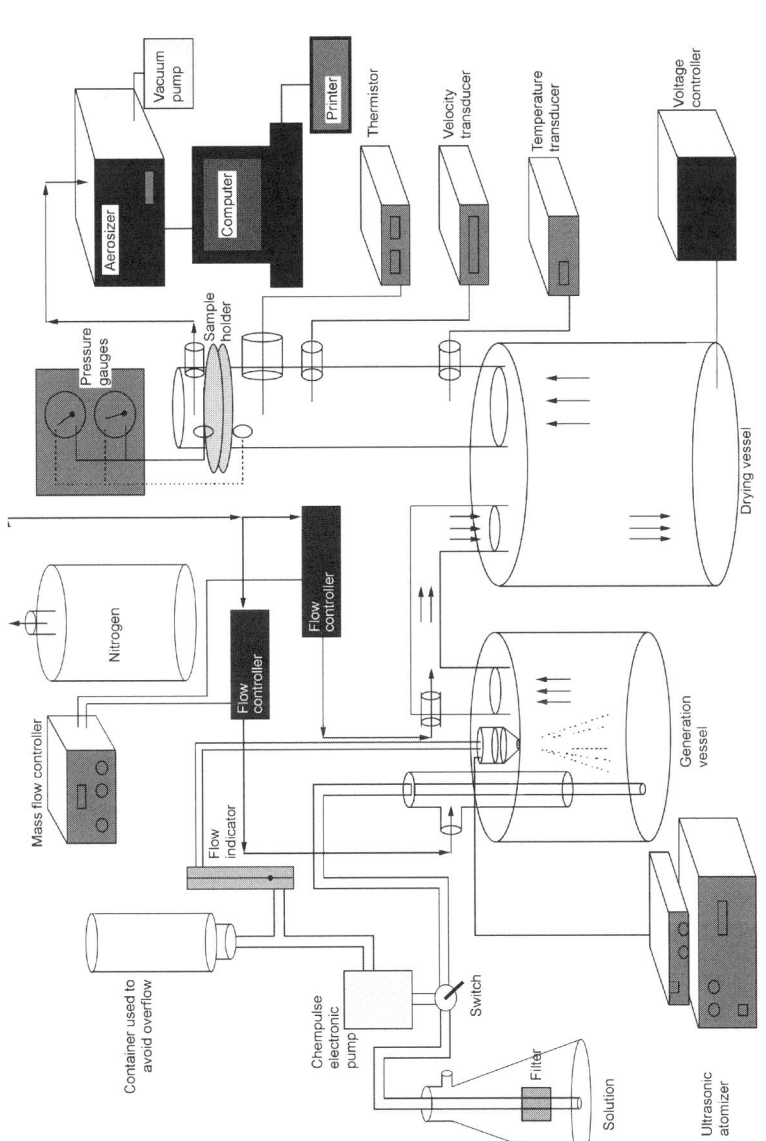

11.6 Aerosol penetration testing apparatus.

Vacuum pump

Printer

Thermistor

Velocity transducer

Temperature transducer

Voltage controller

Aerosizer

Computer

Sample holder

Pressure gauges

Drying vessel

Nitrogen

Flow controller

Flow controller

Flow controller

Mass flow controller

Flow indicator

Generation vessel

Container used to avoid overflow

Chempulse electronic pump

Switch

Filter

Solution

Ultrasonic atomizer

FTMS 191A TM 5514[42] was sometimes used for systems with low-pressure hydrostatic resistance or to test the membranes alone. The membrane faces the water with the fabric reinforcement behind the membrane during testing.

Stiffness[43]

FTMS 191A TM 5202 is designed to determine the directional flex-stiffness of cloth by employing the principle of cantilever bending of the cloth. A Tinius Olsen Stiffness Tester using a 0.46 kg moment, fixed weight is used. The load needed to cause a 60° deflection is measured to calculate the sample's stiffness (flexibility).

Bonding strength[41]

The degree of the cohesion between the fabric and the membrane was measured by the same high-pressure hydrostatic resistance (HPHR) method described above, except that during the tests, water is applied to the shell fabric until the membrane breaks or balloons away from the fabric.

Torsional flexibility[44]

This test is designed to determine the torsional flex–fatigue of cloth by employing twisting and pulling actions to the fabric sample tested. A total of 2000 cycles is used as passing this test. The test is usually conducted at room temperature and at −25 °C for 2000 cycles to measure the effectiveness of the test materials. FTMS 191A TM 5514 is used to measure the integrity of the tested fabric materials for water leakage. If the fabric leaks, it is considered to have failed the torsional flex test.

Scanning electron microscopy[45]

Surface and cross-sections of the membranes were viewed and photographed using an AMRAY Scanning Electron Microscope (SEM) model 1000A. The samples were mounted on aluminum–tin mounts and sputter coated for three five-minute intervals using gold–palladium. The samples were then viewed in the SEM at 10 or 20 kilovolts. Selected SEMs were also taken using an environmental SEM.

Guarded hot plate[46]

The thermal insulative value and the moisture vapor permeability index were measured as outlined in the American Society for Testing Materials

(ASTM) Method D1518-77, 'Thermal Transmittance of Textile Materials Between Guarded Hot Plate and Cool Atmosphere'.

11.6.7 Moisture vapor transport properties

Evaporative cooling potential is measured using the ASTM standard F2298-03.[47,48] This is a test method which determines the amount of water loss over time through a wide range of relative humidity.

11.6.8 Durability testing

System level testing

System level testing of garments as a system is important because these system tests allow the users to see how durably the garments are fabricated, and most importantly how well they protect the wearer. It is also to see how the user is affected by wearing the suit. The experimental suits are usually tested along with commercial off-the-shelf clothing items for test result comparison. Man-in-simulant testing, aerosol testing, physiological testing, rain-court testing, and field exercises are essential and must be performed to find out how well these garments protect the user.

Rain-court testing (NSC test facility)

This test is to see how well a garment resists penetration and determines if there are any leakage points. For statistically valid sampling, eight different suits are required in the rain court, and the testing is performed at the rate of one inch per minute. These tests are performed using manikins that are wearing cotton long underwear, appropriate respirators, and butyl gloves. The manikins are checked from the start of the test at 5, 10, 15 and 30 minutes. Soldier volunteers can also be used, but with a test protocol that has been approved by the Army Research Institute for Environmental Medicine (ARIEM) Human Use Committee. The test lasts for one hour. This will give an indication of any leakage, especially at the sewn seams of the suits and at the interface areas (sleeve-to-glove, trouser-to-jacket, and boot-to-leg). Any sign of leakage at each time period is recorded and reported.

Aerosol system testing (RTI test facility)[49]

This test is to determine how well the chemical protective ensembles protect against penetration by aerosol particulates.

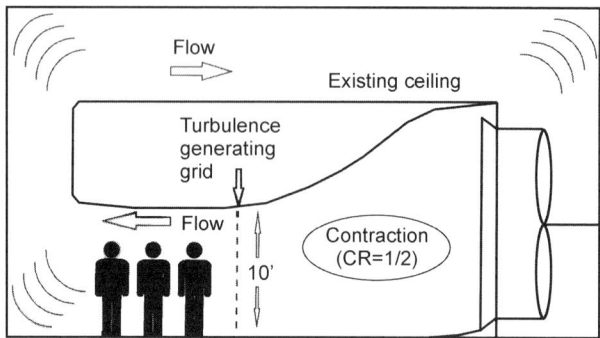

11.7 Schematic of a MIST chamber.

Vapor system testing (man-in-simulant test)

Vapor system testing is a system-level test that measures the amount of vapor that penetrates each suit over a certain period of time. Human test subjects wear each chemical protective garment, along with the appropriate breathing apparatus, and passive adsorption dosimeters (PADs), and enter a man-in-simulant-test (MIST) simulant chamber, and perform a series of physical activities that provide a full range of motion and uniform exposure to a wind stream for two hours. The chamber uses methyl salicylate (MS) as the operative chemical agent simulant. This is used due to its low toxicity and close physical characteristics to those of sulphur mustard (H) vapor. MS is commonly known as oil of wintergreen. The chamber is kept at 27 °C, has a relative humidity of 55%, wind speed of 3–4 mph, and a MS concentration of 85 mg/m^3 throughout the test. The PADs are affixed directly to the skin on the areas of the body shown in Fig. 11.7 to determine how much vapor comes in contact with the body. PADs have the same adsorption rate as human skin to give an accurate measure of the amount of simulant that penetrates the suit. They are removed after the tests and analyzed to determine the protection factor of each suit.[50,51] A manikin has also been used at the Natick Soldier Center (NSC) to test garments and closure designs to cut down actual human based testing cost and time. Figure 11.7 shows a schematic of the NSC MIST chamber.

Physiological testing

The Army Research Institute for Environmental Medicine (ARIEM) conducts physiological testing for NSC using live subjects (soldier volunteers) on each chemical protective suit to determine the effects that wearing the suit has on the user.[52] Initially, each suit is measured on a thermal manikin to get a baseline clo (insulative) value. This baseline measurement also gives us an idea of the degree of heat stress that the live participants will encounter when they don the suits. Heat stress, core temperature, and other physiological signs are measured on each participant wearing the various protective suits. It is hoped that the results of these tests will show significant positive differences in the heat stress levels of the SPM technology over the carbon-based adsorptive technology for CB protection.[53] The clothing system components such as suit (coverall, or jacket and trousers), mask, underwear, socks, gloves, and boots are procured and sized for the subject volunteers prior to the testing to ensure proper fit.

Limited field experiments

Limited field experiments typically range from one week to as long as four weeks with a maximum of two weeks of test time conducted to assess clothing system designs, durability and user comfort while wearing the experimental clothing systems. Limited field experiments are based on ARIEM HUC's approved test protocol, and with structured questionnaires that are used to interview soldier volunteers at the conclusion of the testing. Control garment(s) are used for comparative purposes. Locations are selected by the program managers based on the intended environments and climates. Examples of a few field test locations include: Aberdeen Proving Ground, Aberdeen, MD; Fort Benning, Georgia; Ft. Lewis, Washington; and the Marine Corps Base, Hawaii.

11.7   Future trends

Current and future efforts are concentrated on: (i) novel closure systems for use with carbon-based clothing and SPM-based clothing; (ii) super activated carbon and reactive materials for potential replacement of the current sorptive material that is being used in carbon based fabric systems such as the JSLIST overgarments; (iii) moisture-permeable butyl rubbers for replacement of the current butyl glove to improve comfort through evaporative cooling; (iv) electro-spun nanofiber-based membranes for lighter weight clothing system; (v) nanoscale materials for improved strength and CB protection; (vi) elastomeric SPMs (eSPMs) for minimizing the number of garment sizes and improved CB protection and comfort; (vii) smart materials such as shape memory polymers for allowing greater comfort when used in high-temperature environments; (viii) self-decontaminable materials such as catalytically reactive SPMs for increased safety and protection of wearers as well as support personnel not in CB protective outfits; (ix) biocidal materials for instant-viral/bacterial kill SPMs; (x) TIC resistant SPMs for use in urban warfare environments; and (xi) induction-based fluidic moisture vapor transport facilitated CB protective systems for better and more comfortable protective clothing than that of current fabric systems.

11.8 Acknowledgments

The authors would like to acknowledge the contributions from leaders, scientists, and engineers from the United States and foreign governments, and US Army Natick Soldier Center's industry partners who have been working to provide the individual soldier with comfortable clothing and better protection from toxic war and industrial chemicals, deadly microorganisms, and chemically and biologically derived toxins.

11.9 References

1 http://encyclopedia.fablis.com/index.php/Use_of_poison_gas_in_World_War_I
2 http://www.ndu.edu/WMDCenter/docUploaded/2003%20Report.pdf, At the Crossroads Counterproliferation and National Security Strategy, A Report for the Center of Counterproliferation Research, p. 8. Apr 2004.
3 http://encyclopedia.fablis.com/index.php/Iran-Iraq_War
4 Jane's Information Group, *Jane's Chem-Bio Handbook*, 1340 Braddock Place, Suite 300, Alexandria, VA 22314-1651, Quick Reference.
5 *Jane's NBC Protection Equipment*, 1340 Braddock Place, Suite 300, Alexandria, VA 22314-1651, 1990–1991.
6 http://www.nbcindustrygroup.com/handbook/pdf/AGENT_CHARACTERISTICS.pdf, p. VI.
7 http://encyclopedia.fablis.com/index.php/Biological_warfare
8 http://www.nbcindustrygroup.com/handbook/pdf/AGENT_CHARACTERISTICS.pdf, p. V.
9 Military Specification MIL-DTL-32102, JSLIST Coat and Trouser, Chemical Protective, 3 April 2002.
10 Military Specification MIL-C-43858A, Cloth, Laminated, Nylon Tricot Knit, Polyurethane Foam Laminate, Chemical Protective and Flame Resistant, 17 September 1981.
11 Military Specification MIL-S-43926, Suit, Chemical Protective.
12 http://www.paulboye.com/products_nbcf_1.html
13 Military Specification MIL-U-44435, Undershirt and Drawers, Chemical Protective and Flame Resistant.
14 Military Medical/NBC Technology, *NBC Threat Specialist Q&A – Detection, Protection Rank High on the Army's Medical Technology Agenda*, Vol. 5, Issue 3, 2001, pp. 20–24.
15 http://www.approvedgasmasks.com/suit-rampart.htm
16 http://www.nbcteam.com/products_saratoga.shtml
17 http://www.approvedgasmasks.com/protective-suits.htm
18 http://www.labsafety.com/store/product_group.asp?dept_id=18195&cat_prefix=5WA

19 http://www.frenatus.com/

20 http://www.trelleborg.com/protective/template/T036.asp?id=523&lang=

21 http://www.wolfhazmat.de/hazmat_russia.htm

22 TRUONG Q., M.S. THESIS, Test and Evaluation of Selectively Permeable Materials for Chemical/Biological Protective Clothing, May 1999, University of Massachusetts Lowell, Lowell, Massachusetts.

23 TRUONG, Q., RIVIN, D., *Testing and Evaluation of Waterproof/Breathable Materials for Military Clothing Applications*, NATICK/TR-96/023L, US Army Natick RD&E Center, Natick (1996).

24 W.S. WINSTON HO and KAMALESH, K. SIRCAR, eds., *Membrane Handbook*, Chapter 1, Van Nostrand Reinhold, NewYork, 1992.

25 http://www.gore-tex.com/

26 http://www.diaplex.com/

27 http://www.sympatex.com/

28 WILUSZ, E., in *Polymeric Materials Encyclopedia*, J.C. Salamone, ed., 899, CRC Press, Boca Raton (1996).

29 http://www.bccresearch.com/membrane2003/session4.html

30 TRUONG, Q., U.S. Army Natick RD&E Center, Contract DAAK60-90-C-0105, *Selectively Permeable Materials for Protective Clothing*.

31 KOROS, W. and FLEMING, G., *Journal of Membrane Science*, 83 (1993).

32 HEIDI L. SCHREUDER-GIBSON, QUOC TRUONG, JOHN E. WALKER, JEFFERY R. OWENS, JOSEPH D. WANDER, and WAYNE E. JONES JR., 'Chemical and Biological Protection and Detection in Fabrics for Protective Clothing', *Material Research Society (MRS) Bulletin*, Volume 28, No. 8, Aug 03. *http://www.mrs.org/publications/ bulletin/2003/ aug/aug03_abstract_schreuder-g.html*

33 CIBOROWSKI, S., ERDEC Data Report No. 196, US Army Chemical RD&E Center, Edgewood (1996).

34 WATERS, M.J., *Laboratory Methods for Evaluating Protective Clothing Systems Against Chemical Agents*, US Army CRDC-SP-84010, June 1984.

35 *Chemical Agent Testing*, US Army TOP-8-2-501.

36 RIVIN, D. and KENDRICK, C., *Carbon*, 35, 1295–1305 (1997).

37 National Fire Protection Agency (NFPA) 1994, *Protective Ensembles for Chemical/Biological Terrorism Incidents* (2001 edition). Online reviewis available at: http:// www.nfpa.org/itemDetail.asp?categoryID=279&itemID=18172& URL=Codes%20and%20Standards/Code%20development%20process/ Free%20online%20access&cookie%5Ftest=1

38 PARK, H.B., RIVIN, D., *An Aerosol Challenge Test for Permeable Fabrics*, U.S. Army Natick Research, Development and Engineering Center Technical Report, NATICK/TR-92/039L, July 1992.

39 Federal Test Method Standard No. 191A, Test Method 5030, Thickness of Textile Materials, Determination of, 20 July 1978.

40 Federal Test Method Standard No. 191A, Test Method 5041, Weight of Textile Materials, Determination of, 20 July 1978.

41 Federal Test Method Standard No. 191A, Test Method 5512, Water Resistance of Cloth; High range, Hydrostatic Pressure Method, 20 July 1978.

42 Federal Test Method Standard No. 191A, Test Method 5514, Water Resistance of Cloth; Lowrange, Hydrostatic Pressure Method, 20 July 1978.

43 Federal Test Method Standard No. 191A, Test Method 5202, Stiffness of Cloth, Directional; Cantilever Bending Method, 20 July 1978.

44 Federal Test Method Standard No. 101A, Test Method 2017, Flexing Procedures for Barrier Materials, 13 Mar 1980.
45 Stereoscan 100 SEM, Cambridge Instruments Inc., Eggart and Sugar Roads, Buffalo, New York, NY 14240.
46 American Society for Testing Materials D1518-77, Thermal Transmittance of Textile Materials Between Guarded Hot Plate and Cool Atmosphere.
47 P. GIBSON, C. KENDRICK, D. RIVIN, L. SICURANZA, and M. CHARMCHI, 'An Automated Water Vapor Diffusion Test Method for Fabrics, Laminates, and Films', *Journal of Coated Fabrics*, 24, 322–345, 1995.
48 ASTM Standard Test Methods for Water Vapor Diffusion Resistance and Air Flow Resistance of Clothing Materials Using the Dynamic Moisture Permeation Cell, ASTM F2298-03.
49 Aerosol Protection System Testing. US Army TOP 10-2-022.
50 Royal Military College of Canada, 2002, Canadian Standard Vapour Protection Systems Test Standard Protocol.
51 US Army Standard Vapor Protection Systems Test Standard Protocol. US Army Dugway Proving Ground.
52 US Army Research Institute of Environmental Medicine Standard Test Protocol.
53 US Army Dugway Proving Ground MIST Test Report for the Author (Quoc Truong).

Appendix 11.1 Chemical warfare agent characteristics

				PHYSICAL AND CHEMICAL PROPERTIES				
Agent Type	Chemical Agent; **Symbol** Chemical Structure	Molecular Weight	State @ 20°C	Odor	Vapor Density (Air = 1)	Liquid Density (g/cc)	Freezing/ Melting Point (°C)	Boiling Point (°C)
N E R V E	Tabun; **GA** $C_2H_5OPO(CN)N(CH_3)_2$	162.3	Colorless to brown liquid	Faintly fruity; none when pure	5.63	1.073 at 25°C	−5	240
	Sarin; **GB** $CH_3PO(F)OCH(CH_3)_2$	140.1	Colorless liquid	Almost none when pure	4.86	1.0887 at 25°C	−56	158
	Soman; **GD** $CH_3PO(F)OCH(CH_3)C$ $(CH_3)_3$	182.178	Colorless liquid	Fruity; camphor when impure	6.33	1.0222 at 25°C	−42	198
	(Cyclo-sarin); **GF** $CH_3PO(F)OC_6H_{11}$	180.2	Liquid	Sweet; musty; peaches; shellac	6.2	1.1327 at 20°C	−30	239
	VX $(C_2H_5O)(CH_3O)P(O)S$ $(C_2H_4)N[C_2H_2(CH_3)_2]_2$	267.38	Colorless to amber liquid	None	9.2	1.0083 at 20°C	below −51	298
	V_x ("V sub X")	211.2	Colorless liquid	None	7.29	1.062 at 20°C	−	256
B L I S T E R	Distilled Mustard; **HD** $(ClCH2CH_2)_2S$	159.08	Colorless to pale yellow liquid	Garlic or horseradish	5.4	1.268 @ 25°C; 1.27 @ 20°C	14.45	217
	Nitrogen Mustard; **HN-1** $(ClCH2CH_2)_2NC_2H_5$	170.08	Dark liquid	Fishy or musty	5.9	1.09 @ 20°C	−34	194
	Nitrogen Mustard; **HN-2** $(ClCH2CH_2)_2NCH_3$	156.07	Dark liquid	Soapy (low concentrations); Fruity (high)	5.4	1.15 @ 20°C	−65 to −60	75 at 15 mmHg
	Nitrogen Mustard; **HN-3** $N(CH_2CH_2Cl)_3$	204.54	Dark liquid	None, if pure	7.1	1.24 @ 20°C	−37	256
	Phosgene oximedichloro- foroxime; **CX** CCl_2NOH	113.94	Colorless solid or liquid	Sharp, penetrating	3.9	−	35 to 40	53–54 at 28 mmHg
	Lewisite; **L** $ClCHCHAsCl_2$	207.35	Colorless to brownish	Varies; may resemble geraniums	7.1	1.89 @ 20°C	−18	190
	Mustard-Lewisite mixture; **HL**	186.4	Dark, oily liquid	Garlic	6.5	1.66 @ 20°C	−25.4 (pure)	<190
	Phenyldichlorarsine; **PD** $C_6H_5AsCl_2$	222.91	Colorless liquid	None	7.7	1.65 @ 20°C	−20	252 to 255
	Ethyldichlorarsine; **ED** $C_2H_5AsCl_2$	174.88	Colorless liquid	Fruity, but biting; irritating	6.0	1.66 @ 20°C	−65	156
	Methyldichlorarsine; **MD** CH_3AsCl_2	160.86	Colorless liquid	None	5.5	1.836 @ 20°C	−55	133
B L O O D	Hydrogen cyanide; **AC** HCN	27.02	Colorless gas or liquid	Bitter almonds	0.990 @ 20°C	0.687 @ 20°C	−13.3	25.7
	Cyanogen chloride; **CK** CNCl	61.48	Colorless gas or liquid	Pungent, biting; Can go unnoticed	2.1	1.18 @ 20°C	−6.9	12.8
	Arsine; **SA** AsH_3	77.93	Colorless gas	Mild garlic	2.69	1.34 @ 20°C	−116	−62.5
CHOK- ING	Phosgene; **CG** $COCl_2$	98.92	Colorless gas	New-mown hay; green corn	3.4	1.37 @ 20°C	−128	7.6
	Diphosgene; **DP** $ClCOOCCl_2$	197.85	Colorless gas	New-mown hay; green corn	6.8	1.65 @ 20°C	−57	127–128
V O M I T I N G	Diphenylchloroarsine; **DA** $(C_6H_5)_2AsCl$	264.5	White to brown solid	None	Forms little vapor	1.387 @ 50°C	41 to 44.5	333
	Adamsite; **DM** $C_6H_4(AsCl)-NH)C_6H_4$	277.57	Yellow to green solid	None	Forms little vapor	65 (solid) @ 20°C	195	410
	Diphenylcyanoarsine; **DC** $(C_6H_5)_2AsCN$	255.0	White to pink solid	Bitter almond- garlic mixture	Forms little vapor	1.3338 @ 35°C	31.5 to 35	350
Incapa- citating	**BZ**	337.4	White crystal	None	11.6	Bulk 0.51 solid; Crystal 1.33	167.5	320
T E A R	Chloroacetophenone; **CN** $C_6H_5COCH_3Cl$	154.59	Solid	Apple blossoms	5.3	1.318 (solid) @ 20°C	54	248
	Chloroacetophenone in Chloroform; **CNC**	128.17	Liquid	Chloroform	4.4	1.40 @ 20°C	0.23	Variable, 60 to 247
	Chloroacetophenone and Chloropicrin in Chloroform; **CNS**	141.78	Liquid	Flypaper	−5	1.47 @ 20°C	2	Variable, 60 to 247
	Chloroacetophenone in Benzene and Carbon Tetrachloride; **CNB**	119.7	Liquid	Benzene	−4	1.14 @ 20°C	−7 to −30	Variable 75 to 247
	Bromobenzylcyanide; **CA** $BrC_6H_4CH_2CN$	196	Yellow solid or liquid	Soured fruit	6.7	1.47 @ 25°C	25.5	Decomp- oses at 242
	O-chlorobenzylmalonitrile; **CS** $ClC_6H_4CHC(CN)_2$	188.5	Colorless solid	Pepper	−	1.04 @ 20°C	93 to 95	310 to 315
	CR $(C_6H_4)_2(O)(N)CH$	195.25	Yellow powder in solution	Burning sensation	6.7	−	72	335
	Chloropicrin; **PS** Cl_3CNO_2	164.38	Liquid	Stinging; pungent	5.6	1.66	−69	112

Appendix 11.1 Continued

Agent Type	Vapor Pressure (mm Hg)	Volatility (mg/m³)	Heat of Vaporization (cal/g)	Decomposition Temperature (°C)	Flash Point	Stability
				PHYSICAL AND CHEMICAL PROPERTIES		
NERVE	0.037 @ 20°C	610 @ 25°C	79.56	150	78°C	Stable in steel at normal temperatures
	2.9 @ 25°C; 2.10 @ 20°C	22000 @ 25°C; 16090 @ 20°C	80	150	Non-flammable	Stable when pure
	0.4 @ 25°C	3900 @ 25°C	72.4	130	High enough not to interfere w/ military use	Less stable than GA or GB
	0.044 @ 20°C	438 @ 20°C	90.5	–	94°C	Relatively stable in steel
	0.0007 @ 20°C	10.5 @ 25°C	78.2 @ 25°C	Half-life of 36 hr at 150	159°C	Relatively stable at room temperature
	0.007 @ 25°C; 0.004 @ 20°C	75 @ 25°C; 48 @ 20°C	67.2	–	–	Relatively stable
BLISTER	0.072 @ 20°C	610 @ 20°C	94	149–177	105°C; ignited by large explosive charges	Stable in steel or aluminum
	0.24 @ 25°C	1520 @ 20°C	77	Decomposes before boiling is reached	High enough not to interfere w/ military use	Adequate
	0.29 @ 20°C	3580 @ 25°C	78.8	Below boiling; polymerizes with heat generation	High enough not to interfere w/ military use	Unstable
	0.0109 @ 25°C	121 @ 25°C	74	Below boiling point	High enough not to interfere w/ military use	Stable
	11.2 @ 25°C (solid); 13 @ 40°C (liquid)	1800 @ 20°C	101 at 40°C	Decomposes slowly at normal temperature	–	Decomposes slowly
	0.394 @ 20°C	4480 @ 20°C	58 at 0°C to 190°C	>100	None	Stable in steel and glass
	0.248 @ 20°C	2730 @ 20°C	58 to 94	>100	High enough not to interfere w/ military use	Stable in lacquered steel
	0.033 @ 25°C	390 © 25°C	69	Stable to boiling point	High enough not to interfere w/ military use	Very stable
	2.09 @ 20°	20000 @ 20°C	52.5	Stable to boiling point	High enough not to interfere w/ military use	Stable in steel
	7.76 @ 20°C	74900 @ 20°	49	Stable to boiling point	High enough not to interfere w/ military use	Stable in steel
BLOOD	742 @ 25°C; 612 @ 20°C	1080000 @ 25°C	233	>65.5	0°C; ignited 50% of time when disseminated by artillery shells	Stable if pure; can burn on explosion
	1000 @ 25°C	2600000 @ 20°C	103	100	None	Tends to polymerize; may explode
	11100 @ 20°C	30900000 @ 20°C	53.7 @ -62.5°C	280	Below detonation temp.; mixtures w/ air may explode spontaneously	Not stable in uncoated metal containers
CHOKING	1.173 @ 20°C	4300000 @ 7.6°C	59	800	None	Stable in steel if dry
	4.2 @ 20°C	45.000 @ 20°C	57.4	300 to 350	None	Unstable; tends to convert CG
VOMITING	0.0036 @ 45°C	48 @ 45°C	56.6	300	350	Stable if pure
	Negligible	Negligible	80	>boiling point	None	Stable in glass or steel
	0.0002 @ 20°C	2.8 @ 20°C	71.1	300 (25% decomposed)	Low	Stable at normal temperatures
Incapacitating	0.03 @ 70°C	0.5 @ 70°C	62.9	begins at 170°C	246°C	Adequate
TEAR	0.0041 @ 20°C	34.3 @ 20°C	98	Stable to boiling point	High enough not to interfere w/ military use	Stable
	127 @ 20°C	Indeterminate	n/a	Stable to boiling point	None	Adequate
	78 @ 20°C	610000 @ 20°C (includes solvent)	n/a	Stable to boiling point	None	Adequate
	variable; mostly solvent vapor	Indeterminate	n/a	>247	<4.44°C	Adequate
	0.011 @ 20°C	115 @ 20°C	79.5 @ 20°C	60 to 242	None	Fairly stable in glass, lead, or enamel
	0.00034 @ 20°C	0.71 @ 25°C	53.6	–	197°C	Stable
	0.00059 @ 20°	0.63 @ 25°C	–	–	188°C	Stable
	18.3 @ 20°C	165000 @ 20°C	–	>400	Not flammable	Adequate; unstable in light

Agent Type	PHYSIOLOGICAL ACTION						CWC
	Median Lethal Dose (LD50) (mg-min/m³)	Median Incapacitating Dose (ID50)	Eye & Skin Toxicity	Rate of Action	Physiological Action	Detoxification Rate	CWC Schedule
NERVE	15000 by skin (vapor) or 1500 (liquid); 70 inhaled	<50 inhaled	Very high	Very Rapid	Cessation of breath – death may follow	Slight, but definite	1.A.(2)
	10000 by skin (vapor) or 1700 (liquid); 35 inhaled	25 inhaled	Very high	Very rapid	Cessation of breath – death may follow	Cumulative	1.A.(1)
	2500 by skin (vapor) or 350 (liquid); 35 inhaled	25 inhaled	Very high	Very rapid	Cessation of breath – death may follow	Low, essentially cumulative	1.A.(1)
	2500 by skin (vapor) or 350 (liquid); 35 inhaled	25 inhaled	Very high	Very rapid	Cessation of breath – death may follow	Low	1.A.(1)
	150 by skin or 5 (liquid); 15 inhaled	25 by skin (vapor) or 2.5 (liquid); 10 inhaled	Very high	Very rapid	Produces casualties when inhaled or absorbed	Low, essentially cumulative	1.A.(3)
	–	–	Very high	Rapid	Produces casualties when inhaled or absorbed	Low, essentially cumulative	
BLISTER	900 (inhaled); 5000 by skin (vapor) or 1400 (liquid)	500 (skin); 100 (inhaled); 25 (eyes or nose)	Eyes very susceptible; skin less so	Delayed: hours to days	Blisters; destroys tissue; injures blood cells	Very low – cumulative	1.A.(4)
	1500 (inhaled); 20000 (skin)	200 by eye; 9000 by skin	Eyes susceptible to low concentration; skin less so	Delayed: 12 hours or longer	Blisters; affects respiratory tract; destroys tissue; injures blood cells	Not detoxified; cumulative	1.A.(6)
	3000 (inhaled)	<HN-1 & >HN-3; 100 by eye	Toxic to eyes; blisters skin	Skin – delayed 12 hrs or more; Eyes – faster than HD	Similar to HD; bronchopneumonia possible after 24 hours	Not detoxified; cumulative	1.A.(6)
	1500 (inhaled); 10000 by skin (est.)	200 by eye; 2500 by skin (est.)	Eyes very susceptible; skin less so	Serious effects same as HD; minor effects sooner	Similar to HN-2	Not detoxified – cumulative	1.A.(6)
	3200 (inhaled)	very low	Powerful irritant to eyes and nose; liquid corrosive to skin	Immediate effects on contact	Violently irritates mucous membranes, eyes, and nose; forms wheals rapidly	–	
	1200–1500(inhaled); 100000 (skin)	<300 by eye; >1500 to 2000 by skin	Severe eye damage; skin less so	Rapid	Similar to HD, plus may cause systemic poisoning	Not detoxified	1.A.(5)
	15000 (inhaled); >10000 (skin)	200 by eye; 1500 to 2000 by skin	Very high	Prompt stinging; blistering agent about 13 hours	Similar to HD, plus may cause systemic poisoning	Not detoxified	1.A.(4); 1.A.(5)
	2600 (inhaled)	16 as vomiting agent; 1800 as blister	633 mg-min/m³ produces eye casualty; less toxic to skin	Immediate eye effects; skin effects in 30 to 60 minutes	Irritates; causes nausea, vomiting and blisters	Probably rapid	
	3000–5000 (inhaled); 100000 (skin)	5 to 10 by inhalation	Vapor harmful on long exposure; liquid blisters <L	Immediate irritation; delayed blistering	Damages respiratory tract; effects eyes; blisters; can cause systemic poisoning	Rapid	
	3000–5000 (est.)	25 by inhalation	Eye damage possible; blisters less than HD	Immediate irritation; delayed blistering	Irritates respiratory tract; Injures lungs and eyes; Causes systemic poisoning	Rapid	
BLOOD	Varies widely with concentration	Varies with concentration	Moderate	Very rapid	Interferes with body tissues' oxygen use; accelerates rate of breathing	Rapid: 0.017 mg/kg/min	3.A.(3)
	11000	7000	Low; lacrimatory and irritating	Very rapid	Chokes, irritates, causes slow breathing rate	Rapid: 0.02 to 0.1 mg/kg/min	3.A.(2)
	5000	2500	None	Delayed 2 hours to 11 days	Damages blood, liver, and kidneys	Low	
CHOKING	3200	1600	None	Immediate to 3 hr. depending on conc.	Damages and floods lungs	Not detoxified – cumulative	3.A.(1)
	3200	1600	Slightly lacrimatory	Immediate to 3 hr. depending on conc.	Damages and floods lungs	Not detoxified – cumulative	3.A.(1)
VOMITING	15000 (est.)	12 (>10 minutes	Irritating; not toxic	Very rapid	Like cold symptoms, plus headache, vomiting, nausea	Moderate	
	Variable; avg.: 11000	22 (1 min.); 8 (60 min. exposure)	Irritating; relatively not toxic	Very rapid	Like cold symptoms, plus headache, vomiting, nausea	Rapid in small amounts	
	10000 (est.)	30 (30 sec); 20 (5 min. exposure)	Irritating; not toxic	More rapid than DM or DA	Like cold symptoms, plus headache, vomiting, nausea	Rapid	
Incapacitating	200000 (est.)	112	–	Delayed; 1 to 4 hours depending on exposure	Fast heart beat, vomiting, dry mouth, blurred vision, stupor, increasing random activity	–	2.A.(3)
TEAR	7000 to 14000	80	Temporarily severe eye irritation; mild skin irritation	Instantaneous	Causes tearing; irritates eyes and respiratory tract	Rapid	
	11000 (est.)	80	Temporarily severe eye irritation; mild skin irritation	Instantaneous	Cause tearing; irritates eyes and respiratory tract	Rapid	
	11400	60	Irritating; not toxic	Instantaneous	Vomiting and choking agent as well as a tear agent	Slow because of effect of PS	
	11000 (est.)	80	Temporarily severe eye irritation; mild skin irritation	Instantaneous	Powerfully lacrimatory	Rapid	
	8000 to 11000 (est.)	30	Irritating; not toxic	Instantaneous	Irritates eyes and respiratory passages	Rapid in low dosage	
	61000	10 to 20	Highly irritating; not toxic	Instantaneous	Highly irritating; not toxic	Rapid	
	–	0.15	Highly irritating; not toxic	Instantaneous	Irritates skin, eyes, nose, and throat	Moderate	
	2000	9	Highly irritating	Instantaneous	Acts as tear, vomiting, and choking agent	Slow	3.A.(4)

Appendix 11.2 Selected biological agent characteristics

Agent Type	Disease/Condition *Causative Agent/* *Pathogen*	Description of Agent	Transmissible Person to Person	Infectivity/ Lethality	Incubation Period	Duration of Illness	Persistence/ Stability
B A C T E R I A	Anthrax (inhalation) *Bacillus anthracis*	Rod-shaped, gram-positive, aerobic sporulating micro-organism, individual spores ~(1–1.2) × (3–5)nm	No	Moderate/ High	1–7 days	3–5 days	Spores are highly stable
	Brucellosis *Brucella suis, melitensis & abortus*	All non-motile, non-sporulating, gram negative, aerobic bacterium; ~(0.5–1) × (1–2)rm	No	High/Low	Days to months	Weeks to months	Organisms are stable for several weeks in wet soil and food.
	Cholera *Vibrio cholerae*	Short, curved, motile, gram-negative, non-sporulating rod. Strongly anaerobic, these organisms prefer alkaline and high salt environments.	Negl.	Low/Mode rate-High	1–5 days	1 or more weeks	Unstable in aerosols and pure water, more so in polluted water.
	Glanders *Burkholderia mallei*	Gram-negative bacillus primarily noted for producing disease in horses, mules, and donkeys	Negl.	/Moderate-High	10–14 days	N/A	N/A
	Plague (pneumonic, bubonic) *Yersinia pestis*	Rod-shaped, non-motile, non-sporulating, gram-negative, aerobic bacterium; ~(0.5–1) × (1–2)rm	High	High/Very High in untreated personnel, the mortality is 100%	2 to 6 days for bubonic and 3 to 4 days for pneumonic	1–2 days	Less important because of high transmissibility.
	Shigellosis *Shigella Dysenteriae*	Rod-shaped, gram-negative, non-motile, non-sporulating bacterium	Negl.	High/Low	1–7 days (usually 2–3)	N/A	Unstable in aerosols and pure water, more so in polluted water.
	Tularemia *Francisella tularensis*	Small, aerobic, non-sporulating, non-motile, gram-negative *cocco-bacillus* ~0.2 × (0.2–0.7)μm	No	High/ Moderate if untreated	1–10 days	2 or more weeks	Not very stable
	Typhoid *Salmonella typhi*	Rod-shaped, motile, non-sporulating gram-negative bacterium	Negl.	Moderate/ Moderate if untreated	6–21 days	Several weeks	Stable
R I C K E T T S I A E	Q-Fever *Coxiella burneti*	Bacterium-like, gram-negative organism, pleomorphic 300–700 nm	No	High/Very low	10–20	2 days to 2 weeks	Stable
	Typhus (classic) *Rickettsia prowazeki*	Non-motile, minute, coccoid or rod shaped rickettsiae, in pairs or chains, 300 nm	No	High/High	6–15 days	Weeks to months	Not very stable
V I R U S E S	Encephalitis	Lipid-enveloped virions of 50–60 nm dia., icosohedral nucleocapsid w. 2 glycoproteins					
	-Eastern/Western Equine Encephalitis (EEE, WEE)		Negl.	High/High	5–15 days	1–3 weeks	Relatively unstable
	-Venezuelan Equine Encephalitis		Low	High/Low	1–5 days	Days to weeks	Relatively unstable
	Hemorrhagic Fever						
	-Ebola Fever	Filovirus	Moderate	High/High	7–9 days	5–16 days	Relatively unstable
	-Marburg	Filovirus	Moderate				
	-Yellow Fever	Flavivirus. Isosahedral nucleocapsid 37–50 nm diam., lipoprotein env. w/ short surface spikes	Negl.	High/High	3–6 days	1–2 weeks	Relatively unstable
	Variola Virus (Smallpox)	Asymmetric, brick-shaped, rounded corners; DNA virus	High	High/High	7–17 days	1–2 weeks	Stable
T O X I N	Botulinum Toxin	any of the seven distinct neurotoxins produced by the bacillus, *Clostridium botulinum*	No	NA/High	Variable (hours to days)	24–72 hours/ Months if lethal	Stable
	Ricin	Glycoprotein toxin (66 000 daltons) from the seed of the castor plant	No	NA/High	Hours	Days	Stable
	Staphylococcal enterotoxin B	One of several exotoxins produced by *Staphylococcus aureus*	No	NA/Low	Days to weeks	Days to weeks	Stable
	Trichothecene (T-2) Mycotoxins	A diverse group of more than 40 compounds produced by fungi.	No	NA/High	Hours	Hours	Stable

Appendix 11.2 Continued

Agent Type	Vaccination/ Toxoids	Rate of Action	Symptoms
B A C T E R I A	Yes	Symptoms in 2–3 days: Shock and death occurs with 24–36 hrs after symptoms	Fever, malaise, fatigue, cough and mild chest discomfort, followed by severe respiratory distress with dyspnea, diaphoresis, stridor, and cyanosis
	Yes	Highly variable, usually 6–60 days.	Chills, sweats, headache, fatigue, myalgias, arthralgias, and anorexia. Cough may occur. Complications include sacroiliitis, arthritis, vertebral osteomyelitis, epididymoorchitis, and rarely endocarditis.
	Yes	Sudden onset after 1–5 day incubation period.	Initial vomiting and abdominal distension with little or no fever or abdominal pain. Followed rapidly by diarrhea, which may be either mild or profuse and watery, with fluid losses exceeding 5 to 10 liters or more per day. Without treatment, death may result from severe dehydration, hypovolemia, and shock.
	No	N/A	Inhalational exposure produces fever, rigors, sweats, myalgia, headache, pleuritic chest pain, cervical adenopathy, splenomegaly, and generalized popular/pustular eruptions. Almost always fatal without treatment.
	Yes	Two to three days	High fever, chills, headache, hemoptysis, and toxemia, progressing rapidly to dyspnea, sturdier, and cyanosis. Death results from respiratory failure, circulatory collapse, and a bleeding diathesis.
	No	Symptoms usually within 2–3 days, however, known to demonstrate in as little as 12 hours or as long as 7 days.	Fever, nausea, vomiting, abdominal cramps, watery diarrhea, and occasionally, traces of blood in the feces. Symptoms range from mild to severe with some infected individuals not experiencing any symptoms.
	Yes	Three to five days	Ulceroglandular tularemla with local ulcer and regional lymphadenopathy, fever, chills, headache, and malaise. Typhoidal or septicemic tularemia presents with fever, headache, malaise, substernal discomfort, prostration, weight loss, and non-productive cough.
	Yes	One to three days	Sustained fever, severe headache, malaise, anorexia, a relative bradycardia, splenomegaly, nonproductive cough in the early stage of the illness, and constipation more commonly than diarrhea.
R I C K E T T S I A E	Yes	Onset may be sudden	Chills, retrobulbar headache, weakness, malaise and severe sweats.
	No	Variable onset, often sudden. Terminates by rapid lysis after about 2 weeks of fever	Headache, chills, prostration, fever, and general pain. A macular eruption appears on the fifth to sixth day, initially on the upper trunk, followed by spread to the entire body, but usually not the face, palms, or soles.
V I R U S E S	Yes		Inflammation of the mengies of the brain, headache, fever, dizziness, drowsiness or stupor, tremors or convulsions, muscular incoordination.
	Yes	Sudden	Inflammation of the mengies of the brain, headache, fever, dizziness, drowsiness or stupor, tremors or convulsions, muscular incoordination.
	No		Malaise, myalgias, headache, vomiting, and diarrhea may occur with any of the hemorrhagic fevers May also include a macular dermatologic eruption.
	No Yes	Sudden	May also include a macular dermatologic eruption.
	Yes	2–4 days	Malaise, fever, rigors, vomiting, headache, and backache. 2–3 days later lesions appear which quickly progress from macules to papules, and eventually to pustular vesicles. They are more abundant on the extremities and face, and develop synchronously.
T O X I N	Yes	12–72 hours	Initial signs and symptoms include ptosis, generalized weakness, lassitude, and dizziness. Diminished salivation with extreme dryness of the mouth and throat may cause complaints of a sore throat. Urinary retention or ileus may also occur. Motor symptoms usually are present early in the disease; cranial nerves are affected first with blurred vision, diplopia, ptosis, and photophobia. Bulbar nerve dysfunction causes dysarthria, dysphonia, and dysphagia. This is followed by a symmetrical, descending, progressive weakness of the extremities along with weakness of the respiratory muscles. Development of respiratory failure may be abrupt.
	Not effective	6–72 hours	Rapid onset of nausea, vomiting, abdominal cramps and severe diarrhea with vascular collapse; death has occurred on the third day or later. Following inhalation, one might expect symptoms of weakness, fever, cough, and hypothermia followed by hypotension and cardiovascular collapse.
	Not effective	30 min–6 hours	Fever, chills, headache, myalgia, and nonproductive cough. In more severe cases, dyspnea and retrosternal chest pain may also be present. In many patients nausea, vomiting, and diarrhea will also occur.
	Not effective	Sudden	Victims are reported to have suffered painful skin lesions, lightheadedness, dyspnea, and a rapid onset of hemorrhage, incapacitation and death. Survivors developed a radiation-like sickness including fever, nausea, vomiting, diarrhea, leukopenia, bleeding, and sepsis.

Appendix 11.2 Continued

Agent Type	Treatment	Possible Means of Delivery
BACTERIA	Usually not effective after symptoms are present, high dose antibiotic treatment with penicillin, ciprofloxacin, or doxycycline should be undertaken. Supportive therapy may be necessary.	Aerosol.
	Recommended treatment is doxycycline (200 mg/day) plus rifampin (900 mg/day) for 6 weeks.	Aerosol. Expected to mimic a natural disease.
	Therapy consists of fluid and electrolyte replacement. Antibiotics will shorten the duration of diarrhea and thereby reduce fluid losses. Tetracycline, ampicillin, or trimethoprim-sulfamethoxazole are most commonly used.	1. Sabotage (food/water supply) 2. Aerosol
	Few antibiotics have been evaluated *in vivo*. Sulfadiazine may be effective in some cases. Ciprofloxacin, doxycycline, and rifampin have *in vitro* efficacy. Extrapolating from melioidosis guidelines, a combination of TMP-SMX + ceftazidime ± gentamicin might be considered.	Aerosol.
	Early administration of antibiotics is very effective. Supportive therapy for pneumonic and septicemic forms is required.	May be delivered via contaminated vectors (fleas) causing bubonic type, or, more likely, via aerosol causing pneumonic type.
	The antibiotics commonly used for treatment are ampicillin, trimethoprim/sulfamethoxzole (also known as Bactrim® or Septra®), nalidixic acid, or ciprofloxacin. Persons with mild infections will usually recover quickly without antibiotic treatment. Antidiarrheal agents such as loperamide (Imodium®) or diphenoxylate with atropine (Lomotil®) are likely to make the illness worse and should be avoided.	Contaminated food or water
	Administration of antibiotics with early treatment is very effective. Streptomycin – 1 gm I. M. q. 12 hrs × 10 10–14 d. Gentamicin – 3–5 mg/kg/day × 10–14 d.	Aerosol.
	Chloramphenicol amoxicillin or TMP-SMX. Quilone derivatives and third generation cephalusporins and supportive therapy.	Sabotage of food and water supplies.
RICKETTSIAE	Tetracycline or doxycycline are the treatment of choice and are given orally for 5 to 7 days.	May be a dust cloud either from a line source or a point source (downwind one-half mile or more).
	Tetracyclines or chlormphenical orally in a loading dose of 2–3 g, followed by daily doses of 1–2 g/day in 4 divided doses until ind. becomes afelorite (usually 2 days) plus 1 day.	May be delivered via contaminated vectors (lice or fleas).
VIRUSES	No specific treatment; supportive treatment is essential	Airborne spread possible.
	No specific treatment; supportive treatment is essential	Airborne spread possible.
	No specific treatment; intensive supportive treatment is essential	Airborne spread possible.
	No specific treatment; supportive treatment is essential	Airborne spread possible.
TOXIN	(1) Respiratory failure–tracheostomy and ventilatory assistance, fatalities should be <5%. Intensiva and prolonged nursing care may be required for recovery (which may take several weeks or even months). (2) Food-borne botulism and aerosol exposure–equine antitoxin is probably helpful, sometimes even after onset of signs of intoxication. Administration of antitoxin is reasonable if disease has not progressed to a stable state. Use requires pretesting for sensitivity to horse serum (and desensitization for those allergic). Disadvantages include rapid clearance by immune elimination, as well as a theoretical risk of serum sickness.	1. Sabotage (food/water supply) 2. Aerosol
	Management is supportive and should include maintenance of intravascular volume. Standard management for poison ingestion should be employed if intoxication is by the oral route.	Aerosol
	Treatment is limited to supportive care. No specific antitoxin for human use is available.	1. Sabotage (food/water supply) 2. Aerosol
	General supportive measures are used to alleviate acute T-2 toxicoses. Prompt (within 5–60 min of exposure) soap and water wash significantly reduces the development of the localized destructive, cutaneous effects of the toxin. After oral exposure management should include standard therapy for poison ingestion.	1. Sabotage 2. Aerosol

Appendix 11.3 Protective gloves and shoes

Glove set, chemical protective

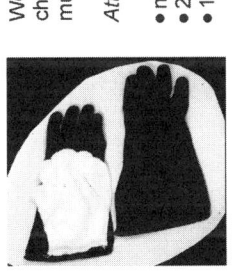

Worn under standard issue handwear

Attributes

- 25 mil. (0.025 inch thick) butyl rubber glove and cotton inner glove
- 24-hour protection
- 14 day field wear

Tactile CB glove

Used in place of the standard CB glove for operations that require greater tactility

Attributes

- lighter and thinner (7 mil and 14 mil)
- 14 mil: 24-hour protection
  - 14 days' wear
- 7 mil: 6-hour protection
  - 14 days' wear

Overshoes, green vinyl or black vinyl (GVOs, BVOs)

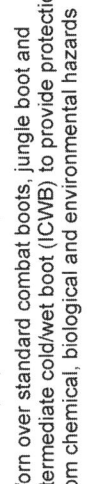

Worn over standard combat boots to provide chemical/biological and wet weather (rain, snow, mud) protection

Attributes

- material: poly vinyl chloride (PVC)
- 24-hour protection
- 14 day field wear

Multipurpose overboots (MULO)

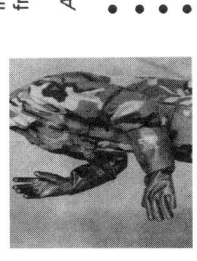
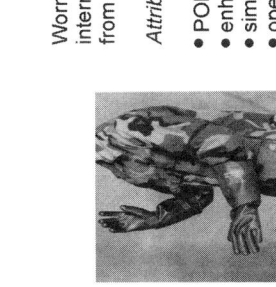

Worn over standard combat boots, jungle boot and intermediate cold/wet boot (ICWB) to provide protection from chemical, biological and environmental hazards

Attributes

- POL, flame and decon solution resistance
- enhanced sole for better traction
- simplified closure system
- operational functional to 0 degrees F
- compatible with standard, jungle and desert combat boots and intermediate cold/wet boot (ICWB)
- 60 days of durability

Appendix 11.4 Overgarment and other chemical protective clothing systems

Appendix 11.5 Improved toxicological agent protective ensemble (ITAP), self-contained, toxic environment protective outfit (STEPO) and other selected civilian emergency response clothing systems

Appendix 11.6 Selected toxic industrial chemicals (TICs)

NFPA 1994 (C/B warfare agents)

1. Ammonia (liquid)
2. Carbonyl chloride (gas)
3. Chlorine (gas)
4. Cyanogen chloride (gas)
5. Dimethyl sulfate (liquid)
6. Lewisite (liquid)
7. Hydrogen cyanide (gas)
8. Sarin (liquid)
9. Distilled sulfur mustard (liquid)
10. V-agent (liquid)

ASTM F1001 (used in ASTM F739-96 (Toxic Industrial Chemical Permeation Test))

1. Acetone
2. Acetonitrile
3. Ammonia (vapor)
4. 1,3 Butadiene (vapor)
5. Carbon disulfide
6. Chlorine (vapor)
7. Dichloromethane
8. Diethylamine
9. Dimethylformamide
10. Ethyl acetate
11. Ethylene oxide (vapor)
12. Hexane
13. Hydrogen chloride (vapor)
14. Methanol
15. Methyl chloride (vapor)
16. Nitrobenzene
17. Sodium hydroxide
18. Sulfuric acid
19. Tetrachloroethylene
20. Tetrahydrofuran
21. Toluene

12

Self-decontaminating materials for chemical biological protective clothing

G. SUN, University of California, USA
S. D. WORLEY, and R. M. BROUGHTON Jr,
Auburn University, USA

12.1 Introduction

Clothing materials are the last line of defense protecting the human body from exposure to any potential hazards, chemical and biological. Military personal are facing even greater challenges than any other professionals. Due to this fact, military uniforms should be multi-functional protective clothing that can prevent wearers from exposure to heat, fire, liquid, biological, and chemical hazards, in addition to protection against bullets and providing sufficient mobility. Military protective clothing currently available can provide different levels of protection against different hazards. For example, chemical and biological protection is mostly achieved by selecting barrier materials to liquid or aerosol agents while sacrificing certain comfort features of the clothing materials. Excellent barrier properties will obviously affect transport performance of the protective clothing, which will result in heat stress to wearers. Therefore, most biological and chemical protective clothing cannot be worn continuously and comfortably for long duration. Moreover, the uniforms after use will become biologically or chemically contaminated on the outside surface. Without proper handling, these uniforms are new sources of contaminants. Many of chemical protective clothing items are only suitable for single use and cannot be reused. For the reusable protective clothing, decontamination after usage is also a challenge. Thus, multi-functional clothing materials that can provide convenient reusable and rechargeable biological and chemical protection to soldiers should be developed.

The ideal self-decontaminating biological and chemical protective clothing should possess the following properties: rapid kill to a broad spectrum of biological agents upon contact; rapid detoxification to most popular chemical warfare agents upon contact; human and environmentally safe; breathable and comfortable to wear for long duration; durable for wearing and storage, and rcusablc; rechargeable for the biological and chemical activities. According to the above prospective requirements, fabric and polymer systems having chemical and biological active functions may meet these challenges and should be investigated extensively. In order to develop materials with multiple active functions against biological and chemical agents, a review of the related chemistry and materials is useful.

12.2   Self-decontaminating materials

Polymer and textile materials that can quickly kill biological agents and detoxify toxic chemical agents upon contact can be defined as self-decontaminating materials. Chemicals that can provide broad antimicrobial functions include disinfectants such as quaternary ammonium salts, alde-hydes, alcohols, halogens, halamine structures, peroxy acids, hydrogen peroxide, and certain heavy metals.[1] These biocides all have limitations in providing biocidal functions. Many of them including alcohols, aldehydes, and heavy metals cannot be safely employed or offer desired functions on textile fabrics. On the other hand, chemical detoxification of warfare agents can be achieved by either enzymatic hydrolysis or oxidative deg-radation.[2–4] However, enzymatic hydrolysis is selective and may not be compatible with biocidal functions on the same materials. Among these chemicals, halogens, halamines, and peroxide compounds are capable of oxidatively detoxifying toxic chemicals.[4] Free halogens cannot be safely employed on materials, and peroxide structures could only provide certain functions at relatively high concentrations, which are not quite safe for human skin.[5,6] Thus, these two groups of chemicals are not preferentially applicable in developing self-decontaminating materials for protective clothing.

Fortunately, halamine structures (N—X, X=Cl, Br, I) are proven safe to humans and have been widely employed as swimming pool disinfectants. The halamine compounds act similarly to chlorine bleach in biological and chemical detoxifying power, but can be handled easily. More recently, hala-mine structures have been incorporated into different polymers and fabrics with different biocidal potentials.[7–13] Fabrics with these structures have demonstrated rapid inactivation against a full spectrum of pathogenic dis-eases and even spores. More interestingly, the halamine fabrics were able to quickly oxidize some carbamate pesticides and other toxic agents and reduce the toxicity to humans. In fact, the halamine fabrics provide the same oxidative function against biological agents and could be defined as disinfectants.

Halamine chemistry can be described by Equations 12.1 and 12.2.

$$\rangle N-Cl + H_2O \rightleftharpoons \rangle N-H + Cl^+ + OH^- \qquad [12.1]$$

$$\underset{/}{\overset{\backslash}{N}}-Cl \underset{\text{Bleach}}{\overset{\text{Kill bacteria}}{\rightleftharpoons}} \underset{/}{\overset{\backslash}{N}}-H \qquad\qquad [12.2]$$

Equation 12.1 represents how halamines serve as water disinfectants in swimming pools. Halamines can release free chlorine in water. But, researchers have also found that the N—Cl bonds (chloramine) can be as biocidal as free chlorine.[14] Most people can have intimate skin contact with the halaimes for long duration in swimming pools. If the halamine itself provides biocidal functions, the overall reaction can be described in Equation 12.2, which serves as the biocidal mechanism of halamine solid materials. The halamine structures will be reduced to the precursor forms after reacting with biological or chemical agents, which is the reversed reaction of the chlorine bleaching of the cyclic amine moieties. N-halamine structures can kill microorganisms directly also without the release of free chlorine, as in Equation 12.2. In fact, N-halamine structures may only release very limited amounts of free chlorine because the dissociation constants of Equation 12.1 are of the order of 10^{-12} to 10^{-4} for imide, amide and amine halamines (Table 12.1).[10] Since N-halamine structures are biocidal, and more importantly quite stable in ambient environments, incorporation of the N-halamine into polymeric and textile materials will bring biocidal functions

Table 12.1 Stability of N-halamine structures.[10] (*Journal of Applied Polymer Science* © 2003)

Dissociation reaction	Dissociation constant for examples
Imide structure	1.6×10^{-2}–8.5×10^{-4} Trichlorocyanuric acid
	2.54×10^{-4} 1,3-dichloro-5,5-dimethylhydantoin
Amide structure	2.6×10^{-8} 1,3-dichloro-2,2,5,5-tetramethyl-4-imidazolidinone
	2.3×10^{-9} 3-chloro-4,4-dimethyl-2-oxazolidinone
Amine structure	$<10^{-12}$

to them. Moreover, since Equation 12.2 is a reversible reaction, the biocidal functions on the materials are rechargeable with a chlorinating agent, such as chlorine bleach. This rechargeable function is primarily suitable for reusable medical textiles and clothing. The latest progress in the application of N-halamine chemistry to textiles and polymers will be presented in this chapter.

12.3   Applications

12.3.1 Graft halamine onto cellulosic materials

Both amide and imide N-halamines have been incorporated into cellulose-containing fabrics by conventional finishing methods with 1,3-dimethylol-5,5-dimethylhydantoin (DMDMH) (Fig. 12.1).[8-10] The DMDMH-treated fabrics exhibit rapid biocidal functions, but the washing durability of the functions requires improvement, due to the dominating imide N-halamine functionality, which is the most reactive but least stable on the fabrics. However, DMDMH fabrics can be employed in personal protection against various biological agents such as bacteria, viruses, fungi, yeasts, and spores. Examples of the treated fabrics demonstrate a complete elimination of pathogens in a contact time as short as two minutes. The biocidal functions can be recharged repeatedly for at least 50 machine washes.

In order to increase washing durability of the N-halamine-treated textiles, the more stable amine N-halamine has been grafted to cellulose in a similar approach by using 3-methylol-2,2,5,5-tetramethylimidazolidin-4-one (MTMIO, Fig. 12.1). The resulting fabrics contain the more stable and less reactive amine N-halamine structure, thus providing slow, but durable, biocidal functions (Table 12.2). The only disadvantage of this process is the use of formaldehyde derivatives in the chemicals. Such a reactive group also limits the applications of DMDMH and MTMIO to cellulosic materials only.

1,3,-dimethylol-5,5-dimethylhydantoin   3-methylol-2,2,5,5-tetramethylimidazolidin-4-one
(DMDMH)   (MTMIO)

12.1 Structures of DMDMH and MTMIO.

Table 12.2 Chlorine loss and antimicrobial effects of MTMIO- and DMDMH-modified cotton samples

Chemical	Washing cycles	Against *E. coli*			Against *S. aureus*		
		Cl (ppm)	Cl loss (%)	Log reduction	Cl (ppm)	Cl loss (%)	Log reduction
MTMIO	0	565	–	6	654	–	6
	2	507	10.2	5	616	6.1	6
	5	498	11.9	4	601	8.4	4
DMDMH	0	863	–	6	934	–	6
	2	218	74.7	1.5	380	59.3	3
	5	157	81	0.9	274	70.7	2

Pure cotton fabric 493#; total finishing bath concentration: 4%. Wet pick-up: 70%. Concentrations of bacteria: *E. coli* 5×10^6 CFU/mL and *S. aureus* 7×10^6 CFU/mL. A six log reduction is equivalent to 99.9999% inactivation. Contact time: 60 min. Machine-washing tests were according to AATCC Standard Test Method 124-1999: Tests 1 and 2. The MTMIO-treated fabric was bleached separately from the DMDMH-treated fabric, with the same concentration of active chlorine (150 ppm) used in each case.[10] (*Journal of Applied Polymer Science* © 2003)

12.2 Structure of ADMH and its grafting reactions on synthetic polymers.

12.3.2 Radical grafting onto polymers

Recently, a hydantoin-containing vinyl monomer, 3-allyl-5,5-dimethylhydantoin (ADMH, as shown in Fig. 12.2), was prepared, which can incorporate amide N-halamine structures into all fibers including synthetic fibers. Due to the amide structure, the thus-produced fabrics could demonstrate both powerful and durable biocidal functions. Synthetic fabrics such as nylon-66, polyester (PET), polypropylene (PP), acrylics, and aramide fibers, as well as pure cotton fabrics, were used in the chemical modification. The ADMH can be incorporated in surfaces of fibers by a controlled radical grafting reaction which can ensure short chain grafts instead of long chain self-polymerization of the monomers (Fig. 12.2).[15,16]

Biocidal properties of the modified fibers could be demonstrated after a chlorination reaction by exposing the grafted fibers to a diluted chlorine solution, with which the grafted hydantoin rings were converted to N-halamine structures. The polymeric N-halamines could provide powerful and rapid antibacterial activities against *E. coli* and *S. aureus*. Most of the fibers could completely inactivate a large number of bacteria (1×10^6 CFU) in a 10–30 minute contact time. In addition, the antibacterial activities of these polymeric N-halamines could be easily recovered after usage by simply exposing the fabrics to chlorine solution again. One of the advantages of this process is that ADMH can be grafted to almost all polymeric structures by a radical polymerization process, which makes this process even more useful (Table 12.3).

12.3.3 Halamine Nomex® fabrics

Chlorination of nitrogen-containing polymers such as polyamides should result in halamine structures on amide groups. However, many similar halamine structures are not stable due to the existence of α-hydrogens next to the amide bond. As an exception, aromatic polyamides, such as poly (*m*-phenylene isophthalamide) (Nomex®) could be easily converted to stable halamine structures because the neighboring carbons have no α-hydrogen atoms. After chlorination, the elimination reaction for the formed N-halamine is not possible. The resultant fabrics provided potent, durable, and refreshable antibacterial activities against both gram-negative and

Table 12.3 Log reduction of *E. coli* after washing[15] (*Journal of Applied Polymer Science* © 2002)

Washing times	Log reduction of *E. coli* (%)					
	Nylon	PET	PP	Acrylic	Cotton	PET/cotton
0	5	5	5	5	5	5
5	5	5	5	5	3	5
15	5	5	5	5	1	5
30	3	3	2	1	UD*	3
50	UD*	1	1	1	UD*	UD*
50**	5	5	5	5	5	5

*no reduction of *E. coli* was detected.
**these samples were re-bleached after 50 times of washing.
Contact time = 30 min (*E. coli* concentration: 10^5~10^6 CFU/mL); all of the samples were tested with machine washing following AATCC Test Method 124. AATCC Standard Reference Detergent 124 was used in all of the machine-washing tests.

Table 12.4 Antibacterial efficiency of selected chlorinated Nomex fabrics* (Bacteria concentration: 10^6~10^7 CFU/mL)[17]

Run	Age (days)	Wash times	[Cl] on fabrics (ppm)	Contact time (min)	Reduction of *E. coli* (%)	Reduction of *S. aureus* (%)
1	2	1	425	2	99	99
2	2	1	425	10	99.9999	99.9999
3	2	1	425	30	99.9999	99.9999
4	15	1	425	10	99.9999	99.9999
5	30	1	425	10	99.9999	99.9999
6	60	1	430	10	99.9999	99.9999
7	90	1	404	10	99.9999	99.9999
8	2	5	420	10	99.9999	99.9999
9	2	10	428	10	99.999	99.999
10	2	15	300	10	99	99.9
11	2	20	157	60	UD**	99
12	2	30	142	60	UD	UD
13	2	1 × 30***	430	10	99.9999	99.9999
14	10	5 × 30****	450	10	99.9999	99.9999

*Chlorination: Room temperature, pH = 11, time = 30 min, active chlorine concentration = 1000 ppm; **No bacterial reduction was detected; ***After 30 washings, this sample was rebleached; ****This sample was treated 5 times with the '30 washing → rebleaching' cycles.

gram-positive bacteria (Table 12.4). In fact, only low crystalline Nomex® fibers, such as Nomex® III, could be chlorinated under the laundering condition.[17,18] The advantages of this development include: (i) such a chlorination reaction has no adverse effect on the original mechanical and thermal properties of the fibers,[17] which provides the direct application to firefighters' and military uniforms; (ii) the biocidal functions can be easily obtained by bleaching and are refreshable,[18] and (iii) there is no additional cost of treatment of the fabrics and garments.

12.3.4 Hydantoinylsiloxane-treated cellulose

Halamine structures can also be introduced into cellulose by condensing hydroxyl groups on 3-trihydroxysilylpropyl-5,5-dimethylhydantoin (SPH) (Fig. 12.3) with those on cellulose followed by chlorination of the amide nitrogen on the hydantoin ring with chlorine bleach.[19] The treated cellulose demonstrated a complete 5.7 log reduction of *S. aureus* in a contact-time interval of 30–60 minutes. A complete 5.9 log reduction of *E. coli* was observed in a contact-time interval of 60–120 minutes on these materials. These results demonstrate conclusively the superiority of cellulose treated with N-halamines in providing biocidal functions. The chlorinated SPH-

3-trihydroxysilylpropyl-5,5-dimethylhydantoin
(SPH)

12.3 Structure of 3-trihydroxysilylpropyl-5,5-dimethylhydantoin.

treated cloth is reasonably stable to loss of chlorine during dry storage. A loss from 0.62% Cl to 0.54% Cl was observed over a 50-day period for the treated cloth stored in a non-airtight plastic bag. In standard washing tests it was found that cotton cloth treated with SPH and chlorinated with an initial chlorine loading of 0.61% retained 0.42% Cl after five machine washings, 0.41% after ten washings, and 0.10% after 50 washings; thus the material still retained some biocidal functionality even after 50 machine washings.[19–21]

The siloxane-based halamine compounds have advantages in broad applications on most textiles. Direct interaction between siloxane hydroxyl groups with cellulose and hydrophilic fibers enhances the binding power of the compounds on surfaces of fibers and other materials.[19–21] The siloxane structure can undergo further polymerization to form a coating of siloxane-based halamine polymers, which can also improve washing durability of the functions on the surfaces.

12.3.5 Chemical detoxifying effects of N-halamine materials

Based on Equations 12.1 and 12.2, halamine structures are oxidative, and could oxidize sulfur to sulfoxide and sulfone structures.[22,23] The oxidative effects were demonstrated in oxidative detoxification of pesticides. Halamine fabrics produced from both MTMIO and DMDMH treatments were exposed to aqueous solutions of aldicarb and methomyl (oxime carbamate pesticides, Fig. 12.4). As expected, the imide halamine structures (fabrics treated with DMDMH) could rapidly react with aldicarb in aqueous solution. In less than two minutes of contact time, over 95% of aldicarb was degraded by the fabric. The amine halamine fabrics (treated with MTMIO) reacted with aldicarb at a much slower rate. After a contact time of 45 minutes, 95% of the pesticide was reacted, which is still quite effective. Both halamine fabrics demonstrated the capability of reacting with aldicarb but at different reaction rates (Fig. 12.5).

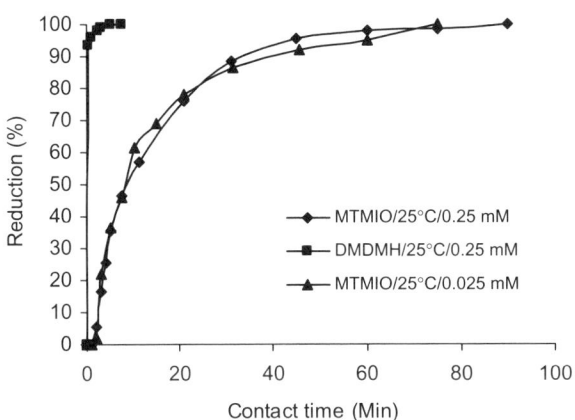

12.4 Structures of methomyl and aldicarb.

12.5 Degradation of aldicarb by DMDMH and MTMIO treated fabrics (5 g of fabrics in 20 mL of 0.25 mMol and 0.025 mMol aldicarb at 25 °C).[24]

The imide-containing fabrics also demonstrated limited detoxifying power against methomyl, while the MTMIO treated (amine halamine) fabrics resulted in almost no reactivity to the pesticide after a contact time of over two hours (Fig. 12.6). The maximum reduction for a low concentration (0.25 mM) of methomyl at even 60 °C was around 60%, much lower than that of aldicarb at a lower temperature for the same duration of reaction, indicating that methomyl had difficulty reacting with halamine.

Both aldicarb and methomyl contain thio (—S—) groups which are more vulnerable to oxidation. Thio groups can be oxidized to sulfoxide (—SO—) or sulfone (—SO$_2$—), which was confirmed by liquid chromatography and using standard aldicarb sulfoxide and aldicarb sulfone as references. The thio group in methomyl is linked to an unsaturated C═N bond, and the double bond structure could conjugate with lone pairs of electrons on the sulfur atom, which will lead to reduced electron density on the sulfur and decreased reactivity with oxidizing agents. These results strongly indicated that halamine structures are capable of detoxifying toxic agents that

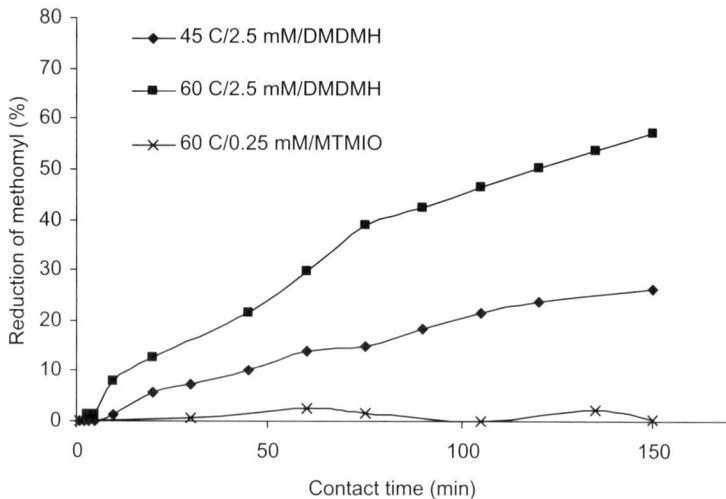

12.6 Degradation of 20 mL methomyl aqueous solutions in different concentrations (2.5 mM and 0.25 mM) by 5 g of chlorinated DMDMH (45 °C and 60 °C) and MTMIO (60 °C) fabrics.[24]

can be oxidatively detoxified. The reaction time was extremely short against certain more labile agents such as aldicarb. Of more importance, the detoxification functions are also refreshable.[25]

12.4   Future trends

Multifunctional textiles will be the new direction for development of fabrics and clothing for military personnel. Self-decontaminating functions, which include biocidal and chemical detoxification, should be required for military uniforms. Although chemically speaking there are only a few options that can provide such functions, research activities are continuing in developing some novel chemistry for such applications. In addition to halamine structures, peroxyl acid moieties have some potential if the reactivity of the groups can be enhanced. Photo-initiated singlet or triplet oxygen species may also become applicable on clothing materials, particularly with the rapid development of nanotechnologies. More environmentally friendly and functional products and processes are another focus of development of new multifunctional materials. Based on such requirements, durable, refreshable, and reusable fabrics with multifunctional properties such as fire resistance, biocidal, anti-chemical, anti-static, and waterproofing should be the future of military protective clothing. Indeed, novel fabrics and clothing should be able to provide protection to professionals on duty. There are always new challenges to textile scientists, and more innovative methods and theories are anticipated for future research in the development of new products.

12.5 Summary

Application of halamine chemistry in polymers and textile materials has resulted in the development of powerful biocidal and self-decontaminating materials for biological and chemical protection. Halamine structures can be incorporated into almost all synthetic and natural fibers and polymers. The thus produced halamine polymers have demonstrated refreshable and durable antibacterial functions.

12.6 Acknowledgments

This research was funded by National Science Foundation (DMI9733981 and DMI0323409), National Textile Center (C02-CD06 and S02-CD01), and the National Personal Protective Technology Laboratory (03NPTTB-041). HaloSource Company (Redmond, WA) and GenTex Corporation (Carbondale, PA) provided partial financial support. The authors would like to thank Drs Yuyu Sun, Xiangjing Xu, Lei Qian, Pengfei Gao, and Song Liu, Ms Anne Sandstrom, and Mr Xin Fei for their contributions to the research.

12.7 References

1 SUN G. and WORLEY S.D. Chemistry of Durable and Regenerable Biocidal Textiles. *Journal of Chemical Education*, 2005 **82** No. 1 pp. 60–64.

2 HARVEY SP, KOLAKOWSKI JE, CHENG TC, RASTOGI VK, REIFF LP, DEFRANK JJ, RAUSHEL FM and HILL C. Stereospecificity in the enzymatic hydrolysis of cyclosarin (GF) *Enzyme and Microbial Technology*, 2005 **37** (5) 547–555.

3 KILBANE JJ and JACKOWSKI K. Biocatalytic Detoxification of 2-Chloroethyl Ethyl Sulfide, *Journal of Chemical Technology and Biotechnology*, 1996, **65**:370–374.

4 YANG YC. Chemical detoxification of nerve agent VX *Accounts of Chemical Research*, 1999 **32** (2) 109–115.

5 HUANG L and SUN G. Durable and Regenerable Antimicrobial Cellulose with Oxygen Bleach: Concept Proofing, *AATCC Review*, 2003 **3** (10) 17–21.

6 HUANG L and SUN G. Durable and Oxygen Bleach Rechargeable Antimicrobial Cellulose: Sodium Perborate as an Activating and Recharging Agent, *Industrial and Engineering Chemistry Research*, 2003 **42** (22) 5417–5422.

7 WORLEY SD and SUN G. Biocidal polymers, *Trends in Polymer Science*, 1996 **4** 364–370.

8 SUN G and XU X. Durable and regenerable antibacterial finishing of fabrics. Biocidal properties. *Textile Chemist and Colorist*, 1998 **30** 26–30.

9 SUN G, XU X, BICKETT JR and WILLIAMS JF. Durable and regenerable antimicrobial finishing of fabrics with a new hydantoin derivative, *Industrial Engineering Chemistry Research*, 2001 **41** 1016–1021.

10 QIAN L and SUN G. Durable and regenerable antimicrobial textiles: Synthesis and applications of 3-methylol-2,2,5,5-tetramethyl-imidazolidin-4-one (MTMIO), *Journal of Applied Polymer Science*, 2003 **89** 2418–2425.

11 SUN YY, CHEN TY, WORLEY SD and SUN G. Novel refreshable N-halamine polymeric biocides containing imidazolidin-4-one derivatives. *Journal of Polymer Science Part A-Polymer Chemistry*, 2001 **39** 3073–3084.

12 CHEN Y, WORLEY SD, KIM J, WEI C-I, CHEN TY, SANTIAGO JI, WILLIAMS JF and SUN G. Biocidal poly(styrenehydantoin) beads for disinfection of water. *Industrial Engineering Chemistry Research*, 2003 **42** 280–284.

13 WORLEY SD and WILLIAMS DE. Halamine water disinfectants. *CRC Critical Reviews Environmental Control*, 1988 **18** 133.

14 WILLIAMS DE, ELDER ED and WORLEY SD. Is free halogen necessary for disinfection? *Applied and Environmental Microbiology*, 1988 **54** 2583–2585.

15 SUN YY and SUN G. Durable and regenerable antimicrobial textile materials prepared by a continuous grafting process, *Journal of Applied Polymer Science*, 2002 **84** 1592–1599.

16 SUN YY and SUN G. Novel refreshable N-halamine polymeric biocides: Grafting hydantoin-containing monomers onto high-performance fibers by a continuous process, *Journal of Applied Polymer Science*, 2003 **88** 1032–1039.

17 SUN YY and SUN G. Novel Refreshable N-Halamine Polymeric Biocides: N-Chlorination of Aromatic Polyamides, *Industrial and Engineering Chemistry Research*, 2004 **43** 5015–5020.

18 SANDSTROM A and SUN G. Durability of Biocidal Nomex Fabrics for Multi-functional Firefighter Uniforms, *Research Journal of Textile and Apparel*, 2006 **10** 13–18.

19 WORLEY SD, CHEN Y, WANG JW, WU R, CHO U, BROUGHTON RM, KIM J, WEI CI, WILLIAMS JF, CHEN J and LI Y. Novel N-Halamine Siloxane Monomers and Polymers for Preparing Biocidal Coatings, *Surf. Coat. Intern. Part B. Coat. Trans.* 2005 **88** 93–99.

20 LIANG J, CHEN Y, BARNES K, WU R, WORLEY SD and HUANG TS. N-Halamine/Quat Siloxane Copolymers for Use in Biocidal Coatings, *Biomaterials*, 2006 **27** 2495–2501.

21 BARNES K, LIANG J, WU R, WORLEY SD, LEE J, BROUGHTON RM and HUANG TS. Synthesis and Antimicrobial Applications of 5,5'-ethylenebis[5-methyl-3-(3-triethoxysilyl-propyl) hydantoin], *Biomaterials*, 2006 **27** 4825–4830.

22 SUN G and WORLEY SD. Oxidation of secondary alcohols and sulfides by halamine polymers, *Chemical Oxidation: Technology for the Nineties*, Edited by W.W. Eckenfelder, A.R. Bowers and J.A. Roth, Technomic Publishing Co. Inc. Lancaster, 1997 V6, 134–144.

23 AKDAG A, WEBB T and WORLEY SD. Oxidation of Thiols to Disulfides with Monochloro Poly(styrenehydantoin) Beads, *Tetrahedral Letters*, 2006 **47** 3509–3510.

24 XIN F, SHIBAMOTO T, GAO P and SUN G. Pesticide Detoxifying Functions and N-Halamine Fabrics, *Archives of Environmental Contamination and Toxicology*, 2006 **51** p509–514.

25 KO LL, SHIBAMOTO T and SUN G. A novel detoxifying pesticide protective clothing for agricultural workers, *Textile Chemist and Colorist & American Dyestuff Reporter*, 2000 **32** (2) p34–38.

Camouflage fabrics for military
protective clothing

P. SUDHAKAR and N. GOBI,
K. S. Rangasamy College of Technology, India, and
M. SENTHILKUMAR, PSG Polytechnic College, India

13.1 Introduction

Fibres and coatings with unique optical, magnetic and electrical properties are being widely researched for both military and commercial applications. New materials are being developed in this research effort, with unique tunable colouration properties across the visible spectrum as well as spanning the infrared and ultraviolet region of the electromagnetic spectrum. These dynamic colour-responsive 'camouflage' fibre systems will have wide application to a variety of new textile products. The new materials will also allow penetration into markets that are normally not dependent on textile materials. New markets, such as the optical communication and electronics market areas, will include biosensors, detector applications for textile materials, and 'smart materials' that can change their hue or depth of shade by applying a static or dynamic electric field for applications such as smart uniforms.

Technology is getting smaller and faster and we all know the speed of this development is increasing. We see an ongoing miniaturization and production of materials equipped with special properties. It is possible to integrate properties of sensitivity, information and intelligence into single materials. Most recently, we have moved towards 'third generation' textiles, enabled by the latest advances in material and biological sciences, Nanotechnology and Intelligent systems. Intelligent textiles represent the next generation of fibres, fabrics and articles produced from them. They can be described as textile materials that think for themselves. Many Intelligent textiles already feature in advanced types of clothing, principally for protection and safety and for added fashion or convenience. One of the main reasons for the fast development of Intelligent textiles is due to the importance given to the military applications. This is because they are used in different atmospheric conditions such as in extreme winter condition jackets or uniforms that change colour so as to improve camouflage effects. Intelligent textiles provide ample evidence of the potential and enormous wealth of opportunities still to be realized in the textile industry, in the fashion and clothing sector, as well as in the technical textiles sector. Moreover, these developments will be the result of active collaboration between people from a whole variety of backgrounds and disciplines: engineering, science, design, process development, and business and marketing. Here are four similar explanations of intelligent textiles:

(i)   They are materials that react to impulses without the need for us to control them.
(ii)   They are able to respond to their environment.
(iii) In garments they react to impulses coming from outside or inside.
(iv)   They react automatically to some kind of stimuli.

Intelligent textiles are fibres and fabrics with a significant and reproducible automatic change of properties due to defined environmental influences. Traditionally, the industry has been concerned not only with hue and intensity, but also with maintaining the colour regardless of environmental influences. Yet there are numerous exciting and important markets which require materials that alter their colour on demand. Camouflage fibres will allow for the creation of value-added products in traditional industry markets, as well as entry into entirely new areas. It is the objective to provide the textile and fibre industry with a series of new 'smart' materials that can quickly change their hue, depth of shade, or optical transparency by the application of an electrical or magnetic field.

Nanta from Toray Industries reported the development of a temperature sensitive fabric with trade name SWAY in 1988 by introducing microcapsules, diameter 3–4 mm, to enclose heat sensitive dyes, which are resin-coated homogeneously, over the fabric surface. The microcapsules were made of glass and contained the dyestuff, the chromophore agent (electron acceptor) and colour-neutralizer (alcohol, etc.) which reacted and exhibited colour/decolour according to the environmental temperature.[2]

Hongu and Phillips for Kanebo Ltd developed a fabric which can change colour from white to blue after irradiating with UV, wavelength range 350–400 nm. The spiropiran type organic compounds used for such photochromic material undergo photolyses and colour change by UV rays. Because of the low stability of spiropiran, Kanebo used a stable compound of spiropiran as the photochromic material and a T-shirt made of photochromic prompted fabric was introduced to the market in 1989.[3] The potential applications range from camouflage fabrics to wall and floor covering. A very interesting effect found from this work is the possibility of information storage and retrieval from organic fibres based on molecular recognition. This may lead to micro-devices encapsulated in fibres for a variety of technical applications.[1]

Photoadaptive fibres and films, which experience photoinduce reversible optical and heat reflectivity changes, are being developed by Mills, Slaten and Broughton et al.[11] Fibres of poly(vinylalcohol) that are insoluble in water have been made by crosslinking the macromolecules with dimethyl-sulfoxide. Incorporation of Ag^+ and $AuCl_4^-$ ions into the fibres was achieved after swelling the polymeric materials with methanolic solutions of the metal ions.

Illumination of the dry fibres containing metal ions yielded nanometre-sized Ag and Au particles only under high intensity artificial light, or direct sunlight, but not under ambient light. Thus, photochemical metallization of the fibres occurs exclusive under high photon fluxes, which is one of the desired properties of the smart fibres that we are attempting to develop. Although a dark oxidation of the metal crystallites was not observed in these simple systems, the metallized fibres are expected to be the basis for flexible materials that act as shields for electromagnetic radiation, where stable metal particles are required. The light-induced changes in the optical properties of these polymeric systems are currently under investigation.

13.2   Methods for production of camouflage textiles

The concept of producing textiles that readily vary in colour has long been an anathema to the textile colourist for whom achieving permanency of colour has been a primary goal stretching back into antiquity. Consequently, colourant manufacturers have striven for many years to develop fast-coloured materials by hunting for dyes and pigments that are chemically inert and physically unresponsive once they have been applied to a substrate.

13.2.1 pH changes

Molecules can change colour dramatically in the presence of acids and bases, but these reagents and the solvents required to transport them make this method extremely difficult to implement in the applications.

13.2.2 Oxidation state changes

This method is also highly effective, but requires the migration of ions. The response time can be fast in solvents, but this complicates the device. Gel-type devices might also be possible, though physical robustness, oxygen stability and response times represent serious engineering challenges. A device built on this principle would be similar to a polymer LED.

13.2.3 Bond breaking/making

There are a number of systems that undergo reversible bond-breaking, bond-forming processes that result in dramatic colour changes. Most commonly, these are light-initiated processes.

13.2.4 Mechanochromism

Certain compounds have been shown to undergo colour changes as a result of applied stress. A mechanochromic system is constructed by surface modification of conducting polymers.

13.2.5 Electric or magnetic field effects

Some highly polaraizable systems have been observed to change colour in the presence of electric or magnetic fields.

13.3   Chromic materials

Chromic materials are also called camouflage fibres, because they can change their colour according to the external conditions. These materials have been used mostly in fashion, to create novel colour changing designs. Because of this, some people fear that the chromic materials will be a short boom, but the accuracy and endurance of the materials are being improved. Chromic materials refer to materials that radiate the colour, erase the colour or just change it because its induction is caused by the external stimuli (as 'chromic' is a suffix that means colour). So the chromic materials can be classified depending on the stimuli affecting them:

- Photochromic: external stimuli energy is light.
- Thermochromic: external stimuli energy is heat.
- Electrochromic: external stimuli energy is electricity.
- Piezorochromic: external stimuli energy is pressure.
- Solvatochromic: external stimuli energy is liquid.
- Carsolchromic: external stimuli energy is electron beam.

13.3.1 Photochromic materials

The phenomenon produced in photochromic materials is called photochromism, where the change in colour is due to incident light. However, to date, photochromism is most important for optical switching data and imaging systems, rather than for textile applications. It is possible to classify photochromic fibres into different groups: those which emit the colour when activated by visible light and those which emit the colour when activated by ultraviolet radiation. Although in the first group we can find both organic and inorganic materials, the most studied are the former because they are colourful, have high density and a wide range of application. Organic photochromic materials generally do not show this phenomenon in crystal form, they show their photochromism only after dissolving in some solvent. The problem is that material behaviour (such as colour emitted, reaction speed, resistance, density) is largely affected in a positive or negative way by the solvent nature. For this reason, in order to apply these materials to fibres, it is important to consider which solvent needs to be used.

There are fibres that emit fluorescent colour, for example red, green or blue, under ultraviolet radiation in a dark place, though they maintain their original colour when exposed to natural light. The inorganic fluorescent paints used for this purpose are mixed at an approximate rate of 10% in the liquid during the spinning operation. It is important to note that the colour can be freely controlled by mixing various inorganic paints together or by adding the paints to the natural colour of the threads.

13.3.2 Thermochromic materials

Thermochromic organic colourants have increasingly been the subject of investigations in academia and industry over the past few decades, for use in producing novel colouration effects in textiles as well as other applications. Aesthetics has been the focus in the textile sector, with the emphasis firmly on novelty items, for example ski-wear or promotional T-shirts. This situation is paralleled in the textile field and is due, in part, to the inability of known thermochromic colourants to meet performance requirements for functional textile applications such as camouflage 'smart' fabrics, as exemplified by adaptive camouflage materials, which seem some way off becoming a reality.

The majority of thermochromic systems are unacceptable simply because the changes in colour require relatively large amounts of energy, as intramolecular transformations are involved. The molecular structure of the colourant is altered, either through the breaking of covalent bonds or the flipping of the molecule into a different conformation. Consequently, the colour changes lie well above room temperature, typically 100–200 °C, and so are not of much use for producing thermochromic effects on textiles. Interestingly, systems that depend on less drastic changes in structure are typically intermolecular in nature and require less energy to drive them, so that the colourants can be persuaded to alter their colour at and around room and body temperatures. One such type of material that has been explored as a means of conferring thermochromic properties to textiles is based on a special class of liquid crystalline substances that are referred to as cholesteric or chiral nematic systems. Using this variety of liquid crystal, it is possible to achieve significant changes in appearance over narrow temperature ranges (5–15 °C) and to detect small variations in temperature (<1 °C). Two types of thermochromic systems used in textiles are: the liquid crystal type and the molecular rearrangement type. In both cases, the dyes are entrapped in microcapsules and are applied to the garment fabric like a pigment in a resin binder. The most important type of liquid crystal for thermochromic systems is the so-called cholesteric type, where adjacent molecules are arranged so that they form helices. Thermochromism results from the selective reflection of light by the liquid crystal. The wavelength of the light reflected is governed by the refractive index of the liquid crystal and by the pitch of the helical arrangement of its molecules. Since the length of the pitch varies with temperature, the wavelength of the reflected light is also altered, and results in colour changes.

An alternative method of inducing thermochromism is by means of a rearrangement of the molecular structure of the dye, as a result of a change in temperature. The most common type of dye which exhibit thermochromism through molecular rearrangement is the spirolactone, although other types have also been identified. A colourless dye precursor and a colour developer are both dissolved in an organic solvent. The solution is then microencapsulated, and is solid at lower temperatures. Upon heating, the system becomes coloured or loses colour at the melting point of the mixture. The reverse change occurs at this temperature and the mixture is then cooled. However, although thermochromism through molecular rearrangement in dyes has aroused a degree of commercial interest, the overall mechanism underlying the changes in colour is far from clear-cut and is still very much open to speculation. The source of the thermochromic behaviour lies in the way that the highly ordered structure of the liquid crystal interacts with light and responds to variations in temperature. All liquid crystal phases have structures that fall somewhere in between those of the rigid regular lattices of crystalline solids and of relatively structureless isotropic liquids. Since the thermochromism involves reflected light, the effects are best seen against a dark background which soaks up the light transmitted by the liquid crystal, so that only those wavelengths that are reflected by the liquid crystal phase are observed, and not the substrate. While the changes in colour may be striking and relatively sensitive to temperature, opportunities to take advantage of this technology for textile colouration are limited by the necessity of using dark substrates as well as the requirement that the liquid crystal systems must be microencapsulated. This process is necessary since these thermochromic materials exist as liquids that must be enclosed in order to preserve their thermochromic behaviour. While the resultant capsules can be treated as conventional pigments and printed on fabrics, there are drawbacks associated with microencapsulation concerning colour strength reduction and durability. In addition, the process adds to the expense of these liquid crystalline systems; their relatively high cost is the main reason behind the lack of commercial exploitation of these materials for textile colouration.

pH changes

A cheaper, albeit less sensitive, thermochromic alternative that has been applied to textiles does not actually employ chromophores that are themselves thermochromic, but ones which are instead sensitive to some variable other than temperature. For example, pH sensitive colourants (see Fig. 13.1) known as colour formers, such as Crystal Violet lactone, can be used as the basis of such a system in which the variable is acidity. Functional dyes of this type tend to ring open from a colourless to a coloured state when placed in contact with acid. For example, Crystal Violet lactone turns from colourless to a deep blue upon protonation. The trick for inducing thermochromic effects using colour formers is to design a system in which the acidity experienced by the colourant varies with heating and cooling, which in turn leads to the colour of the system changing in response to the differences in temperature. One way of achieving this requires the combined use of three components: the colour former, an acidic colour developer (typically a phenolic material) and a non-polar co-solvent medium (often a low-melting point, long-chain alkyl compound) that controls the interaction between the first two ingredients of the formulation. When the components are heated and mixed together in the correct proportions so that the colour former and developer are dissolved in the co-solvent and the solution is then cooled, the solid composite formed is intensely coloured. Heating the composition above its melting point (determined largely by that of the co-solvent) results in complete colour loss. The colour change is reversible; the point at which it occurs corresponds closely to the range of temperatures at which the formulation melts. Since these systems involve changes in phase between coloured solid and colourless liquid states, applications must

13.1 pH sensitive colourants.

generally employ microencapsulation or lamination to protect the composites and safeguard their thermochromic properties. For textile applications, the former approach is used. This restricts the techniques available for applying the thermochromic material to textiles to that of pigment printing of fabrics or incorporation into synthetic fibres during their manufacture.

13.3.3 Electrochromic materials

Electrochromism is the reversible change in optical properties that can occur when a material is electrochemically oxidized, and is of great academic and commercial interest. Traditionally, materials have been considered as being electrochromic when they display distinct visible colour changes, with the colour change commonly being between a transparent ('bleached') state and a coloured state, or between two coloured states. In cases where more than two redox states are electrochemically available, the electrochromic material may exhibit several colours and can be described as polyelectrochromic. However, the working definition of electrochromism has now been extended to include devices for modulation of radiation in the near infrared, thermal infrared and microwave regions, so 'colour' can now mean a response by detectors at these wavelengths, and not just by the human eye.

13.4   Identification of chromophores

Coloured substances owe their colour to the presence of one or more unsaturated groups responsible for electronic absorption. Examples: $C=C$, $C\equiv C$, $C=N$, $C\equiv N$, $C=O$, $N\equiv N$. They all absorb intensely at the short wave length end of the spectrum but some of them (example carbonyl) have less intense bands at higher wavelength owing to the participation of electrons.

  A chromophore is that part (atom or group of atoms) of a molecular entity in which the electronic transition responsible for a given spectral band is approximately localized. The term is derived from the dyestuff industry, referring originally to the groupings in the molecule that are responsible for the dye's colour. The molecular entity is any constitutionally or isotopically distinct atom, molecule, ion, ion pair, radical, radical ion, complex, conformer, etc., identifiable as a separately distinguishable entity (see Table 13.1). Molecular entity is used in this compendium as a general term for singular entities, irrespective of their nature, while chemical species stands for sets or ensembles of molecular entities. Note that the name of a compound may refer to the respective molecular entity or to the chemical species, e.g. methane, may mean a single molecule of CH_4 (molecular entity) or a molar amount, specified or not (chemical species), participating in a

Table 13.1 Molecular entity

Chromophore	Example	λ_{max}, nm	Solvent
C=C	Ethene	171	Hexane
C≡C	1-Hexyne	180	Hexane
C=O	Ethanal	290	Hexane
		180	Hexane
N=O	Nitromethane	275	Ethanol
		200	Ethanol
CB$_r$	Methyl bromide	205	Hexane
Cl	Methyl Iodide	255	Hexane

13.2 Highly polarizable, highly conjugated compounds.

reaction. The degree of precision is necessary to describe a molecular entity depends on the context. For example 'hydrogen molecule' is an adequate definition of a certain molecular entity for some purposes, whereas for others it is necessary to distinguish the electronic state and/or vibrational state and/or nuclear spin, etc. of the hydrogen molecule.

13.5   Synthesis of new polymers

13.5.1 Highly polarizable, highly conjugated compounds

Several intensely coloured compounds that are highly polarizable, have been prepared. The compounds were designed to undergo a dramatic change in their molecular dipole moment upon application of an electric field. Each can be dissolved within a polymer, or covalently bound to it (see Fig. 13.2). These complexes are designed to work with conventional fibre systems, such as polyesters or polyamides. In addition to compounds 1–4, we are also examining several less exotic chromophores as well as a por-phyrin-based system. Working with the Clemson laboratory, we will soon prepare films containing known concentrations of chromophore and monitor the effects of fields on the electronic spectra. Structure/activity relationships will be noted and used to design more efficient, second-generation compounds.[12]

13.5.2 Orientation of chromophores by surface attachment to polymers

It is suspected that the strategy outlined in Section 13.5.1 will run into problems due to the fact that the chromophores are randomly oriented in the film matrix. This will mean that the effect of the field on the net dipole moment will vary from molecule to molecule, which will lead to an averag-ing of the colour change as well as to a reduction in the maximum electro-chromic shift observed. To this end, we are exploring a method of attaching chromophores to the surface of films in a highly oriented fashion. Our first candidate is Compound 3, which will be modified with a long alkyl chain terminated by a thiol. This nucleophilic end-group will, in turn will be attached to the surface of a polymer. Finally, Compound 4 can be attached to the surface-bound 3 in a sequential manner to make a multi-layered electrochromic device of controlled colour intensity. (see Fig. 13.3).

13.3 Orientation of chromophores by surface attachment to polymers.

13.5.3 Colour change by 2 + 2 cycloaddition reactions

We are also preparing a series of diolefins, which can undergo a solid state, photoinduced 2 + 2 cycloaddition reaction. The diolefins have been chosen for their colour, colour intensity and because their conjugation length is critically dependent upon the olefin. Thus, the cycloaddition reaction completely shifts the compounds electronic absorption into the UV, giving a light yellow or colourless material. We hope to demonstrate a selective method of reversing the cycloaddition (and restoring the colour) by application of a high frequency electric field. At this point, we have prepared a number of chromophores and demonstrated their ability to photocyclize either as neat microcrystalline powders or dispersed in a polymer matrix. A few representative compounds are shown in Fig. 13.4.

13.5.4 Redox polymers as chromophores

Many electrically conducting polymers, and even smaller oligomers built from the same monomers, undergo distinct colour changes when oxidized

13.4 2 + 2 Cycloaddition reactions.

or reduced. By tuning the substitutents on the polymer backbone, we can adjust the colours that a particular system displays. The polymer is designed to correlate structure with chromatic properties, switching speed, oxygen stability, etc. Poly(p-phenylenevinglenes) (PPV, see Section 13.7.1) molecules are 'tunable' by small structural variations

- PPV oligomers emit at different wavelengths for different molecular weights.
- PPV's with kinked chains have different emission spectra from those with non-kinked chains
- PPV's with kinked chains have enhanced solubility.

The construction and development of synthetic routes to oligomers of PPV with well-defined length and structure is shown in Fig. 13.5.

13.5 Synthetic routes to oligomers of PPV.

13.6 Synthesis of monomeric and oligomeric chromophores

In addition to polymeric chromophores, we can synthesize molecular chromophores. These are highly conjugated species that should be particularly sensitive to external fields. In some cases, the chromophores are designed to be added to a fibre much like traditional dyestuffs. In others, chromophores can be attached by the surface of films and fibres in a highly organized manner. An example of the first type of chromophore is shown in Fig. 13.6.

We can also investigate porphyrins and porphyrin arrays as field-sensitive chromophores. An example is the synthesis of a thiol-linked octaethyl porphyrin that will self-assemble into well-defined layers on the surface of electronically conducting polymers. We can also construct acetylene and diacetylene-linked arrays of porphyrins, with or without metals (see Fig. 13.7).

13.7 Conductive/conjugated polymers

Conjugated polymers have a framework of alternating single and double carbon–carbon (sometimes carbon–nitrogen) bonds. These conductive polymers, especially those with a conjugated π-bond system, often yield higher conductivity once having undergone a doping process. Single bonds are referred to as σ-bonds, and double bonds contain a σ-bond and a π-bond. All conjugated polymers have a σ-bond backbone of overlapping sp^2

13.6 Synthesis of molecular chromophores.

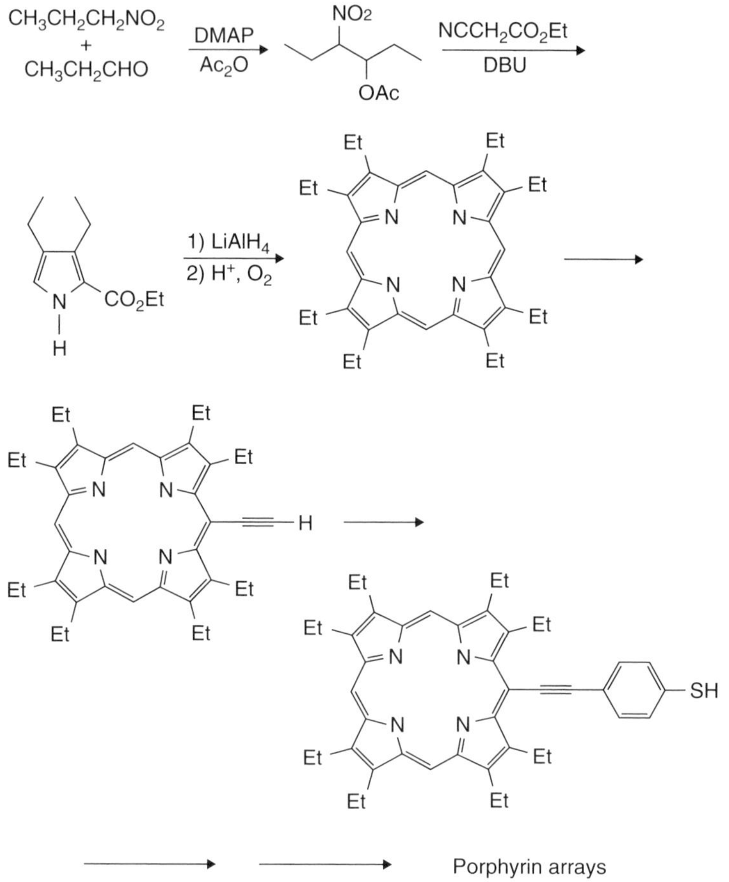

13.7 Porphyrins and porphyrins arrays.

hybrid orbitals. The remaining out-of-plane p_z orbitals on the carbon (or nitrogen) atoms overlap with neighbouring p_z orbitals to give π-bonds.[7]

Although the chemical structures of these materials are represented by alternating single and double bonds, in reality, the electrons that constitute the π-bonds are delocalized over the entire molecule (see Fig. 13.8). For this reason, polyaniline (PAn) and poly (N-vinylcarbazole) (PVCZ) are considered to be conjugated polymers, with the nitrogen p_z orbital assisting the delocalization of the π-electrons. In some conjugated polymers, such as polyacetylene (PA) and PAn, delocalization results in a single (degenerate) ground state, whereas in other polymers the alternating single and double bonds lead to electronic structures of varying energy levels. The behaviour of conjugated polymers is dramatically altered with chemical doping. Generally, polymers such as polypyrrole (PPy) are partially oxidized to produce p-doped materials (Fig. 13.9).

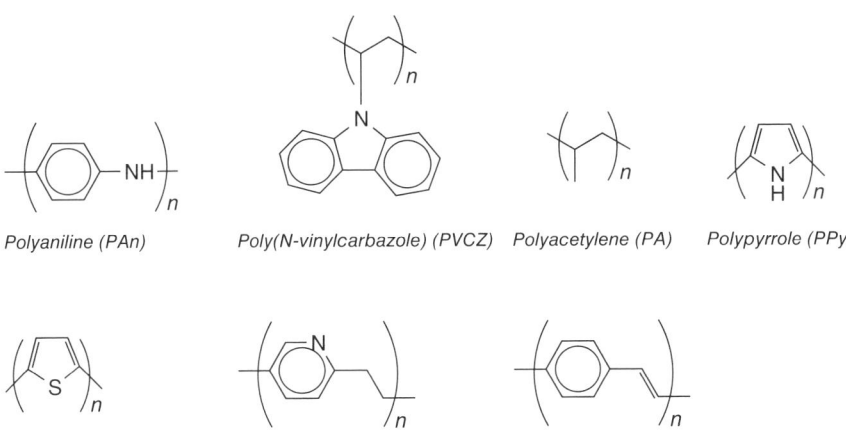

13.8 Conductive/conjugated polymers.

Polyaniline (PAn)   Poly(N-vinylcarbazole) (PVCZ)   Polyacetylene (PA)   Polypyrrole (PPy)

Polythiophene (PTh)   Poly(2-vinylpyridine) (P2VP)   Poly(p-phenyleneyinylenes) (PPV)

13.9 p-Doped polymers.

p-Doped polymers have wide application – for example, electrochronic devices, rechargeable batteries, capacitors, membranes, charge dissipation, and electromagnetic shielding. Less effort has gone into synthesizing and characterizing n-doped materials.

13.7.1 Poly (*p*-phenylenevinylenes) (PPV)

So far, most success has been achieved by using photovoltaic devices containing PPVs. The fabrication of PPV-containing photodiodes with a structure has been described. The PPV layer was obtained by spin coating the sulfonium salt precursor and then heating the polymer to 250 °C in vacuo. These devices were capable of generating open-circuit voltages of ~1.2 V when aluminum and magnesium electrodes were used or ~1.7 V when calcium electrodes were used. Quantum efficiencies of ~1% were obtained at low-light intensity (0.1 mW/cm^2).

The efficiency of photovoltaic devices containing conjugated polymers is determined by the ability to generate excitons from incoming radiation, and then to separate these excitons at appropriate interfaces before they recombine. Given that typical exciton capture zones are limited to 10 nm or less, more efficient structures are needed. This need has led several workers to the idea that interpenetrating networks of donor (electron donating–hole accepting) and acceptor (electron accepting–hole donating) polymers should give better results. The addition of cyano groups to a dialkoxy derivative of PPV forms the CN–PPV, making it a better electron acceptor. Underivatized PPV is a good hole-transporting material. Using blends of MEH–PPV (a soluble PPV derivative) as a hole transporter and CN–PPV as an electron transporter, results in quantum efficiencies of up to 6%. More recently, even higher quantum efficiencies (up to 29%) with overall power conversion of ~2% (using a simulated solar spectrum) were obtained using a modified organic solvent-soluble polythiophene as the hole acceptor and a cyano derivative of PPV (MEH–CN–PPV) as electron acceptor. Perylene is another electron acceptor that increases the quantum efficiency to 6%. An alternative approach uses the fullerene C_{60} as the electron acceptor, giving a quantum efficiency of ~29% and an energy conversion efficiency of 2.9%.

Structural characterization of PPV films with extension

The structure/property of fibre and film processing on the development of anisotropic molecular structure and its influence on the chromic, electrical and mechanical properties of electrical conducting and electric field modifiable polymers has been studied. With increasing interest in the structure and electrical properties of three-phase fibres as a camouflage fibre architecture, the effect of processing on the structure and electrical properties of different intrinsically conducting polymers has been compared. Two polymers, polyaniline (PAn) and poly (phenylenevinylene), PPV, were prepared and were investigated.[6]

The three different processes used to prepare the oriented films were spin coating and drawing to prepare one set of oriented emeraldine base polyaniline (PAn EB) films (see Section 13.7.2); casting and drawing another set of PAn EB films; and synthesis of a PPV precursor with drawing and the conversion to PPV while restrained. A modified prism waveguide was used to obtain the three-dimensional refractive index of PPV and otherwise immeasurable PAn. The three-dimensional refractive indices coupled with the intrinsic optical parameters allowed the absolute orientation functions about the three principal axes to be calculated and plotted on a triangular plot for comparison. From Fig. 13.10 we see that unstretched PPV film (squares) has a planar structure and one-way stretching converts the planar structure to a uniaxial structure. In contrast, unstretched PAn EB film has close to a random structure and one-way stretching converts the random structure to a uniaxial structure (circles and triangles follow the dashed fz

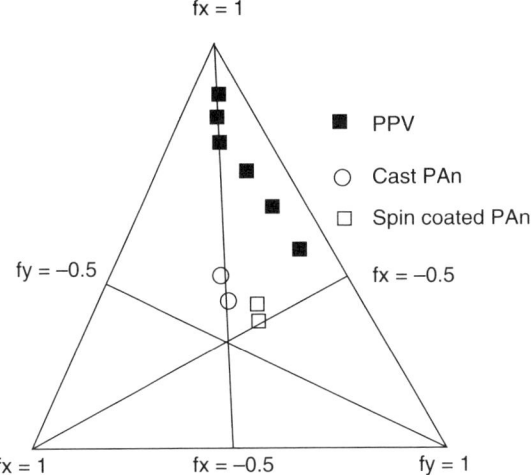

fx = 1

PPV

Cast PAn

Spin coated PAn

fy = −0.5

fx = −0.5

fx = 1

fx = −0.5

fy = 1

13.10 Orientation triangle plot for the PAn (EB) and PPV systems.

line). It can also be seen that PPV films reach high orientation states easily while PAn EB cannot be stretched to high orientation states based on the present stretching procedure.

The three-dimensional conductivity can be determined from impedance spectroscopy measurements of both HCl-doped PAn film and undoped PPV film as a function of orientation function. Note the much higher orientation achieved by the PPV films. The orientation function for PAn is the orientation function of the EB measured before doping.

The first thing that was noticed is that HCl-doped PAn films have higher conductivities than those of undoped PPV films. That is to be expected because conducting polymers are insulators in the neutral state and become conductive upon doping. Secondly, for both polymers, the in-plane conductivity is of magnitudes higher than the through-plane conductivity. Similar conductivity anisotropy has been reported for polypyrrole films where the in-plane conductivity is greater than the through-plane conductivity by four orders of magnitude. With respect to the effect of orientation, the conductivity along the stretch direction of the HCl-doped PAn increases with orientation, while the conductivity along the stretch direction of the undoped PPV is insensitive to orientation. This can be attributed to the fact that the main charge carrier in doped PAn is electronic in nature, while it is not in undoped PPV. Since the electronic charge carriers move along the polymer chain, it makes sense that the conductivity along the stretch direction increases with orientation function as more and more chains align toward the stretch direction. PPV samples are not doped and therefore whatever small conductivity that is detected must have come from residual impurity species in the PPV film. These are probably ionic in nature and therefore do not necessarily move along the polymer chain.

13.7.2 Polyanilines

Polyaniline is one of the oldest conductive polymers known. Leitch *et al.* first prepared it, in 1862, by anodic oxidation of aniline in sulphuric acid. It has been known as an Electrically Conductive Polymer (ECP) for the past thirty years. ECPs are able to conduct electricity because of their conjugated π-bond system, which is formed by the overlapping of carbon *p* orbitals and alternating carbon–carbon bond lengths. In some systems (such as polyaniline) nitrogen p_z orbitals and C rings are also part of the conjugation system. The conjugated double bonds permit easy electron mobility throughout the molecule because the electrons are delocalized. Delocalization is the condition in which π-bonding electrons are spread over a number of atoms rather than localized between two atoms. This condition allows electrons to move more easily, thus making the polymer electrically conductive.

PAn offers the possibility of providing properties to end products that are very difficult to obtain by existing commercial methods. In addition, it has a conjugated double bond structure, the benzenoid ring, between the quinoid imine and the benzenoid amine structures, which render the polymer a candidate as an ECP (see Fig. 13.11). PAn exists in three forms of oxidation state: leucoemeraldine (fully reduced or only benzenoid amine structures), emeraldine (neutral or partially reduced and partially oxidized), and pernigraniline (fully oxidized or only quinoid imine structures). The emeraldine-based (EB) form of PAn was used for this research because only doped EB PAn is conductive among the three oxidation states. The emeraldine-based form of PAn is also the most stable of the three states because leucoemeraldine is easily oxidized when exposed to air and pernigraniline is easily degraded. Some barrier devices containing polyanilines have also been produced. They can be made using either chemical or electrochemical oxidation (see Fig. 13.12).

13.11 Structural formula of undoped PAn (EB).

13.12 Electrochemical oxidation.

13.13 Introduction of electron acceptors to the PTh chain.

The electrochemical method can be used to produce thin films directly on conductive substrates such as indium tin oxide ITO. The chemical method can be used to produce a material with the de-doped emeraldine base (EB) form soluble in solvents such as 1-methyl-2-pyrrolidinone (NMP) and some doped forms. The materials are doped with appropriate surfactants such as dodecyl benzenesulfonic acid (DBSA), camphor sulfonic acid (CSA), or *p*-toluenesulfonic acid (pTS), all of which are soluble in common organic solvents. Polyaniline has been widely used in photoelectrochemical cells. Early researchers investigated the photoelectrochemical reduction of chloral (CCl_3CHO) to trichloroethanol (CCl_3CH_2OH). Shen and Tian[10] used PANi electrodes to induce the photoelectrochemical reduction of peroxidisulfate. Photocurrents generated at polyaniline are potential- and electrolyte-dependent.

13.7.3 Polythiophenes

The photoelectrochemical properties of polythiophenes (PThs) have also been of interest for some time. Their ability to electrodeposit regular structures with minimal impurities makes it possible to attain high photocurrents. The photocurrents that can be attained using PTh-based electrodes have been enhanced by using conjugated linkers to introduce electron acceptors to the PTh chain (see Fig. 13.13). The electron-accepting NO_2 group facilitates charge separation upon irradiation, resulting in sustained photocurrent. More recently, a photoelectrochemical cell that uses a solid polymer electrolyte based on poly(ethylene oxide) (PEO) has been described. Quantum efficiencies of up to 0.6% could be achieved.

A series of urethane-based diacetylenes have been prepared which are known to undergo a unique solid state photopolymerization to give highly

13.14 Photopolymerization of urethane-based diacetylenes.

coloured, highly conjugated polymer crystals (see Fig. 13.14). These materials are solvatochromic, thermochromic and mechanochromic. The presence of a long chain urethane on one or both sides of the diacetylene dramatically increases the solubility of the polymer in organic solvents. Thus, we can blend polydiacetylenes with both conventional textile polymers and certain piezoelectric materials, which results in a mechanical stress upon application of an electric field. As the conjugated backbone of the polydiacetylene is stressed, its colour changes. These molecules formed the initial basis of our investigations into the production of true camouflage fibres capable of changing their adsorption characteristics, and therefore their colour, under applied electrical or magnetic stress.[5,9]

13.8   Emissive polymers

Fibres which can quickly change their hue, depth of shade or optical transparency by application of an electrical or magnetic field could have applications in coatings, additives or stand alone fibres. Varying the electrical or magnetic field changes the optical properties of certain oligomeric and molecular moieties by altering their absorption coefficients in the visible spectrum as a result of changes in their molecular structure. These materials produce electroactive and magneto active oligomeric molecules with unique abilities to change their absorption and/or reflection of electromagnetic radiation in the infrared, visible, and ultraviolet frequency ranges. Molecules and oligomers identified in the initial phase of this research that possess these 'tunable' properties are being introduced into fibre-forming polymeric matrices. These are being formed into fibres and films (films provide ease of study by a variety of spectroscopic methods) and their optical properties evaluated under differing electrical, magnetic and thermal stress. These new emissive compounds, when polymerized or attached to oligomers, are highly polarizable and have the correct electronic structure to render significant changes in absorption of electromagnetic energy in the visible spectrum. The change in colour is due to the absence of specific wavelengths of light that will vary with the application of an electromagnetic field, due to structural changes.

New synthetic routes for emissive compounds have been developed and enable the construction of monomeric materials that have unique emissive properties in the visible spectrum. These molecules undergo significant changes in their polarizabilities, microstructure, or conjugation lengths dependent on the strength of the applied field. Some of the work carried out incorporated these materials into fibres and films, imparting to them dynamic colour change ability when exposed to either a static or dynamic electric field. The resultant colour may be changed by application of either a static field, or in this particular case, interrogating radiation, which may, in fact, make the fibre containing the material an ideal sensor. This molecule and other closely associated derivatives are easily polymerized at the alpha carbon position on the poly-ethylenedioxythiophene (PEDOT) end units. This ensures linearity in the polymerization of the monomer due to the low threshold potential of the alpha carbon which prevents branching. Such molecules as these PEDOT moieties and others show significant promise for development of spectral tunable materials as either neat fibres, fibre and film coatings, or as additives to fibre and film matrices. These polymers can be easily tuned to different colours and/or extend the emissions into the infrared or ultraviolet by simply changing the R group either on the benzoid di -oxy group or on the vinylene linkage. These new emissive compounds, when polymerized or attached to oligomers, are highly polarizable and have the correct electronic structure to render significant changes in absorption of electro magnetic energy in the visible spectrum. The change in color is due to the absence of specific wavelengths of light that will vary with the application of an electromagnetic field due to structural changes. Figure 13.15 illustrates highly emissive bis cyano monomer capped with PEDOT (di-oxy-polypara phenylene vinylene) monomer for subsequent polymerization.

13.15 Emissive polymer for camouflage fibres.

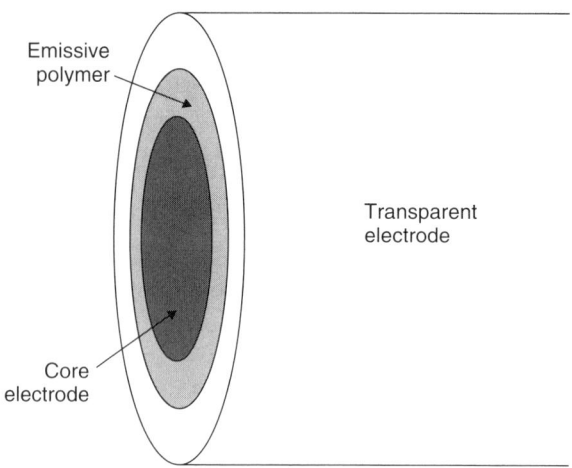

13.16 Camouflage fibre architecture.

Conductive polymers have been the subjects of study for many decades as possible synthetic metals. Many of these polymers, especially those with a conjugated π-bond system, often yield higher conductivity once they have undergone the doping process. However, practical uses of conducting polymers are not very likely because of their poor mechanical properties (such as strength and processibility), which rarely meet industrial requirements. Thus, the unique combination of electronic and mechanical properties of blends of conductive polymers with conventional polymers seems to have great promise for many applications. Since the conducting polyblends are stable and retain the mechanical properties of the host polymer, films, fibres, and coatings can be fabricated by solvent evaporation or by melt-processing for use in anti-static applications, for electromagnetic shielding and/or absorption, for transparent conducting films, and so on (see Fig. 13.16).

13.9   Surface attachment of chromophores to conducting polymers

A major thrust is the development of methods to attach chromophores to the surface of films and fibres, and particularly interesting is attachment to electrically conducting polymers, as this gives us a method to alter the chromophore's oxidation state or to subject it to an electrical field. The optical response of a chromophore to an electric field is highly dependent upon its molecular orientation. One way to ensure consistent sharp colour change is to assemble the chromophores on or very near the surface of the

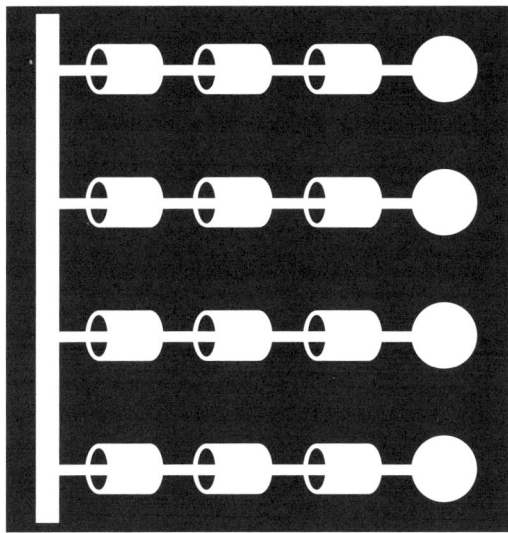

13.17 Chromophoric array.

film or fibre. Thiols are powerful nucleophiles that might serve as anchors for attaching molecular devices to conducting polymers. The mechanism and extent of the nucleophilic attack will determine how useful this process will be in a given system. The chromophoric array (see Fig. 13.17), might be assembled into an organized 'near-monolayer' on the surface of a polymer film or fibre.

Films of four conducting polymers (while doped to their cationic, conducting forms) were immersed for one hour in a THF solution of an alkanethiol or a fluoroalkanethiol. The films were then rinsed thoroughly to ensure that any adsorbed thiol was removed. Some ongoing research is focused on the attachment of chromophoric species to the surfaces of conducting polymers. For example, porphyrins and porphyrin arrays have been attached and their spectroelectrochemistry (see Fig. 13.18) has been explored.

13.10 Processing of electrically conducting polymers

Electrically conducting organic polymers are required to generate the electric fields for camouflage materials. Stable, homogeneous fields can be achieved only from high-quality, homogeneous polymers. Unfortunately, the development of processing technology in this field has lagged far behind the exploratory synthesis. Polyaniline is being explored as electrode materials. Solution processing of PANi is well established, but the possibility of thermal processing has yet to be fully explored.

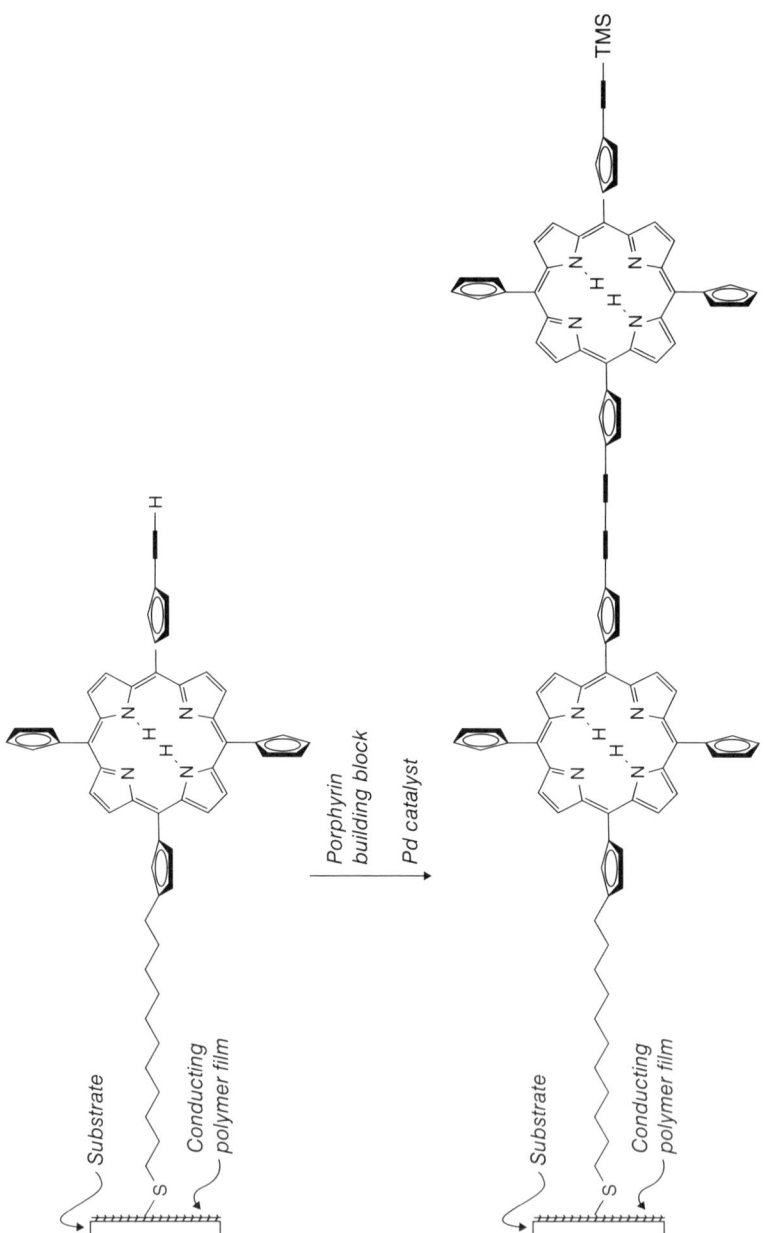

13.18 Spectroelectrochemistry of porphyrin.

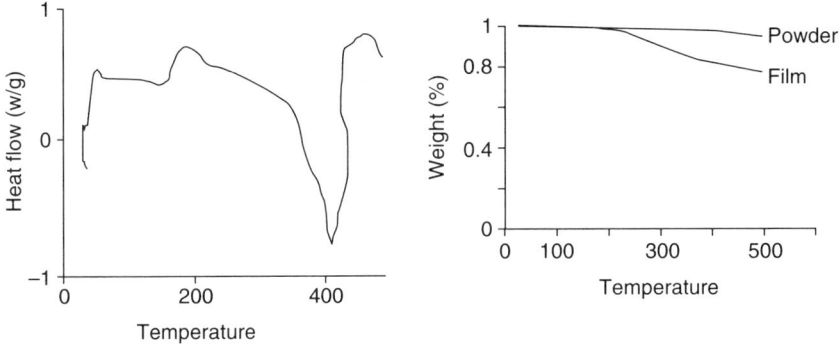

13.19 Thermal gravimetric analysis (TGA).

The leucoemeraldine base (LEB) form of polyaniline is the fully reduced form. In order to become conducting, the polymer must be partially oxidized and protonated. The LEB form is somewhat oxygen sensitive, particularly at higher temperatures, so these studies were performed either under nitrogen or vacuum. Powders of LEB were prepared by deprotonation. The material shows a Tg of 170–200 °C and a strong endotherm at 365–395 °C. As can be seen from the TGA (thermal gravimetric analysis, see Fig. 13.19), this event is not associated with a mass loss. Heating the sample below the melting transition results in a dramatic increase in the crystallinity of the powder, as evidenced by X-ray diffraction. The conductivity of the sample is not adversely affected by the annealing process and the LEB can be converted to its conducting form either before or after heat treatment.

13.11 Assembling of gold nanoparticles

The technology exists for assembling gold nanoparticles onto conducting polymer films with self-assembled monolayers (SAMs) built from long-chain alkyl thiols. These SAMs feature a functional end (the thiol) designed to interact with the nanoparticle surface and a long alkane chain that encourages self-assembly into highly ordered, close-packed arrays through van der Waals interactions. The preparation of gold nanoparticles is coating with a monolayer of a long chain thiol featuring a diacetylene moiety midway along a long hydrocarbon tail. Conjugated diacetylenes can undergo a fascinating photopolymerization in the solid-state if the proper arrangement of the reacting centers can be achieved. The reaction has been observed in crystals, Langmuir–Blodgett films, liposomes and in self-assembled monolayers on metal and oxide surfaces.

13.12 Conclusions

The creation of field-responsive fibres, camouflage fibres, is a multi-disciplinary endeavour. In addition to chromophores, polymeric materials may be able to generate a uniform, stable field for excitation of the colour change processes. This may lead to microdevices encapsulated in fibres for a variety of technical applications. Such intelligent clothes are worn like ordinary clothing, providing help in various situations according to the designed application. Although lots of new products have come into existence, there is still vast scope to utilize new technologies such as camouflage textiles for protective wear.

13.13 Acknowledgment

We thank all the authors we quoted in the references and also our institution which encouraged us to produce this material.

13.14 References

1 R. GREGORY, R. SAMUELS and T. HANKS, 'Chameleon Fibres' *US National Textile Centre Annual Report* M-98 C01.
2 S. NANTA, Temperature-Sensitive Colouring Materials SWAY, *Sen-1 Kikai Gakkaishi* (1989) 42 (9) 435–439.
3 T. HONGU, G.O. PHILLIPS, Kanebo Ltd, *New Fibres*, New York, Ellis Horwood, 1990.
4 T. LIU and R.J. SAMUELS, 'Novel Method for Optical Characterization of Films', Tao Liu and Robert Samuels, *Proceedings of Interpak 2001*, Kauai, Hawaii, July 8–13, (2001) 1–5.
5 P. LEITCH and T.H. TASSINARI, 'New Material in New Millenium', *Journal of Industrial Textile* (2000) 29 (3) 173–191.
6 R. OU, X. WANG, R. GREGORY and R.J. SAMUELS, 'Evolution of the Anisotropic Structure of Poly(Phenylene Vinylene) Films with Stretching', *SPE ANTEC* (2000) 46 1449.
7 R. GREGORY, *Handbook of Conductive Polymers*, Ch. 18; 2nd ed., T. Skotheim, R. Elsenbaumer and J. Reynolds (eds) Marcel Dekker. 1997.
8 S.S. HARDAKER, C.Y. CHA, S. MOGHAZY and R.J. SAMUELS, *Advances in Polyimide Science and Technology*, C. Feger, M. Khojasteh and M. Htoo (eds), Lancaster, PA, Technomic Pub. Co. (1993) 571.
9 C. CHA, S.S. HARDAKER, R.V. GREGORY and R.J. SAMUELS, 'New Approaches to the Study of Polyaniline', *J. Synthetic Metals* (1997) 84 (1) 743.
10 P.K. SHEN and Z.Q. TIAN, *Electrochim. Acta* (1989) 34 1611–1613.
11 G. MILLS, L. SLATEN, R. BROUGHTON, Photoadptive Fibers for Textile Materials', *National Textile Center Annual Report*, November, 2001.
12 R.V. GREGORY, R.J. SAMUELS, T. HANKS, 'Chameleon Fibers: Dynamic Color Change from Tunable Molecular and Oligomeric Devices, *National Textile Center Annual Report*, November 2001.

14

New developments in coatings and fibers for military applications

P. SUDHAKAR, S. KRISHNARAMESH
and D. BRIGHTLIVINGSTONE,
K. S. Rangasamy College of Technology, India

14.1   Introduction

The US Army uses a wide variety of equipment for tactical applications, notably tanks, trucks, missiles, aircraft, artillery, and shelters. Less obvious equipment includes water purification units, generators, and high-mobility forklifts. This tactical equipment is usually painted during its original manufacture, then repainted as necessary at facilities ranging from large-scale Army depots and small 'touch-up' operations to unit-level organizations. The same coating systems, defined by their military specifications, are used at all facilities, but in many different environmental conditions. The coating systems used to paint Army equipment are highly engineered materials, formulated to meet multiple performance requirements. The selection of coating ingredients, formulation conditions, application methods, and curing conditions can influence the final properties of the coating, see Fig. 14.1. Generally, Army coatings are formulated using polymer binders, a variety of solvents, pigments for tinting, extender pigments for the control of gloss, and functional additives. In this chapter, we will discuss recent research findings and product developments that have led to the next generation of coating systems designed for Army applications.

14.2   Chemical agent resistant coatings

Chemical Agent Resistant Coating (CARC) systems form an important category of products currently being researched, developed, and implemented. Finishes included in this group resist penetration by chemical warfare agents and can be readily decontaminated.

In the camouflage coatings initially developed in the 1970s and 1980s, the chemical components of the CARC finishes were alkyd enamels modified

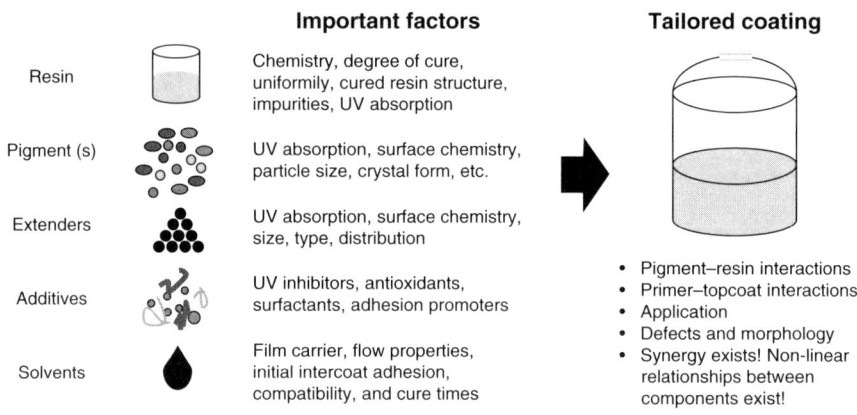

Important factors

Resin — Chemistry, degree of cure, uniformily, cured resin structure, impurities, UV absorption

Pigment (s) — UV absorption, surface chemistry, particle size, crystal form, etc.

Extenders — UV absorption, surface chemistry, size, type, distribution

Additives — UV inhibitors, antioxidants, surfactants, adhesion promoters

Solvents — Film carrier, flow properties, initial intercoat adhesion, compatibility, and cure times

Tailored coating

- Pigment–resin interactions
- Primer–topcoat interactions
- Application
- Defects and morphology
- Synergy exists! Non-linear relationships between components exist!

14.1 Formulation of highly engineered materials.

according to their application. The primers utilized heavy metal pigments, in particular, lead, hexavalent chromium and inert iron oxides, to arrest and prevent corrosion. CARC topcoats employed a variety of organic and inorganic pigments to mimic woodland or desert surroundings; and silica and talc were used to reduce the gloss of the coatings. The first-generation polyurethane-based CARC topcoat proved to be a tremendous improvement over the enamel finish it replaced. However, while the high pigment content and low resin volume met the Army's low gloss requirement, it did so at the expense of long-term durability, resistance, and flexibility. The Army's current topcoats are based on aliphatic polyurethanes that meet the requirements of military specifications. These coatings provide excellent performance characteristics, such as exterior durability and excellent chemical resistance. They also withstand decontamination procedures and provide camouflage properties near the visible and near infrared regions of the spectrum.

When assessing the inherent balance of properties found in a CARC coating system, there are three principles to guide and assist in the formulation, scale-up, and final implementation of new coating products to the field. The three primary principles are Durability, Environmental Compliance, and Survivability, see Fig. 14.2. This approach has produced products that ensure the best durability, meet or exceed current or proposed environmental and user safety laws, and meet survivability requirements for camouflage coatings.

Survivability requirements drive two unique military and technical challenges: camouflage that blends with a given background while also providing resistance to live chemical warfare agents.

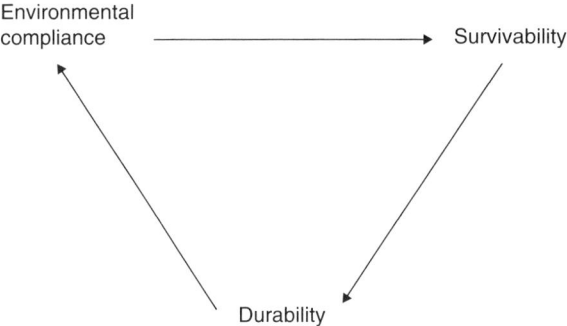

14.2 Primary principles for camouflage coatings.

14.3   Influence of environmental regulations

A variety of environmental regulations and worker safety issues have motivated the development of environmentally compliant CARC coatings. For example, Federal and local regulations resulting from the Clean Air Act and its amendments restrict the amount of volatile organic compounds (VOCs) that can be emitted during the application of protective coatings, and Occupational Safety and Health Administration (OSHA) regulations on worker safety restrict exposure to many of the materials used in their manufacture. To address these regulations, the coating has been reformulated to be lead- and hexavalent-chromate-free, low in VOC content, and in many instances free of hazardous air pollutants (HAPs).

From a historical perspective, the two main avenues to VOC reduction in coatings were formulations with lower molecular weight polymers and the use of waterborne, water-dispersible, or water-reducible polymers. Reducing a polymer's molecular weight usually reduces its viscosity and reduces the need for solvents to control the system viscosity during application. This is the traditional high solids solution to VOC problems. But the high solids versions of CARC topcoats did not meet VOC regulations in certain localities, and new technology was needed to develop a CARC that would solve current and anticipated VOC problems. Alternatively, systems in which water can be used for this viscosity control can greatly alleviate the need for solvents. Until recently, water-compatible coatings did not match the performance of solvent-based analogs, but recent developments in polymer technology have enabled the development of high performance polyurethane systems with excellent performance and chemical agent resistance.

14.4 Water-reducible, two-component polyurethane, chemical agent-resistant coating (CARC) topcoat

The recent development of a water-reducible, two-component polyure-thane CARC topcoat has been a significant achievement. This product uses water-dispersible hydroxy-functional polyesters and water-dispersible poly-isocyanates. Traditionally, the presence of water must be eliminated in non-aqueous two-component polyurethane formulations due to the unfavorable reaction with isocyanate. The reaction forms an unstable carbamic acid, which quickly decomposes to generate carbon dioxide and an amine (Equation 14.1). The amine then reacts with further isocyanate to yield the substituted urea (Equation 14.2).

$$RNCO + H_2O \rightarrow [RNHCOOH] \rightarrow CO_2\uparrow + RNH_2 \qquad [14.1]$$

$$RNH_2 + RNCO \rightarrow RNHCONHR \qquad [14.2]$$

This reaction may inhibit or adversely affect the stoichiometry and develop-ment of cross-linking that is crucial to the integrity and performance typical of two-component polyurethanes.[1,2] Developments in waterborne polyure-thane technology have enabled high performance coatings to be formulated using water-dispersible polyisocyanates and hydroxyl-functional polyure-thane dispersions.[7] While there is a competing reaction occurring with water, the kinetics, raw materials selection, and proper indexing of isocya-nate (NCO) to hydroxyl (OH) groups can ensure that sufficient crosslink density is established. Dynamic Mechanical Analysis, Fourier-transform infrared spectroscopy and desorption-gas chromatography/mass spectros-copy have shown the role of indexing on the final physical properties of these polyurethane materials.

14.5 Contribution of binders and pigments

Research into water-dispersible urethanes has resulted in a new CARC coating system. The VOC of the regulated solvent is less than 180 g/L (a 60% reduction) and it contains zero HAPs. At the same time, through careful selection of pigmentation, the survivability and durability of this environmentally compliant coating are outstanding. Chemical agent resis-tance is provided by the aliphatic polyurethane binder, and camouflage properties are provided by the appropriate selection of tinting pigments for visual color and near-infrared reflectance, plus extender pigments for gloss control. Camouflage requirements complicate the development of Army coatings because of the need for low gloss, which leads to proportionally higher pigment to binder ratios. Higher binder content in the formulation enhances chemical agent resistance while high pigment content reduces gloss. The Army Research Laboratory (ARL) has replaced the solvent-borne polyester and polyisocyanate binder components of the current CARC with the water-dispersible system. At the same time, an improved pigment package has been developed to reduce problems (such as marring and reduced flexibility) resulting from the high pigment content required for low gloss. The resulting coating formulation survives decontamination and provides chemical agent resistance. The use of novel pigmentation has improved the low-temperature flexibility, mar resistance, and weathering durability of this coating. The selection of binders and pigments also affects the degradation of CARC coatings.[4] Research has demonstrated that replacing diatomaceous silica and talc with polymeric-based extender pigments enhances the durability of CARC systems. Alternative pigment materials, such as a blend of urea resin containing a negligible quantity of free methylol groups, have also been explored. The primary particles, with an average grain diameter of 0.1–0.5 µm, form agglomerates of approximately 4–5 µm. This results in a high pore volume and a steep grain distribution, two important factors providing preconditions for an excellent matting effect. As a result of the almost ideal spherical shape of the particles, the coating rheology remains unaffected, as opposed to the more needle-shaped silica particles. The other active flattening agents used have a polyurethane composition with a medium particle size of 18 µm. Most are spherical vesiculated-type materials that have enhanced matting properties and are also extremely chemical resistant. The combination of these materials provides a very dynamic and active flattening package.

CARC materials technologies are developing coating formulations to meet both performance requirements and environmental regulations. In addition, today's CARCs function as camouflage and have increased environmental degradation resistance. These newer CARC systems meet the current design strategies of providing durability, survivability, and environmental compliance.

14.6   Functional garments for soldiers

The issue of thermal protection has always been a dilemma: if a protective garment insulates the wearer from heat and flames, it also insulates in the reverse direction. It can trap body heat and moisture inside the garment, leading to fatigue, discomfort and moisture build-up, and it can compromise the body's ability to maintain proper body temperature (heat stress). The result is the so-called 'plastic wrap effect'. Synthetic fibers can be used to produce functional flame-resistant fabrics, but these are unsatisfactory with respect to the management of heat and moisture. Research has been undertaken into various treatments to make synthetics more comfortable, but the improvements have not been adequate. An alternative approach is to start with a natural, comfortable fiber, such as cotton, and treat it topically with FR chemicals to make it flame resistant. This approach results in fabrics that have not only lost mechanical strength, but more importantly, have lost most of the comfort characteristics of cotton. Furthermore, the flame retardance is not permanent, but is lost during laundering. An alternative solution has to be found.

14.7   New-generation fibers for military applications

Kermel, a high performance *meta*-aramid technical fiber, is naturally non-flammable and provides excellent thermal insulation as well as good resistance to mechanical stress and chemicals. Its circular cross-section and low modulus make it comfortable. In addition, it is spun-dyed during the manufacturing process, which gives it lifelong color fastness and resistance to UV light and washing.

Viscose, a natural, regenerated cellulosic fiber, was also chosen for research. An ambitious development program resulted in a revolutionary, inherently flame-resistant, halogen-free, comfortable, breathable, anti-static viscose fiber, which has facilitated the next advances in thermal protection fabrics and apparel.

In addition to providing comfortable protection against heat, the new-generation fabrics also have enhanced infrared (IR)-reflectance characteristics and excellent printability properties, thereby making them desirable for camouflage and battle-dress garments. The blending of solution-dyed *meta*-aramids, such as Kermel, and printable, inherently flame-resistant viscose fibers, such as Lenzing FR, and the careful selection of specific dyes, enable the production of the desired IR-reflection parameters. Of course, camouflage prints have to meet extremely high standards of color fastness, IR-reflectance and consistency of shade. With the right dyes and pigment preparation, it is indeed possible to achieve color matching of shades in the visual range, to meet the defined fastness requirements and produce the required IR-reflectance value.

14.8   Acknowledgment

The authors would like to thank the coatings and corrosion team, and various other armament and research development centers, for their provision of ideas through their journals. We thank Mr John A. Escarsega, research chemist and leader of the Coatings and Corrosion Team in the Materials Application Branch of the US Army Research Laboratory, and also Mr Kestutis Chesonis of the Weapons and Materials Research Division at Aberdeen. We also thank Dr Dawn M. Crawford, Mr Jeffrey L. Duncan and Ms Pauline M. Smith for their valuable contribution. We also thank all those who have been involved, either directly or indirectly, in the preparation of this chapter.

14.9   References

1 US Department of Defense. *Coatings, Water Dispersible, Aliphatic Polyurethane, Chemical Agent Resistant*. MIL-DTL- 64159, 30, January 2002.
2 G. WOODS, *The ICI Polyurethane Book*. New York: John Wiley, 1990.
3 G. OERTEL, *Polyurethane Handbook*. New York: Hanser, 1985.
4 J.A. ESCARSEGA and K. CHESONIS, United States Patent, 5 569 410, *Water Dispersible Low-reflectance Chemical Resistance Coating Composition*, November 25, 1997.

14.10 Bibliography

http://amptiac.alionscience.com/quarterly
J.A. ESCARSEGA, D.M. CRAWFORD, P.J. KASTE, *Analysis and Performance Evaluations of Chemical Agent Resistant Coating Systems*, ARL-TR-2545, July 2001.
J.A. ESCARSEGA, *Multifunctional Protective Coatings For Weapon Systems: Chemical Agent Resistant Coatings (CARC) Technical Exchange, TACOM Briefing*, September 5, 2002, (http://www.arl.army.mil/wmrd/coatings/Current/TECHNIC ALEX090502.ppt)
P. SMITH, K. CHESONIS, C. MILLER, J.A. ESCARSEGA, *Replacement Alternatives to the Chromate Wash Primer* DOD-P- 15328D, ARL-TR-3220, June 2004.
P.B. JACOBS and P.C. YU, *Coatings Technology*. Vol. 65, p. 45, 1993.
US Department of Defense. *Coatings, Aliphatic Polyurethane, Chemical Agent Resistant*. MIL-C-46168, Amendment 3, May 1993.
US Department of Defense. *Coatings, Aliphatic Polyurethane, Single Component, Chemical Agent Resistant*. MIL-C-53039, 28, November 1988.
W. KOSIK, *et al. Mechanism of Military Coatings Degradation*, SERDP Pollution Prevention Project 1133, FY99-02.

15
Military fabrics for flame protection

C. WINTERHALTER, US Army Natick Soldier Research,
Development and Engineering Center, USA

15.1　Introduction

Military burn injuries on the battlefield can be attributed to three causes: flame and thermal threats, incidental or secondary hazards, and accidents. Threat-generated burns result from the direct employment of a flame and thermal weapon. Incidental or secondary burns result from flame and thermal weapons or other threats (i.e. ballistic, blast, directed energy) that ignite battlefield combustibles, including clothing or equipment that can also present a burn hazard. Accidents comprise the remaining burn hazard. Historically, even in combat, burn injuries more often result from the ignition of battlefield combustibles due to accidents or secondary flame effects in the environment, than from direct flame and thermal threats.[1] In addition, the specific cause of casualties and fatalities has been difficult to characterize in terms of burn injuries because many were documented as involving multiple trauma and explosions. Until recently the flame threat had generally been categorized as relatively low for the individual soldier, and due to the increased use and exposure to fuels, burn injuries had been more prevalent for both Navy and Air Force personnel. However, during Operation Iraqi Freedom, the use of Improvised Explosive Devices (IEDs) against United States ground forces has become prevalent. The IEDs serve not only as ignition points for battlefield combustibles, but have been enhanced to include fuels making them a direct battlefield threat to the soldier and increasing the likelihood of burn injuries.

Kim has reviewed the characteristics of several battlefield threats and combustibles. He estimated that the values of the heat flux of thermobaric, incendiary and flame weapons, and JP-8 fuel converge to a common heat flux of about 2.0 cal/(cm²·sec) over time.[2] Since the likelihood of survival of a direct hit from a flame and thermal weapon is low based on the initial ignition temperatures and known blast effects, and the fact that many burn injuries result from the secondary effects of the environment, a thermal flux of 2.0 cal/(cm²·sec) was selected as most representative of a military fire hazard that is survivable if exposure time is short. Appropriately developed protective clothing should provide critical seconds of increased escape time to distance oneself from the flame hazard and thereby provide a lesser, but survivable exposure time.

15.2   Types of fabrics and their performance

Flame protection is one of several desired performance characteristics in a combat uniform. Due to costs and logistics issues, fabrics are desired that are multifunctional, and clothing systems that are modular and lightweight. The ideal fabric would provide flame protection, visual and near infrared camouflage protection, and comfort in climatic extremes such as the hot and humid tropics and hot and dry desert. In addition, it would be lightweight, low cost, durable to extreme wear and tear, and have the ability to maintain a sharp military appearance with low maintenance. The standard fabric used in United States combat uniforms is made from a nylon and cotton fiber blend that is dyed and camouflage printed to provide visual and near infrared camouflage protection, and is constructed in either a ripstop poplin[3] or twill[4] weave construction. This fiber blend fabric is used by every branch of the United States military, as well as the United States Coast Guard. The physical property requirements for these fabrics are listed in Table 15.1. While these fabrics provide excellent durability and serviceability, they are flammable. Ground forces desire flame-resistant combat uniforms to protect against secondary and accidental flame hazards but they are too expensive to issue to every soldier. However, United States Army tankers and aviators of all services, which account for a smaller percentage of the force, are required to wear flame-resistant clothing systems made from Nomex® and Kevlar® fiber blend fabrics. As described below, alternative materials have been developed and investigated, as well as clothing system protection strategies that are affordable and provide flame protection for the individual soldier.

As a guide, the military standard camouflage printed Nomex® and Kevlar® fabric[6] has been used for general performance and flame-resistance goals. This fabric is composed of 92% Nomex®, five percent Kevlar® to prevent break through when the fabric is exposed to flame, and three percent of a carbon core nylon sheathed fiber for electrostatic dissipation. Fibers, fiber blends and functional finishes investigated include: Basofil, flame retardant treated (FRT) cotton/nylon, FRT Tencel®, FRT cotton/ Kevlar®/nylon, carbonized rayon/Nomex®, Kevlar®/flame-resistant (FR) rayon, Nomex®/FR rayon, FRT cotton/PBI, and wool/Nomex®. Weight and strength data for these fabrics are listed (Table 15.2) with those for

Table 15.1 Physical property requirements for cotton, and nylon/cotton combat uniform fabrics

Property	MIL-C-43468[5] 100% Cotton Ripstop poplin	MIL-DTL 44436 50% Nylon 50% Cotton Ripstop poplin	MIL-C-44031 50% Nylon 50% Cotton Left hand twill
Weight, grams/meter^2 ASTM D 3776	193–237	203–237	230 minimum
Yarns per centimeter, minimum, warp × filling ASTM D 3775	41 × 20	41 × 20	34 × 21
Breaking strength, newtons, minimum, warp × filling ASTM D 5034	490 × 303	890 × 400	890 × 556
Tearing strength, newtons, minimum, warp × filling ASTM D 1424	18 × 18	31 × 22	50 × 36
Air permeability, maximum cubic feet/ minute/foot squared ASTM D 737	9.1	7.6	12.7

the military solid and camouflage printed Nomex® fabrics. Many of the inherently flame-resistant fibers were eliminated for use in a homogeneous fabric due to their high cost and the requirement for visual and near infrared camouflage. The high polymer orientation of the aramids and PBI, for example, contributes to their flame resistance, but also reduces or eliminates their ability to be dyed with traditional dyestuffs due to the lack of chemical dye sites. Some of these materials may achieve coloration by pigment injection in solution form, but their versatility is limited. The aramid blends used today are dyed and camouflage printed using proprietary technology that adds significantly to the final cost of the finished fabric. Still prized for their inherent flame resistance, some of these fibers were blended with low-cost fibers to enhance the overall flame resistance and strength of the fabric. Flame-retardant rayon, which is inherently flame-resistant rather than flame-retardant treated, was blended with the aramids in 60/40 and 35/65% blend ratios, but these materials fell short of the desired fabric strength and the camouflage print demonstrated poor colorfastness.

Table 15.2 Performance characteristics of developmental and commercial flame-resistant fabrics

Fiber blend		Weight, (g/m²)	Tearing strength, (N)	Breaking strength, (N)	Flame resistant
MIL-C-83429 92% Nomex® 5% Kevlar® 3% P140	Class 5 Camouflage Printed	159–203 ASTM D 3776	40 × 31 (min.) ASTM D 1424	801 × 445 (min.) ASTM D 5034	Yes
	Class 6 Solid Color	145–169	53 × 36 (min.)	801 × 445 (min.)	Yes
93% Nomex® 5% Kevlar® 2% P140 (Spun Laced)		105	Does not tear	485 × 445	Yes
58% FR cotton (sheath) 2% Mannacryl (sheath) 40% Kevlar® (core)		152	36 × 36	479 × 271	Yes
80% FR cotton 20% PBI		170	40 × 31	369 × 214	Yes
50% Wool 50% Nomex®		203	49 × 36	712 × 520	Yes
58% FR cotton 27% Kevlar® 15% Nylon		214	13 × 13	680 × 712	Yes
58% FR rayon 40% Kevlar® 2% Conductive fiber		217	44 × 58	668 × 445	Yes
65% Nomex® 35% Rayon		220	40 × 27	583 × 454	Yes
80% FR cotton 20% PBI		224	49 × 27	618 × 334	Yes
Carbonized rayon Nomex®		234	44 × 36	881 × 556	Yes
100% FR Lyocell (Tencel®)		251	27 × 31	801 × 520	Yes
88% FR cotton 12% Nylon		261	27 × 31	441 × 320	Yes
88% FR cotton 12% Nylon		373	22 × 40	716 × 683	Yes

Flame-retardant treated cotton has long been the industry standard for use in low-cost flame-resistant industrial work wear. However, the most commonly used treatment 'Indura' adds 20% to the weight of the fabric. Flame-retardant treated cotton was blended with nylon in 88/12% blend ratios, where the nylon was added to improve strength. While the addition of the nylon did not negatively impact the flame resistance, a heavier weight fabric of 373 grams/meter^2 was required to achieve relatively acceptable breaking and tearing strengths.

Flame-retardant treated cotton was also blended with Kevlar® and nylon to enhance flame resistance and improve abrasion resistance in a 58/27/15% blend ratio. While the Kevlar® was the strongest of the three fibers, it occupied less than 50% of the total material composition to keep costs down. The strength of the fabric is dictated by the dominant fiber, which in this case, was the lower strength cotton. In addition, the high end and pick count required to anchor the Kevlar® fiber detrimentally reduced the fabric tearing strength to three pounds in the warp and filling directions.

PBI was blended with flame-retardant treated cotton in a 20/80% blend ratio in 170 and 224 grams/meter^2 weights; however, the performance characteristics of the predominate fiber, cotton, prevailed.

Basofil fiber demonstrated low fiber tenacity and developmental efforts were directed toward insulation, knitted underwear, headwear and handwear applications, where high strength is not a critical factor.

Blends of carbonized rayon and Nomex® were investigated and while they demonstrated good strength performance they could not be dyed and camouflage printed.

Flame-retardant treated Tencel® demonstrated good strength but the camouflage demonstrated poor colorfastness performance.[7] Core spun yarns were investigated and developed with the primary intent of manufacturing a yarn with a high strength, inherently flame-resistant core, and low cost, readily camouflage printable sheath fiber. The best performing material combination was a cotton sheath, Kevlar® core yarn. These materials also fell short on strength because only the Kevlar®-based core and not the sheath contributed to the fabric strength.

While all of the developmental materials investigated and described above met the fabric[6] bench scale flame performance goals for vertical flame, which are 2.0 seconds maximum after flame; 25.0 seconds maximum after glow; 4.0 inches maximum char length, when tested in accordance with ASTM D 6413, 'Flame Resistance of Textiles (Vertical Test)[8], strength and other performance requirements such as colorfastness of the camouflage fell short.

The most promising of the developmental fabrics was the wool and Nomex® blend because it met the desired performance requirements of MIL-C-83429, for flame resistance, weight, tearing strength, and was only

11% short on breaking strength. In addition, the fabric met performance goals for near-infrared reflectance and colorfastness. Only the wool fiber was dyed and camouflage printed. The nylon/cotton blend fabric,[3] which is flammable, is the standard for near-infrared reflectance and colorfastness, and the wool and Nomex® blend fabric met those performance characteristics.

15.3   Measuring flame and thermal performance

Bench-scale testing, such as the vertical flame test, is an excellent screening tool for initial evaluations of fabrics and materials. However, the instrumented manikin test, ASTM F 1930, 'Evaluation of Flame Resistant Clothing for Protection Against Flash Fire Simulation Using an Instrumented Manikin[9] provides information about the flame protection provided by the entire clothing system. This test provides information not only about fabric performance in a clothing system, but also about the effects of clothing configurations (system layering), garment design (coverall, or shirt and trousers), closure systems (buttons, zippers, Velcro), effects at the component interfaces (shirt to pants, glove to sleeve cuff), garment fit (silhouette, ease, drape, air gaps), and system behavior in a flash fire exposure. The test, which is a research tool and is not recommended for quality assurance testing, quantifies the thermal protection of the garment or ensemble and the predicted burn injury by the measured sensor response that indicates how well the garment or protective clothing configuration blocks heat from the manikin surface.[9] In addition, the test targets a thermal flux of 2.0 cal/ (cm^2·sec), which has been identified as representative of a military fire scenario. The test method requires an upright adult male manikin, with a minimum of 100 heat sensors uniformly distributed with the exception of the hands and feet, and a data acquisition system and burn assessment program that has the capability of acquiring and storing the output of the sensors, calculating the heat flux, and predicting the burn injury level at each sensor, and the total predicted burn injury area as a result of the thermal exposure.[9] Due to differences between sensors, data acquisition methods, and burn assessment programs among different testing facilities the results have not been statistically correlated in a round robin study, and therefore are not recommended for quality assurance testing associated with procurement actions.

Military medical doctrine was used to establish performance guidelines because the test method does not provide pass/fail criteria. A partial listing of the accepted criteria for transferring patients to a burn center include: (a) partial-thickness burns of 20% total body surface area burned (TBSAB) or greater in adults, (b) full-thickness burns exceeding five percent TBSAB, and (c) burns involving the face, feet, hands, perineum, or major joints.

During an armed conflict, however, additional criteria include casualties with small burns involving the hand, foot, or perineum that may be considered for early evacuation since their activity will be disproportionately limited.[10,11,12] In addition, during combat, available resources must be expended on those individuals with the greatest chance for survival. Casualties at the extremes of age (that is, those under ten or over 60 years old) or those with 70% of the total body surface area burned will, in general, do poorly, and should be allocated a lesser share of the resources. The care of casualties with less than a 20% TBSAB can safely be delayed pending either their evacuation to a higher-echelon medical treatment facility or the availability of more resources.[10,11,12] Pass/fail criteria for the instrumented manikin test was established as no more than a 20% second-degree body burn and no more than a 5% third-degree body burn based on the likelihood of the availability of treatment during combat.

15.4 Clothing system configurations and their performance

Instrumented manikin testing was conducted on both aviator and tanker protective clothing systems beginning with summer weight and adding clothing layers up to the full winter weight configuration (Table 15.3). The clothing systems were tested from three to ten seconds of exposure (Table 15.4).[26] Exposure limits were established for each clothing configuration (Table 15.5) based on the protection provided by each system. This approach was selected for two reasons: (i) many military protective clothing systems provide protection against multiple battlefield threats and hazards and environmental conditions, and the protection mechanisms for each may be mutually exclusive, and (ii) the flame and thermal incident exposure on the battlefield is random and therefore difficult to specify.

The first line of defense against the flame assault is the clothing system outer-layer, which in every configuration tested is flame resistant. Each additional clothing under-layer adds insulation and increases protection time. Tests performed with both cotton long underwear and Nomex® long underwear show no difference in performance. These findings corroborate another study[27] demonstrating that the entire clothing system, especially the under-layers, does not necessarily need to be made from flame-resistant materials. The aviator and tanker systems performed similarly at various exposures with the exception of the summer weight configuration. The tanker coverall demonstrated a second-degree body burn of eight versus 18% for the two-piece aviator system. The differences are likely due to differences in material type and garment construction. The tanker coverall was made from a 152 grams/meter^2 producer colored Nomex® fabric composed

Table 15.3 Aviator and tanker clothing configurations

Configuration	Aviator	Tanker
1	T-shirt[13] and briefs[14] Aircrew coat[15] and trouser[16] Lightweight balaclava[18] Nomex® gloves[19] Combat boots[20]	T-shirt and briefs Combat vehicle crewman coverall[17] Lightweight balaclava Nomex® gloves Combat boots
2	Cotton long underwear Aircrew coat and trouser Lightweight balaclava Nomex® gloves Combat boots	Cotton long underwear Combat vehicle crewman coverall Lightweight balaclava Nomex® gloves Combat boots
3	Nomex® long underwear[21] Aircrew coat and trouser Lightweight balaclava Nomex® gloves Combat boots	Nomex® long underwear Combat vehicle crewman coverall Lightweight balaclava Nomex® gloves Combat boots
4	Nomex® long underwear Aircrew coat and trouser Aircrew jacket and liner[22] Lightweight balaclava Nomex® gloves Combat boots	Nomex® long underwear Combat vehicle crewman coverall Combat vehicle crewman jacket[23] Lightweight balaclava Nomex® gloves Combat boots
5	Nomex® long underwear Aircrew coat and trouser Bib overall[24] Aircrew jacket and liner Cold weather balaclava[25] Nomex® gloves Combat boots	Nomex® long underwear Combat vehicle crewman coverall Bib overall Combat vehicle crewman jacket Cold weather balaclava Nomex® gloves Combat boots

of fully crystallized fiber, and the aviator coat and trousers were made from a 186 grams/meter² camouflage printed Nomex® fabric composed of partially crystallized fiber. According to the fiber manufacturer, the terms 'fully' and 'partially' crystallized are relative to the level of crystallinity achievable in Nomex® fiber. The fully crystallized producer colored fabric shows superior performance compared to the partially crystallized camouflage printed

Table 15.4 Summary of predicted percent burn injury for aviator and tanker clothing

Configuration	Exposure time											
	3 seconds			4 seconds			6 seconds			10 seconds		
	% Predicted burn*			% Predicted burn*			% Predicted burn*			% Predicted burn*		
	Second degree	Third degree	Total	Second degree	Third degree	Total	Second degree	Third degree	Total	Second degree	Third degree	Total
1A	18	2	20	23	7	30	25[a]	31[a]	56[a]			
2A	0	2	2	2	2	4	17	3	19			
3A	0	2	2	2	2	4	26	2	28			
4A				2	2	4	9	2	11			
5A				0[a]	2[a]	2[a]	0	2	2	1	4	5
1B	8	2	10	13	16	29	22[a]	43[a]	65[a]			
2B	1	2	3	3	3	6	18	14	32			
3B	0	2	2	2	2	4	21	9	30			
4B				2	2	4	8	4	13			
5B				0[a]	2[a]	2[a]	0	2	2	2	3	5

* Average of replicate measurements.
[a] Result of one measurement.

Table 15.5 Exposure limits for aviator and tanker clothing

Configuration Layer	1	2	3	4
One	T-shirt, briefs cotton	Long underwear Cotton or Nomex®	Long underwear Nomex®	Long underwear Nomex®
Two	Coverall or ABDU Nomex®	Coverall or ABDU Nomex®	Coverall or ABDU Nomex®	Coverall or ABDU Nomex®
Three			Jacket Nomex®	Bib-overalls Nomex®
Four				Jacket Nomex®
% Body burn	Less than 20% at 3 seconds	Less than 20% at 4 seconds	Less than 20% at 6 seconds	Less than 20% at 10 seconds

fabric, even at a lower weight. When initially developed, the weight of the camouflage printed fabric was increased from 152 to 186 grams/meter^2 to compensate for the lower strength of the partially crystallized fiber. In addition, two-piece clothing systems such as a coat and trousers usually don't demonstrate as high a level of protection as a one-piece coverall because of the chimney effect observed with some coats. Hot air and gas from the test propane torches rises and travels under the coat, closer to the manikin sensors, and potentially out through the neckline. However, when initially designed, a fabric-encased elastic band was incorporated at the back waist of the aviator coat specifically to prevent hot air from rising through the bottom back of the coat. This feature is not present on the front of the garment.

A comparison of garments made from the same fabric weave and fiber type demonstrates that fabric weight contributes to the clothing system's flame protection.[28] Flame-retardant treated cotton and nylon twill fabric in an 88/12% blend level made in eight and eleven ounce/yard^2 fabrics were evaluated in a combat uniform configuration of coat[29] and trousers,[30] cotton t-shirt and briefs, balaclava, gloves, and boots. In Table 15.6, at a four-second exposure, the configuration made from the heavier weight fabric provided greater protection than the lighter weight fabric, and is acceptable based on military pass/fail criteria. When used in combination, however, where the coat is fabricated from the lighter weight fabric for comfort purposes, and trousers fabricated from the heavier weight fabric for improved durability, the body burns exceeded those of the configuration fabricated completely of the lighter weight fabric and demonstrated the greatest % body burn of the three configurations.

Table 15.6 Summary of predicted percent burn injury for nylon/cotton, and lightweight and heavy weight FRT cotton/nylon clothing

Configuration	3 seconds			4 seconds		
	Second	Third	Total	Second	Third	Total
Hot weather nylon/cotton combat coat and trouser, cotton T-shirt and briefs, balaclava, gloves, boots	9	23	32*			
Lightweight FRT cotton/nylon combat coat and trouser, cotton T-shirt and briefs, balaclava, gloves, boots				26	7	33
Heavyweight FRT cotton/nylon combat coat and trouser, cotton T-shirt and briefs, balaclava, gloves, boots				13	3	16
Lightweight FRT cotton/nylon combat coat, heavyweight FRT cotton/nylon combat trouser, cotton T-shirt and briefs, balaclava, gloves, boots				45	19	64

*All garments in this configuration were doused with water one minute after the start of the test to prevent damage to the manikin. The predicted burn levels would have been much higher if the after-flame was allowed to continue burning. The results in this table are only relevant when the assumption is made that the garment will be doused with water 60 seconds after being initially exposed.

The amount of coverage provided by each component in a clothing system contributes to its flame protection. Flame-resistant rayon and Kevlar® fabric in a 58/40% blend level with 2% conductive fiber in a 203 grams/meter^2 fabric were evaluated in a combat uniform configuration of coat and trousers, cotton short sleeve T-shirt and briefs, balaclava, gloves, and boots, and a second configuration of long cotton, and long Nomex® underwear.[28] In Table 15.7, at three seconds, while the T-shirt and briefs configuration slightly exceeds the military pass/fail criteria, the addition of long underwear substantially reduces the percent body burn. At four seconds both ensembles with long underwear, i.e. the Nomex® long underwear, and the cotton long underwear continue to provide substantially lower percent body burns than the ensemble with T-shirt and briefs.

An ensemble consisting of flame-resistant clothing layers topped with a flammable outer garment provides interesting results.[28] In Table 15.8, the United States aviator clothing system fabricated from the standard Nomex®/Kevlar® and conductive fiber blend fabric was evaluated with both the long Nomex® underwear and commercial polyester/cotton blend

Table 15.7 Summary of predicted percent burn injury for nylon/cotton, and FR rayon/Kevlar® clothing

Configuration	3 seconds			4 seconds		
	Second	Third	Total	Second	Third	Total
Hot weather nylon/cotton combat coat and trouser, cotton T-shirt and briefs, balaclava, gloves, boots	9	23	32*			
FR rayon/Kevlar® combat coat and trouser, cotton T-shirt and briefs, balaclava, gloves, boots	19	4	23	36	6	42
FR rayon/Kevlar® combat coat and trouser, Nomex® long underwear, balaclava, gloves, boots	4	2	6	6	2	8
FR rayon/Kevlar® combat coat and trouser, cotton long underwear, balaclava, gloves, boots				7	3	10

*All garments in this configuration were doused with water one minute after the start of the test to prevent damage to the manikin. The predicted burn levels would have been much higher if the after-flame was allowed to continue burning. The results in this table are only relevant when the assumption is made that the garment will be doused with water 60 seconds after being initially exposed.

Table 15.8 Summary of predicted percent burn injury for aviator clothing with ECWCS outerwear

Configuration	4 seconds			6 seconds		
	Second	Third	Total	Second	Third	Total
Nomex® aviator coat and trousers, Nomex® long underwear, balaclava, gloves, boots	4	1	5	26	2	28
Nomex® aviator coat and trousers, polyester/cotton long underwear, balaclava, gloves, boots	5	2	7	35	3	38
Nomex® aviator coat and trousers, Nomex® long underwear, ECWCS parka and trousers, balaclava, gloves, boots	2	3	5*	6	8	14*

*All garments in this configuration were doused with water one minute after the start of the test to prevent damage to the manikin. The predicted burn levels would have been much higher if the after-flame was allowed to continue burning. The results in this table are only relevant when the assumption is made that the garment will be doused with water 60 seconds after being initially exposed.

underwear and each provided significant protection at four seconds of exposure with total percent body burns of five and seven percent, respectively. At six seconds, both ensembles exceeded the performance goal of no greater than a 20% second-degree body burn. Additional ensembles were tested but with an added flammable outer garment known as the Extended Cold Weather Clothing System (ECWCS) parka[31] and trousers.[32] The ECWCS parka and trouser are made of a polytetrafluoroethylene film sandwiched between two nylon fabrics.[33] The total laminated fabric weighs no more than 200 grams/meter^2. Usually, layered clothing ensembles with a flame-resistant garment as the outermost will eventually self-extinguish when tested in accordance with ASTM F 1930. However, when the flammable ECWCS parka and trousers was added to the aviator configuration and exposed for four and six seconds the configuration had to be doused with water one minute after the start of the test to prevent damage to the manikin. While the burn injury prediction model calculated burns of 5 and 14% at four and six seconds respectively, the predicted burn injury levels would have been much higher if the after-flame had been allowed to continue burning.

Table 15.9 Summary of predicted percent burn injury for tanker clothing with JSLIST, and ECWCS outerwear

Configuration	4 seconds			6 seconds		
	2nd	3rd	Total	2nd	3rd	Total
Nomex® tanker coveralls, Nomex® long underwear, balaclava, gloves, boots	1	2	3	24	1	25
Nomex® tanker coveralls, polyester and cotton long underwear, balaclava, gloves, boots	5	2	7	25	3	28
Nomex® tanker coveralls, Nomex® long underwear, JSLIST, gloves, boots	<1	7	7*	8	10	18*
Nomex® tanker coveralls, Nomex® long underwear, ECWCS parka and trousers, balaclava, gloves, boots	2	2	4*	10	7	17*

* All garments in this configuration were doused with water one minute after the start of the test to prevent damage to the manikin. The predicted burn levels would have been much higher if the after-flame was allowed to continue burning. The results in this table are only relevant when the assumption is made that the garment will be doused with water 60 seconds after being initially exposed.

A similar ensemble was evaluated consisting of the United States tanker's coveralls, long underwear, balaclava, gloves and boots.[28] As listed in Table 15.9, the tanker's coveralls are fabricated from the standard Nomex®/Kevlar® and conductive fiber blend fabric and was evaluated with both the Nomex® long underwear and commercial polyester/cotton long underwear. As seen in the aviator configuration, at four seconds both the long Nomex® underwear and commercial polyester/cotton blend underwear provided significant protection with total percent body burns of three and seven percent respectively. At six seconds both configurations exceeded the performance goal of no greater than a 20% second-egree body burn. In a subsequent test, the tanker clothing system was also topped with the ECWCS parka and trousers and had to be doused with water one minute after the start of the test to prevent damage to the manikin. While the burn injury prediction model calculated burns of four and 17 percent, at four and six seconds respectively, the predicted burn injury levels would have been much higher if the after-flame had been allowed to continue burning. The tanker configuration was also tested when topped with the flammable Joint Service Lightweight Integrated Suit Technology (JSLIST) chemical protective garment.[34] This garment is fabricated from a nylon/cotton outershell that has been treated with a water- and oil-repellent treatment.[3] The active ingredient in the vapor sorptive liner is carbon spheres. This configuration, as with the ECWCS outer garment, and the aviator and ECWCS configuration, also had to be doused with water one minute after the start of the test, and the predicted burn injury levels of seven and 18%, at four and six seconds respectively, would have been much higher if the after-flame had been allowed to continue burning.

In summary, the data demonstrates that the first line of defense against the flame assault is the clothing system outer layer. When the outer layer is flame resistant, each additional clothing under layer adds insulation and increases protection time. Tests performed with cotton, polyester/cotton, and Nomex long underwear showed little difference in performance when exposure time is short. As the exposure time is increased the configuration with the heaviest weight underwear fabric demonstrated the lowest-percent body burn. Greater clothing component coverage, such as long-sleeved undershirts and long-legged underpants versus short-sleeved T-shirt and briefs provides greater protection. Heavier weight fabrics made from the same fabric weave and fiber blend provide greater protection than those that are lighter weight. When a flame-resistant ensemble is topped with a flammable outer garment, such as the ECWCS or JSLIST described here, the predicted burn injury levels may be misleading if after-flame behavior and configuration dousing, if necessitated, are not taken into consideration.

15.5   Future trends

Since cost is a significant driver in soldier flame protection, non-woven fabric structures have also been investigated. As an alternative to woven fabrics made from low-cost fiber blend yarns, direct fiber to fabric manufacturing appear to provide potential savings on yarn spinning, fabric weaving, and finishing, which are eliminated in the spun-laced non-woven fabric manufacturing process. A spun-laced non-woven Nomex® and Kevlar® blend fabric developed by DuPont demonstrated strength that was equal to or greater than the former all-cotton Hot Weather Battledress Uniform (BDU) fabric[5] and was half the weight. Due to its lightweight, open, air-permeable construction, it could be worn over the existing BDU and together this configuration provides a 40% cost savings over the camou-flage printed Nomex®/Kevlar® Aircrew BDU used by Army aviators. However, garments made from spun-laced non-woven fabrics are not designed for long-term durability. Instrumented manikin testing (Table 15.10) demonstrated that when the Nomex® coverall (DuPont™ Nomex® Limitedwear) was worn over the BDU, the total predicted body burn was reduced from 84 to 7%. Permissible exposure limits up to five seconds were established. The non-woven Nomex®/Kevlar® outershell provided ignition resistance, and the under layers of the BDU, T-shirt and briefs provided thermal insulation.

This approach of augmenting the existing combat uniform with a quick don coverall rather than replacing the uniform with a flame-resistant version was further investigated. Military field evaluations of the DuPont commer-cial coverall were conducted at the Ranger Training School in Dahlonega, Georgia; National Training Center, Fort Irwin, California; and a vehicle and machinery maintenance unit at Fort Lewis, Washington. The results indi-cated that it could not be worn 'as is'. It needed to be re-sized to fit the military population, re-designed to support military use, and hardened. A

Table 15.10 Summary of predicted percent burn injury for hot weather BDU and Limitedwear*

Configuration	3 seconds			4 seconds			5 seconds			6 seconds		
	2nd	3rd	Total	2nd	3rd	Total	2nd	3rd	Total	2nd	3rd	Total
Cotton T-shirt, briefs, HWBDU	12	72	84									
Cotton T-shirt, briefs, HWBDU, Limitedwear	1	6	7	4	7	11	5	10	15	8	13	21

*The manikin head, which was not protected during this test, accounts for 7% of the predicted burn injury.

new garment sizing system and patterns were developed. The garment design was modified to include a center front two-way zipper with flap, back elasticized waist, wrists and ankles, and knee, seat and elbow patches. DuPont developed a novel pigment-based camouflage print formulation which was tested and approved for use. The new military version of the coverall was tested in accordance with ASTM F 1930 and Table 15.11 lists the data which demonstrates that the changes to the garment did not negatively impact its flame protection. Additional configuration testing was

Table 15.11 Summary of predicted percent burn injury for hot weather BDU, and non-woven coverall*

Configuration	3 seconds			4 seconds		
	Second	Third	Total	Second	Third	Total
Cotton T-shirt and briefs, Hot weather combat uniform	12	72	84			
Cotton T-shirt and briefs, Hot weather combat uniform, Military non-woven coverall				1	7	8

*The manikin head, which was not protected during this test, accounts for 7% of the predicted burn injury.

Table 15.12 Summary of predicted percent burn injury for non-woven coverall and underwear*

Configuration	3 seconds			4 seconds			5 seconds		
	Second	Third	Total	Second	Third	Total	Second	Third	Total
Cotton T-shirt and briefs, non-woven coverall	15	7	22	18	17	35	22	34	56
Polyester/cotton long underwear, non-woven coverall	0	7	7	4	7	11	16	11	27
Cotton long underwear, non-woven coverall	0	7	7	1	7	8	3	7	10
Nomex® long underwear, non-woven coverall	0	7	7	2	7	8	10	7	17

*The manikin head, which was not protected in this test, accounts for 7% of the predicted burn injury.

conducted to determine the effect of layering and material on predicted burn injury. In hot and dry, and hot and humid climates, it may be desirable to wear the coverall directly over underwear without the added insulation of the combat uniform. It is noted that the configurations listed in Table 15.12 did not include headwear, handwear, or boots. The manikin head, which was not protected, accounts for seven percent of the predicted body burn and must be taken into consideration. There are no sensors in the manikin's hands or feet. At three seconds of exposure, the data demonstrates that the coverall, which is made from a very lightweight 102 grams/meter^2 Nomex$^®$/Kevlar$^®$ non-woven fabric, is barely sufficient to be worn over 152 grams/meter^2 T-shirt and briefs, and at four and five seconds, the protection decreases with predicted burn injuries of 35 and 56%, respectively. At three seconds, all of the long underwear ensembles, i.e. polyester/cotton, cotton, and Nomex$^®$, provide the same level of predicted burn injury, which is zero if the seven percent third-degree burn to the head is negated. The difference in predicted burn injury between the T-shirt and briefs over the other ensembles is due to the insulation provided by the long sleeves and trouser legs and these are significant contributors to the overall burn protection of the configuration. At four seconds, the differences between the long underwear material types begin to emerge. The 458 grams/meter^2 cotton, and 295 grams/meter^2 Nomex$^®$ underwear provide greater protection over the 203 grams/meter^2 polyester/cotton underwear, while all are still within the desired 20% second-degree range. At five seconds, the configuration with polyester/cotton long underwear begins to approach the maximum second-degree performance goal of 20% and the maximum third-degree performance goal of five percent. The ensemble with cotton long underwear, which is the heaviest weight underwear fabric of the three, provides the most protection with a three percent predicted burn injury.

Table 15.13 Summary of predicted percent burn injury for non-woven coverall and ballistic armor*

Configuration[a]	4 seconds			5 seconds		
	Second	Third	Total	Second	Third	Total
Ballistic armor under non-woven coverall	3	7	10	7	7	14
Ballistic armor over non-woven coverall	2	7	8	7	7	14

*The manikin head, which was not protected during this study, accounts for 7% of the predicted burn injury.
[a] Each reported value is the result of one measurement.

The final flame protection configuration study was conducted with the coverall worn over or under the Interceptor Body Armor (IBA). As demonstrated in Table 15.13, the coverall can be worn over or under the IBA with no loss in flame protection.

After the final burn study was completed, field evaluations of the garment design, fit, comfort, and durability were conducted at Fort Benning, Georgia; Fort Bragg, North Carolina; Fort Lee, Virginia; and Fort Leonard Wood, Missouri.[35] The soldier's work assignments included vehicle maintenance (e.g. change tires and oil, refueling), driving (e.g. Bradley Fighting Vehicle), maintenance work, office work, welding, conducting smoke missions, handling of petroleum, oils and lubricants, washing equipment, and construction. The findings indicate that the garment was comfortable and breathable, and the fit was just right. Durability concerns were reported regarding fuzz and pilling that occurred at the knees, seat and elbows; they are a common characteristic of clothing made from non-woven fabric.

15.6   References

1 TUCKER, D.W., REI, S.A., *Soldier Flame/Thermal Hazard Assessment, Natick/TR-99/039L*, U.S. Army Soldier and Biological Chemical Command, Natick, MA, 01760-5020.

2 KIM, I.L., *Battlefield Flame/Thermal Threats or Hazards and Thermal Performance Criteria, TR-00/015L*, U.S. Army Soldier and Biological Chemical Command, Natick, MA, 01760-5020.

3 MIL-DTL-44436, *Cloth, Camouflage Pattern, Wind Resistant Poplin, Nylon/Cotton Blend*, Defense Support Center Philadelphia, 700 Robbins Avenue, Philadelphia, PA 19111-5096.

4 MIL-C-44031, *Cloth, Camouflage Pattern: Woodland, Cotton and Nylon*, Defense Support Center Philadelphia, 700 Robbins Avenue, Philadelphia, PA 19111-5096.

5 MIL-C-43468, *Cloth, Camouflage Pattern, Wind Resistant Poplin, Cotton*, Defense Support Center Philadelphia, 700 Robbins Avenue, Philadelphia, PA 19111-5096.

6 MIL-C-83429, *Cloth, Plain and Basket Weave, Aramid*, Defense Support Center Philadelphia, 700 Robbins Avenue, Philadelphia, PA 19111-5096.

7 WINTERHALTER, C.A., *Experimental Battledress Uniform Fabrics Made From Amine Oxide Solvent Spun Cellulosic Fibers, Natick/TR-02/007*, US Army Soldier and Biological Chemical Command, Natick, MA, 01760-5020.

8 ASTM D 6413, *Flame Resistance of Textiles (Vertical Test)*, ASTM International, West Conshohocken, PA, 19428-2959.

9 ASTM F 1930, *Evaluation of Flame Resistant Clothing for Protection Against Flash Fire Simulation Using an Instrumented Manikin*, ASTM International, West Conshohocken, PA, 19428-2959.

10 Committee on Trauma, 'Resources for Optimal Care of Patients with Burn Injury,' *Resources for Optimal Care of the Injured Patient*, American College of Surgeons, Chicago, 1990, 57–60 pp.

11 *The Military Medicine Series, Part I. Warfare, Weaponry, and the Casualty, Volume 5: Conventional Warfare: Ballistic, Blast, and Burn*, Office of the Surgeon General, Department of the Army, United States of America, Washington, DC, 1990.

12 'Chap. 3 Burn Injury', *Emergency War Surgery, 2nd US Rev. of The Emergency War Surgery NATO Handbook*, U.S. Department of Defense, Bowen, T.E., Bellamy, R.F., Eds, Washington, DC, 1989, 35–56 pp.

13 MIL-U-44096, *Undershirt, Cotton, Brown 436*, Defense Support Center Philadelphia, 700 Robbins Avenue, Philadelphia, PA 19111-5096.

14 MIL-D-43783, *Brief Drawers, Cotton, Brown 436*, Defense Support Center Philadelphia, 700 Robbins Avenue, Philadelphia, PA 19111-5096.

15 MIL-C-44371, *Coat, Aircrew, Combat*, Defense Support Center Philadelphia, 700 Robbins Avenue, Philadelphia, PA 19111-5096.

16 MIL-T-44372, *Trousers, Aircrew, Combat*, Defense Support Center Philadelphia, 700 Robbins Avenue, Philadelphia, PA 19111-5096.

17 MIL-C-44077, *Coveralls, Combat Vehicle Crewman's*, Defense Support Center Philadelphia, 700 Robbins Avenue, Philadelphia, PA, 19111-5096.

18 MIL-H-24936, *Hood, Anti-flash, Flame Resistant*, Defense Support Center Philadelphia, 700 Robbins Avenue, Philadelphia, PA, 19111-5096.

19 MIL-DTL-81188, *Gloves, Flyer's Summer, Type GS/FRP-2*, Defense Support Center Philadelphia, 700 Robbins Avenue, Philadelphia, PA, 19111-5096.

20 MIL-B-44152, *Boots, Combat, Mildew and Water Resistant*, Defense Support Center Philadelphia, 700 Robbins Avenue, Philadelphia, PA, 19111-5096.

21 MIL-D-85040, *Drawers and Undershirts, Flyer's, Anti-Exposure, Aramid, High Temperature Resistant*, Defense Support Center Philadelphia, 700 Robbins Avenue, Philadelphia, PA 19111-5096.

22 MIL-DTL-31010, *Jacket, Aircrew, Cold Weather, With Removable, Liner, and Hood*, Defense Support Center Philadelphia, 700 Robbins Avenue, Philadelphia, PA, 19111-5096.

23 MIL-J-43924, *Jacket, Cold Weather, High Temperature Resistant*, Defense Support Center Philadelphia, 700 Robbins Avenue, Philadelphia, PA, 19111-5096.

24 MIL-O-44109, *Overall, Combat Vehicle Crewman's*, Defense Support Center Philadelphia, 700 Robbins Avenue, Philadelphia, PA, 19111-5096.

25 MIL-H-44265, *Hood, Combat Vehicle Crewman's (Balaclava)*, Defense Support Center Philadelphia, 700 Robbins Avenue, Philadelphia, PA, 19111-5096.

26 *A Report on the Thermal Protective Performance (TPP) and the Pyroman Evaluation for the U.S. Army Soldier Systems Command*, Center for Research on Textile Protection and Comfort (T-PACC), College of Textiles, North Carolina State University, Raleigh, NC, 1999.

27 CAPECCI, T.M., SWAVERLY, C.B., 'Flammability and Thermal Protection Testing of Long Underwear for Navy and Marine Corps Aircrew Use', *Performance of Protective Clothing: Fifth Volume, ASTM STP 1237*, ASTM, 1995, 625–634 pp.

28 BARKER, R., DEATON, A.S., FOWLER, J., *A Report to U.S. Army Soldier Systems Command, Characterization of Test Garments Using the PyroMan System*, Center for Research on Textile Protection and Comfort (T-PACC), College of Textiles, North Carolina State University, Raleigh, NC, March 2003.

29 MIL-DTL-44048, *Coats, Camouflage Pattern, Combat*, Defense Support Center Philadelphia, 700 Robbins Avenue, Philadelphia, PA, 19111-5096.

30 MIL-T-44047, *Trousers, Camouflage Pattern*, Defense Support Center Philadelphia, 700 Robbins Avenue, Philadelphia, PA 19111-5096.

31 MIL-P-44188, *Parka, Cold Weather, Camouflage*, Defense Support Center Philadelphia, 700 Robbins Avenue, Philadelphia, PA 19111-5096.

32 MIL-T-44189, *Trousers, Cold Weather, Camouflage*, Defense Support Center Philadelphia, 700 Robbins Avenue, Philadelphia, PA 19111-5096.

33 MIL-C-44187, *Cloth, Laminated, Waterproof and Moisture Vapor Permeable*, Defense Support Center Philadelphia, 700 Robbins Avenue, Philadelphia, PA 19111-5096.

34 *Joint Service Lightweight Integrated Suit Technology (JSLIST) Coat and Trouser, Chemical Protective*, Defense Support Center Philadelphia, 700 Robbins Avenue, Philadelphia, PA 19111-5096.

35 PACITTO, S., *Field Evaluation of Flame Resistant Coveralls, Fort Benning, GA, Fort Bragg, NC, Fort Lee, VA, and Fort Leonard Wood, MI, Phase II, December 2006*, US Army, Natick Soldier Center, Kansas St., AMSRD-NSC-SS-P, Natick, MA 01760-5020.

Index